Surface-Water-Quality Assessment of the Yakima River Basin, Washington

Distribution of Pesticides and Other Organic Compounds in Water, Sediment, and Aquatic Biota, 1987–91

By JOSEPH F. RINELLA, STUART W. MCKENZIE, J. KENT CRAWFORD, WILLIAM T. FOREMAN, GREGORY J. FUHRER, and JENNIFER L. MORACE

with a section on

Dissolved Organic Carbon in the Yakima River Basin

By GEORGE R. AIKEN

U.S. GEOLOGICAL SURVEY WATER-SUPPLY PAPER 2354–B

U.S. DEPARTMENT OF THE INTERIOR
BRUCE BABBITT, Secretary

U.S. GEOLOGICAL SURVEY
Charles G. Groat, Director

UNITED STATES GOVERNMENT PRINTING OFFICE: 1999

For sale by the
U.S. Geological Survey
Information Services
Box 25286
Federal Center
Denver, CO 80225

Library of Congress Cataloging in Publication Data

Surface-water-quality assessment of the Yakima River Basin, Washington : distribution of pesticides and other organic compounds in water, sediment, and aquatic biota, 1987-91 / by Joseph F. Rinella ... [et al.] : with a section on dissolved organic carbon in the Yakima River Basin by George R. Aiken.

p. cm. -- (U.S. Geological Survey water-supply paper ; 2354-B)

Includes bibliographical references.

Supt. of Docs. no.: I 19.13:2354-B

ISBN 0-607-89533-0 (alk. paper)

1. Water chemistry. 2. Water quality--Washington (State)--Yakima River Watershed. 3. Trace elements in water--Washington (State)--Yakima River Watershed. I. Rinella, Joseph F. II. Geological Survey (U.S.) III. Series.

GB857.2.W2S86 1999 99-31199

363.739'42'0979755--dc21 CIP

FOREWORD

The mission of the U.S. Geological Survey (USGS) is to assess the quantity and quality of the earth resources of the Nation and to provide information that will assist resource managers and policymakers at Federal, Tribal, State, and local levels in making sound decisions. Assessment of water-quality conditions and trends is an important part of this overall mission.

One of the greatest challenges faced by water-resources scientists is acquiring reliable information that will guide the use and protection of the Nation's water resources. That challenge is being addressed by Federal, Tribal, State, interstate, and local water-resource agencies and by many academic institutions. These organizations are collecting water-quality data for a host of purposes that include: compliance with permits and water-supply standards; development of remediation plans for a specific contamination problem; operational decisions on industrial, wastewater, or water-supply facilities; and research on factors that affect water quality. An additional need for water-quality information is to provide a basis on which regional and national-level policy decisions can be based. Wise decisions must be based on sound information. As a society we need to know whether certain types of water-quality problems are isolated or ubiquitous, whether there are significant differences in conditions among regions, whether the conditions are changing over time, and why these conditions change from place to place and over time. The information can be used to help determine the efficacy of existing water-quality policies and to help analysts determine the need for and likely consequences of new policies.

To address these needs, the Congress appropriated funds in 1986 for the USGS to begin a pilot program in seven project areas to develop and refine the National Water-Quality Assessment (NAWQA) Program. In 1991, the USGS began full implementation of the program. The NAWQA Program builds upon an existing base of water-quality studies of the USGS, as well as those of other Federal, Tribal, State, and local agencies. The objectives of the NAWQA Program are to:

- Describe current water-quality conditions for a large part of the Nation's freshwater streams, rivers, and aquifers.
- Describe how water quality is changing over time.
- Improve understanding of the primary natural and human factors that affect water-quality conditions.

This information will help support the development and evaluation of management, regulatory, and monitoring decisions by other Federal, Tribal, State, and local agencies to protect, use, and enhance water resources.

The goals of the NAWQA Program are being achieved through ongoing and proposed investigations of about 50 of the Nation's most important river basins and aquifer systems, which are referred to as study units. These study units are distributed throughout the Nation and cover a diversity of hydrogeologic settings. About two-thirds of the Nation's freshwater use occurs within the study units and more than one-half of the people served by public water-supply systems live within their boundaries.

National synthesis of data analysis, based on aggregation of comparable information obtained from the study units, is a major component of the program. This effort focuses on selected water-quality topics using nationally consistent information. Comparative studies will explain differences and similarities in observed water-quality conditions among study areas and will identify changes and trends and their causes. The first topics addressed by the national synthesis are pesticides, nutrients, volatile organic compounds, and aquatic biology. Discussions on these and other water-quality topics will be published in periodic summaries of the quality of the Nation's ground and surface water as the information becomes available.

This report is an element of the comprehensive body of information developed as part of the NAWQA Program. The program depends heavily on the advice, cooperation, and information from many Federal, State, interstate, Tribal, and local agencies and the public. The assistance and suggestions of all are greatly appreciated.

Robert M. Hirsch

Robert M. Hirsch
Chief Hydrologist

CONTENTS

FIGURES

TABLES

CONVERSION FACTORS

[SI = International System of units, a modernized metric system of measurement]

Multiply	By	To obtain
A. Factors for converting SI metric units to inch/pound units		
	Length	
micrometer (µm)	0.00003937	inch (in)
millimeter (mm)	0.03937	inch
meter (m)	3.281	foot (ft)
	Volume	
milliliter (mL)	0.001057	quart (qt)
liter (L)	1.057	quart
liter	0.2642	gallon (gal)
	Mass	
gram (g)	0.03527	ounce (oz avoirdupois)
kilogram (kg)	2.205	pound (lb avoirdupois)
	Temperature	
degree Celsius (°C)	Temp degree F = 1.8 (Temp degree C) + 32	degree Fahrenheit (°F)
B. Factors for converting inch/pound units to SI metric units.		
	Volume per unit time (flow)	
cubic feet per second (ft^3/s)	0.02832	cubic meters per second (m^3/s)
	Distance	
mile (mi)	1.609	kilometer (km)
	Area	
acre	4,047	square meters (m^2)
acre	.4047	hectare (ha)
square miles (mi^2)	2.590	square kilometers (km^2)
	Volume	
acre-feet	1,233	cubic meters (m^3)
C. Factors for converting SI metric units to other miscellaneous units		
	Concentration, in water	
milligrams per liter (mg/L)	1	parts per million (ppm)
nanograms per liter (ηg/L)	1	parts per trillion (ppt)
	Concentration, in bed sediment and soil	
micrograms per kilogram (µg/kg)	1	parts per billion (ppb)
	Concentration, in aquatic biota	
micrograms per gram (µg/g)	1	parts per million
nanograms per gram (ηg/g)	1	parts per billion
micrograms per kilogram (µg/kg)	1	parts per billion

Electrical conductivity is measured as specific electrical conductance, in units of microsiemens per centimeter (µS/cm) at 25 degrees Celsius.

Surface-Water-Quality Assessment of the Yakima River Basin, Washington

Distribution of Pesticides and Other Organic Compounds in Water, Sediment, and Aquatic Biota, 1987–91

By Joseph F. Rinella, Stuart W. McKenzie, J. Kent Crawford, William T. Foreman, Gregory J. Fuhrer, *and* Jennifer L. Morace

With a section on Dissolved Organic Carbon in the Yakima River Basin

By George R. Aiken

Abstract

This report presents an interpretation of chemical data from an assessment of pesticides and other organic compounds in samples of surface water, suspended sediment, streambed sediment, agricultural soil, and aquatic biota (fish, crayfish, mollusk, and plant) from the Yakima River Basin in south-central Washington. During 1987–91, data were collected from about 100 stations to determine the occurrence, distribution, transport, and fate of organic compounds in the basin. The assessment is a component of the U.S. Geological Survey's National Water-Quality Assessment (NAWQA) Program, which was designed to describe the status and trends in the quality of the Nation's ground- and surface-water resources, and to determine factors that affect water quality. In 1986, the Yakima River Basin was selected for one of the Nation's four surface-water pilot studies.

In 1989, about 180 pesticides were applied in the Yakima River Basin. Fifty-four of the compounds were analyzed in this study, and 43 of the 54 compounds (80 percent) were detected in water, sediment, soil, and (or) aquatic biota. Including other organic compounds associated with industrial and urban activities, as well as persistent pesticides that were used historically, more than 110 organic compounds were detected in the basin from 1987–91.

In 1988, concentrations of hydrophilic and hydrophobic pesticides generally began to increase in agricultural runoff in June in response to increased irrigation following springtime pesticide applications. Irrigation at or near peak water use in the basin flushed relatively high pesticide loads to streams in June and July 1988. In March 1989, high pesticide loads also were flushed to the streams from agricultural fields during a period of storm runoff.

In late June 1989, when pesticide concentrations were expected to be high, water-quality data were collected synoptically from 29 stations in the Yakima River Basin. The most frequently detected compounds in the water column were organochlorine compounds (chlordane, DDT+DDE+DDD, dieldrin, and endosulfan I), organophosphorus compounds (chlorpyrifos, diazinon, dimethoate, malathion, parathion, phorate, phosphamidon), thiocarbamate and sulfite compounds (EPTC and propargite), acetamide compounds (alachlor and metolachlor), triazine compounds (atrazine, prometon, and simazine), and chlorophenoxy-acetic acid and benzoic compounds (2,4-D and dicamba). Quantifiable concentrations of these compounds generally ranged from 1 to 100 nanograms per liter; however, maximum concentrations of dicamba, atrazine, phorate, simazine, parathion, propargite, 2,4-D, dimetoate, and diazinon were higher, ranging from 100 to 410 nanograms per liter. With the exception of

DDT+DDE+DDD, dieldrin, chlordane, and prometon, these compounds were among the most abundantly used compounds in the basin in 1989. The abundance probably accounts for their frequent detections. Although the use or sale of DDT, dieldrin, and chlordane were banned or restricted in 1973, 1974, and 1987, respectively, these persistent compounds were detected frequently. The widespread occurrence of these persistent pesticides suggests broad historical usage throughout the basin. Although prometon use was minimal, its long half-life (up to 500 days) in soils probably accounts for its increased number of detections.

The pesticides that most frequently exceeded chronic-toxicity water-quality criteria or guidelines for the protection of freshwater aquatic life in June 1989 include DDT+DDE+DDD, dieldrin, diazinon, and parathion. Most of the exceedances occurred in agricultural return flows and Yakima River stations downstream from the city of Yakima. Reductions in pesticide concentrations that would be needed to meet criteria or guidelines for DDT+DDE+DDD, dieldrin, diazinon, and parathion ranged from 29 to 99 percent at 19 of 29 stations, 47 to 95 percent at 7 stations, 31 to 98 percent at 14 stations, and 19 to 93 percent at 4 stations, respectively. Assuming that instream pesticide concentrations could be reduced to the concentrations that occurred in November 1988 during a period of minimal overland runoff, parathion and diazinon would meet guidelines. Dieldrin would meet the criteria in the main stem and Moxee Drain, and DDT+DDE+DDD would meet the criteria in the main stem. In the other agricultural return flows, however, the criteria would be exceeded by as much as 3 ηg/L for dieldrin and 18 ηg/L for DDT+DDE+DDD.

In 1989–90, samples of fish, mollusks, and aquatic plants were collected from 33 stations for analyses of organic compounds. One DDT metabolite, 4,4'-DDE, was the most widely occurring organic compound detected in aquatic biota. This organochlorine compound was detected in fish samples at all stations sampled in 1989 and 68 percent of the stations in 1990. Other organochlorine compounds, including DDT, DDD, cis-chlordane, trans-nonachlor, and dieldrin, were detected frequently, with each compound being detected at 32 percent or more of the stations. Dicofol, PCBs, toxaphene, and other chlordane-related compounds were detected less frequently. Similar to water-column and bed-sediment results, the highest organochlorine-compound concentrations generally were detected at main-stem and tributary stations downstream from the city of Yakima, where agriculture is the primary land use.

In the Yakima River Basin, most samples of whole resident fish collected downstream from the city of Yakima had concentrations of DDT+DDE+DDD, PCBs, chlordane-related compounds, dieldrin, toxaphene, and to a lessor extent, dicofol that were higher than those concentrations expected to result in an increased lifetime cancer risk of 1:1,000,000. The highest increased cancer risk for an individual compound was computed to be 640:1,000,000 and was based on the detection of a high PCB concentration (0.90 micrograms per gram, wet weight) in a composited sample of largescale suckers from the Yakima River at Kiona, Washington. The increased cancer risk associated with ingestion of whole resident fish from the lower Yakima River Basin generally averaged less than 50:1,000,000 for each of these individual compounds. In the lower Yakima River Basin, concentrations of DDT+DDE+DDD, toxaphene, and dieldrin in whole fish exceeded guidelines recommended by the National Academy of Sciences-National Academy of Engineering Committee on Water Quality for the protection of fish-eating predators. Seventeen, 12, and 1 whole-fish samples had concentrations of DDT+DDE+DDD, toxaphene, and dieldrin, respectively, that exceeded their guidelines of 1.0, 0.1, and 0.1 micrograms per gram, wet weight, respectively.

Major pathways for pesticide transport from agricultural fields to surface water include (1) the erosion of soil and sorbed contaminants from the fields, (2) flushing of pesticides from soil-pore water, and (3) dissolution of pesticides from soils and sediment. This transport may be reduced by controlling excessive applications of irrigation water, which will help to reduce overland runoff. To further minimize the likelihood of transport, pesticide characteristics should include (1) soil half-lives less than 3 weeks, so compounds are more likely to degrade, (2) water solubilities less than 30 milligrams per liter, to reduce the flushing effect of pesticides from soils, and (3) sediment-water partition coefficients higher than 300 to 500 milliliters per gram, to increase the propensity of the pesticide to remain sorbed to the soils in the fields.

INTRODUCTION

Thousands of man-made organic compounds are used annually in industry, agriculture, and forestry, and an unknown additional number of impurities and degradation products result from the production and use of these compounds. Some of these compounds are hydrophobic and tend to sorb onto particles of soil and stream sediment, whereas other compounds are hydrophilic and tend to readily dissolve in water.

Erosion runoff and effluent discharges from agricultural, urban, and industrial sources can result in direct transport of these compounds into surface water. Hydrophilic compounds also can percolate down to ground-water supplies and seep into surface water. The ease of transport of a great number of these toxic compounds into surface and ground water makes knowledge of their occurrence and behavior essential for management of the Nation's water resources. Even in small concentrations in surface water, many of these compounds are harmful to aquatic organisms. Some organisms bioaccumulate these compounds and can attain concentrations in tissues that are several orders of magnitude greater than their concentrations in water. Bioaccumulation of contaminants, as well as acute exposure to nonbioaccumulating contaminants, poses a threat to both aquatic biota and their predators, including humans.

Beginning in 1986, Congress appropriated funds for the U.S. Geological Survey (USGS) to test and refine concepts and protocols for the National Water-Quality Assessment (NAWQA) program. Long-term goals of the NAWQA program are to:

(1) provide a nationally consistent description of current water-quality conditions for a large part of the Nation's water resources;

(2) define long-term trends (or lack of trends) in water quality;

(3) identify, describe, and explain, as possible, the major factors that affect observed water-quality conditions and trends; and

(4) explain the implications of major findings from the assessment as they relate to water-resource monitoring, management, and regulation.

This information, which will be obtained on a continuing basis, will be made available to water managers, policy makers, and the public to provide an improved scientific basis for evaluating the effectiveness of water-quality management programs (Hirsch and others, 1988).

In 1992, the NAWQA program completed a pilot phase that lasted about 6 years. The Yakima River Basin in Washington was one of four areas in the Nation selected for testing and developing assessment concepts for the surface-water component of the program. The other surface-water, pilot-project areas were the Lower Kansas River Basin in Kansas and Nebraska; the Upper Illinois River Basin in Illinois, Indiana, and Wisconsin; and the Kentucky River Basin in Kentucky (Hirsch and others, 1988; Leahy and others, 1990).

Even though pesticide use has been extensive in the Yakima River Basin, relatively few historical samples have been collected to determine the magnitude, spatial distribution, and seasonal variability of compound concentrations in the aquatic environment. The limited amount of historical data that are available, however, show that some pesticide concentrations in water have exceeded U.S. Environmental Protection Agency (EPA) chronic-toxicity water-quality criteria for the protection of freshwater aquatic life (Rinella, McKenzie, and Fuhrer, 1992). Currently, minimum reporting levels for organic compounds are 10 or more times lower than those concentrations reported in the historical data prior to 1986. Accordingly, these low-level analyses provide water managers with better information for determining the occurrence and transport of these compounds in surface water.

Purpose and Scope

This report contains an assessment of pesticides and other organic compounds in the surface water of the Yakima River Basin during 1987–91. Objectives of this assessment are to describe for pesticides and other organic compounds:

(1) the spatial distribution of concentrations in water, sediment, and aquatic biota;

(2) temporal variability (short-term and long-term trends) in concentrations;

(3) the suitability of surface water for designated beneficial uses, including aquatic-life and domestic uses; and

(4) major natural and human factors that affect the spatial and temporal distribution of concentrations.

Because of intense agricultural activities in the basin, the main focus of this study is on the occurrence of agricultural pesticides. Water, suspended sediment, bed sediment, aquatic biota, and a few soil samples were collected and analyzed for a variety of pesticides and other organic compounds that have been, and (or) continue to be used, in the basin. Sampling stations were located throughout the basin to examine the effects of human activity on the distribution of organic compounds. Stations also were located along the main stem and at the mouths of major tributaries to better understand the sources, transport, and fate of these compounds. The study was performed in several phases, with each phase involving a specific combination of data-collection activities for different sampling stations, times, and media (water, sediment, aquatic biota, and soils).

Acknowledgments

The authors wish to acknowledge the aid and advice provided by members of the Yakima NAWQA Liaison Committee. During 1992, this committee included:

David W. Zimmer	Bureau of Reclamation
Don Schram	Bureau of Reclamation
Richard Albright	U.S. Environmental Protection Agency
William Garrigues	U.S. Forest Service
Terry W. Berkompas	U.S. Bureau of Indian Affairs
Kate Benkert	U.S. Fish and Wildlife Service
Jannine Jennings	Yakama Indian Nation
Bob Barwin	Washington State Department of Ecology
Perry Harvester	Washington State Department of Fisheries
Brent Renfrow	Washington State Department of Wildlife
Glen Patrick	Washington State Department of Health
Dr. L. Clint Duncan	Washington Water Research Center, Central Washington State University
Skip Steinmetz	Yakima County Health Department
Ronald L. Van Gundy	Yakima River Basin Association of Irrigation Districts
Ray L. Wondercheck	U.S. Soil Conservation Service
Elaine Taylor	Yakima Valley Conference of Governments
Don Chaplin	Yakima County Extension Service
Mike Tobin	North Yakima Conservation District
Richard C. Bain, Jr.	Kittitas County Conservation District

Special thanks to (1) the Yakama Indian Nation for their cooperation in providing staff time for electrofishing, reviewing reports, and giving USGS permission to collect water-quality samples on the Yakima Indian Reservation, (2) U.S. Fish and Wildlife Service, Washington State Department of Fisheries, Washington State Department of Wildlife, U.S. Forest Service, and Jean-Pierre Wilson (Heritage College, Toppenish, Washington) for assistance and cooperation in electrofishing, (3) Bureau of Reclamation and the Prosser Experimental Station (Washington State University and U.S. Department of Agriculture) for providing laboratory facilities in the basin, (4) Gregory D. Foster, Frank A. Rinella, and Paul M. Gates (USGS) for developmental work using the Goulden large-sample extractor to reduce minimum analytical reporting levels for selected organic compounds, (5) Washington State Department of Wildlife for providing refrigeration and freezer space at the Naches Fish Hatchery, (6) Mary Janet (USGS) for technical and editorial assistance, and (7) Shen Xianchen (Institute of Water Conservancy and Hydroelectric Power Research, Chinese Academy of Sciences, Ministry of Water Resources and Electric Power, Beijing, China) for assistance in sampling and developing quality-assurance procedures for spiking bed-sediment samples.

THE YAKIMA RIVER BASIN

The Yakima River is 214 miles long and drains 6,155 square miles in south-central Washington (fig. 1); it flows southeastward from the eastern slopes of the Cascades to the Columbia River (Columbia Basin Inter-Agency Committee, 1964). A schematic of selected inflows and outflows to the Yakima River is shown in figure 2. Many of the tributaries shown in figure 2 are used to convey agricultural return flow to the Yakima River.

Rinella, McKenzie, and Fuhrer (1992) provide a detailed description of the characteristics of the Yakima River Basin, including topography, geology, climate, and hydrology. The basin contains a variety of landforms, including high peaks and deep valleys of the Cascade Range, broad river valleys, and lowlands of the Columbia River. Altitude in the basin ranges from 8,184 feet in the Cascade Range to about 340 feet at the Columbia River. Present-day geomorphic changes in the basin are controlled by precipitation, snowmelt, wind (erosion of top soil), steep gradients, and the slow but continuing erosion of the high-mountain terrain by streams and small glaciers. Mean-annual precipitation ranges from 140 inches in the higher mountains to less than 10 inches in the agricultural areas and in the Kennewick area near the mouth of the basin. In the mountains, winters are cold and summers are warm and dry. The lower valleys are arid throughout most of the summer, requiring extensive irrigation to support agriculture.

The Yakima River Basin is one of the most intensively irrigated areas in the United States. The Yakima River and its major tributary, the Naches River, have perennial streamflow. Peak runoff occurs during peak snowmelt, usually in May and (or) June. For irrigation, the basin has five large reservoirs (total storage capacity of 1,065,400 acre-feet) used to augment summer flows from June through October (fig. 1). The Yakima irrigation system consists of more than 1,900 miles of canals and laterals, three hydroelectric power plants, six major irrigation projects, and numerous smaller irrigation systems to support about 450,000 acres of irrigated agriculture. Surface-water drains and wasteways, most of which are natural streams, convey agricultural return flow, livestock wastes, and sewage-treatment-plant effluent to the Yakima River. Annual surface-water diversions from the Yakima River system for irrigation are equivalent to about 60 percent of the mean annual streamflow leaving the basin. During the summer, the quality of the agricultural return flow determines water quality in the lower Yakima River downstream from the city of Yakima, because return flows contribute as much as 80 to 90 percent of the flow in the lower main stem during the irrigation season.

Prior to 1880, anadromous fish runs were estimated to be more than one-half million fish annually in the Yakima River Basin. By 1905, construction of large storage reservoirs and other water-resource developments for irrigation (for example, small dams without fish ladders and irrigation diversions without screens) had seriously affected fish habitat and fish migrations. By 1920, anadromous fish runs had declined almost 98 percent to 12,000 per year and have remained at approximately this level for 70 years (Rinella, McKenzie, and Fuhrer, 1992). Major factors that are suspected of affecting fish in the basin are degraded habitat and loss of smolts as they migrate downstream in the Yakima and Columbia Rivers, excessive fishing pressure in the Columbia River and Pacific Ocean, low streamflows, degraded water quality, and predation by fish species that thrive in warmer water (Perry Harvester, Washington Department of Fisheries, written commun., January 1993).

Water-quality issues related to fisheries in the Yakima River Basin include (1) erosion of fine-grained soils that deposit in low- to moderate-gradient spawning beds, (2) high stream temperatures in slow-flowing, pooled reaches (low flows caused by large irrigation diversions), and (3) extensive pesticide usage that has resulted in concentrations above safe chronic-exposure levels for fish and fish predators in the main stem and agricultural return flows (Johnson and others, 1986; Confederated Tribes and Bands of the Yakama Indian Nation and others, 1990; Rinella, McKenzie, and Fuhrer, 1992).

Land Use and Sources of Organic Compounds

Major land uses in the basin include growing and harvesting timber, grazing on nonirrigated land,

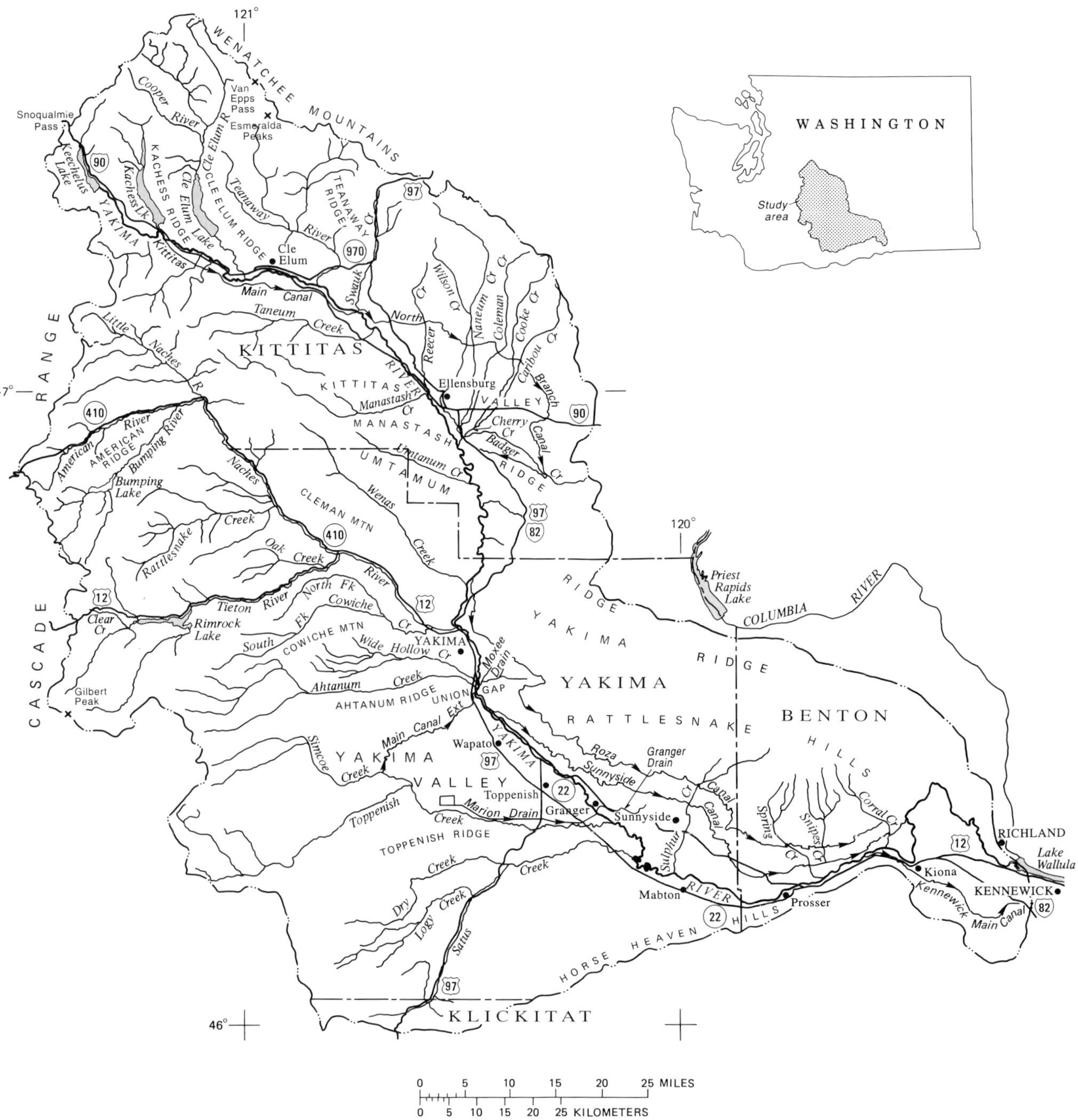

Base modified from U.S. Geological Survey 1:250,000 quadrangles

Figure 1. Drainage and topographic features in the Yakima River Basin in south-central Washington.

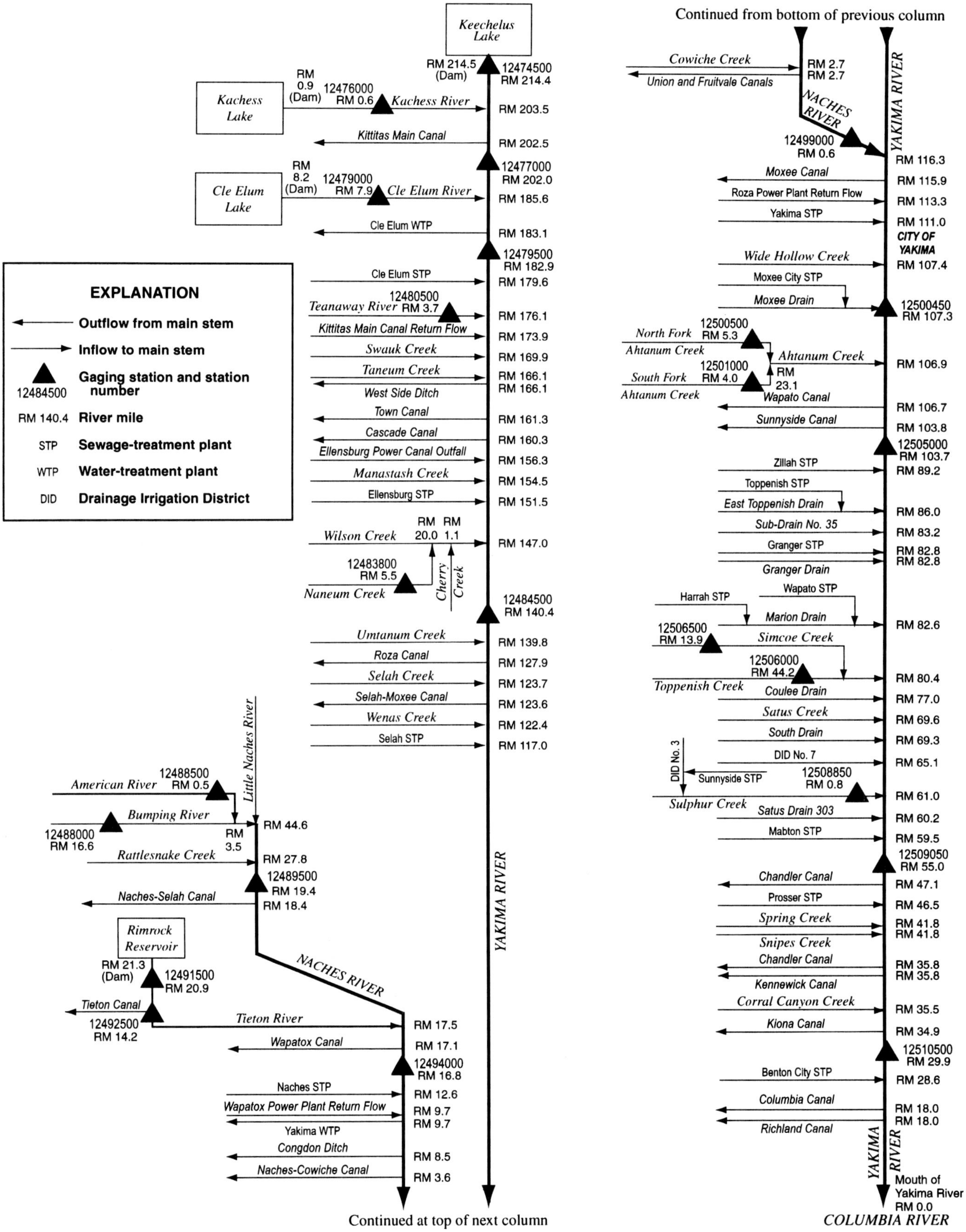

Figure 2. Schematic diagram showing selected tributaries, diversion canals, return flows, and stream-gaging stations, Yakima River Basin, Washington.

intense agriculture that requires irrigation, and urbanization (fig. 3).Even though the area covered by irrigated land (700 square miles) is smaller than the areas covered by forest (2,200 square miles) or pasture land (2,900 square miles), intense water and pesticide use for farming greatly influences water-quality conditions and beneficial uses.

The forested northern and western areas of the basin lie within the Wenatchee and Snoqualamie National Forests along the eastern slopes of the Cascade Range. Forest lands are used for wildlife habitat, timber harvesting, and recreation. About one-fourth of this area is designated as wilderness land where only nonmotorized recreation is permitted. Prior to 1985, pesticides were applied to only several hundred acres of National Forest in the Yakima River Basin. Since 1985, however, pesticides have not been applied to the federally owned land (Bill Garrigues, U.S. Forest Service, Naches, Washington, oral commun., October 1992). Some of the forested area, however, is owned by private timber companies and is receiving pesticide applications. For example, (2,4-dichlorophenoxy) acetic acid [2,4-D], carbaryl, and *Bacillus thuringiensis* commonly are used (Perry Harvester, Washington Department of Fisheries, written commun., January 1993). Rangeland is used for cattle grazing, wildlife habitat, and military training.

Urbanization/industrial and agricultural activities are sources of nonpesticide semivolatile organic compounds (including polycyclic aromatic hydrocarbons [PAHs] and phthalate esters) in the Yakima River Basin. PAHs are formed during the combustion of hydrocarbons (for example, burning of fossil fuels, forest fires—both lightning-ignited and controlled burning, burning of weeds and brush in irrigation ditches, and use of smudge pots for heating orchards) and also may be released from oil spills. Phthalate esters are plasticizers, which can occur in urban runoff, as well as in municipal- and industrial-waste discharges (table 1). Some of these esters also are used as solvents in pesticide sprays. Major population areas in the Yakima River Basin, which are potential sources of organic compounds, including semivolatile compounds, are shown in figure 3. The Yakima River Basin has a population of about 250,000 (1990); the area with the largest population is Wide Hollow Creek Subbasin with more than 80,000 people (Elaine Taylor, Yakima Valley Conference of Governments, written commun., October 1992). Virtually all of the city of Yakima is located in the Wide Hollow Creek Subbasin (fig. 3).

Crop types and acreages for agricultural areas are listed in table 2, and large areas of irrigation are shown in figure 4. In general, irrigated pasture and alfalfa-timothy hay dominate irrigated areas in Kittitas Valley, whereas orchards, pasture, corn, hops, grapes, and mint dominate the downstream areas.

Crop acreage for most of the Yakima River Basin (table 2) was determined using 1990 data compiled by irrigation districts on 41 crops (Shirley Mangold, Bureau of Reclamation's [BOR's] Crop Production Report for 1990, written commun., June 23, 1992). Data for small and older irrigation systems and for dryland agricultural activities were not included in BOR's data compilation.

Crop acreages, listed in table 2, were calculated as follows:

(1) Kittitas area—Tom Hoffmann, Kittitas County Extension Agent, provided a 1991 summary for Kittitas County (written commun., September 4, 1992);

(2) Tieton area—sum of crop acreage in the Yakima Valley Canal District and Yakima-Tieton Irrigation District for 1990 (BOR's Crop Production Report for 1990), the crop acreage in the Ahtanum Irrigation District for 1990 (Mike Tobin, North Yakima Conservation District, Yakima, Washington, written commun., April 1993), and the crop acreage in the Ahtanum unit of the Wapato Irrigation Project for 1990 (Linda Wahsise, Wapato Irrigation Project, Wapato, Washington, written commun., April 1993);

(3) East side—sum of crop acreage in the Moxee Ditch, Selah-Moxee, Terrace Heights, Union Gap, Grandview, Granger, Outlook, Snipes Mountain, Roza, and Sunnyside Valley Irrigation Districts (BOR's Crop Production Report for 1990);

(4)West side—sum of crop acreage in the Wapato Irrigation Division (BOR's Crop Production Report for 1990) and 10,000 acres of pasture in Toppenish Creek and Satus Creek Subbasins;

(5) Other areas—sum of crop acreage in the Naches-Selah Irrigation District and 1.05 times the acreage established under the "Special and Warren Act Contracts" (BOR's Crop Production Report for 1990);

(6) Total basin—sum of crop acreage in the Kittitas, Tieton, East side, West side, and other areas; and

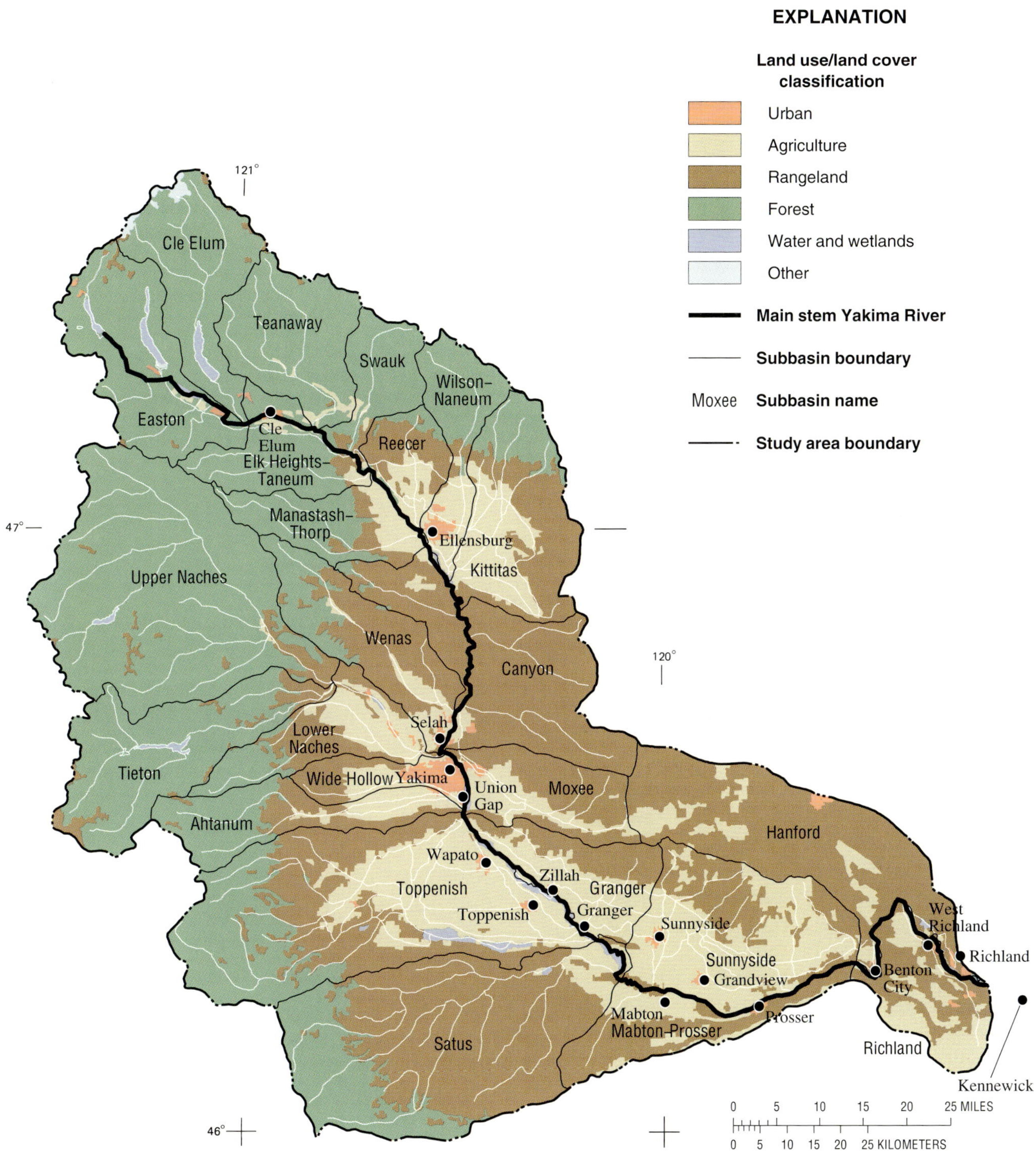

Base modified from U.S. Geological Survey 1:250,000 quadrangles

Figure 3. Land use and land cover for subbasins in the Yakima River Basin, Washington, 1981. (Source: U.S. Geological Survey, 1986; population data from Elaine Taylor, Yakima Valley Conference of Governments, written commun., October 1992.)

Subbasin	Population (1990)	Subbasin	Population (1990)	City	Population (1990)
Ahtanum	5,585	Moxee	9,630	Yakima	54,827
Canyon	2,180	Reecer	952	Kennewick	42,152
Cle Elum	197	Richland	14,759	Richland	32,315
Easton	670	Satus	865	Ellensburg	12,361
Elk Heights-Taneum	4,249	Sunnyside	32,350	Sunnyside	11,238
Granger	11,602	Swauk	181	Toppenish	7,419
Hanford	8,382	Teanaway	500	Grandview	7,169
Kittitas Wilson-Naneum	18,390 (Total)	Tieton	300	Selah	5,113
		Toppenish	24,848	Prosser	4,476
Lower Naches	10,157	Upper Naches	885	West Richland	3,962
Mabton-Prosser	7,290	Wenas	11,654	Wapato	3,795
Manastash-Thorp	1,430	Wide Hollow	80,593	Union Gap	3,120
				Granger	2,053
				Zillah	1,911
				Benton City	1,806
				Cle Elum	1,778
				Mabton	1,482

Figure 3. Land use and land cover for subbasins in the Yakima River Basin, Washington, 1981. (Source: U.S. Geological Survey, 1986; population data from Elaine Taylor, Yakima Valley Conference of Governments, written commun., October 1992.)—Continued

Table 1. Major municipal point-source dischargers, Yakima River Basin, Washington, 1987
[Effluent discharges are 1987 mean daily discharges, in million gallons per day, from files of Washington State Department of Ecology in Olympia, Washington; listing excludes several small dischargers; RM, river mile]

Yakima river mile location	Municipal waste effluent: Name	Municipal waste effluent: Effluent discharge
RM 179.6	Cle Elum	0.609
RM 151.5	Ellensburg	3.26
RM 117	Selah	.976
Naches River (RM 116.3)	Naches	.074
RM 111	Yakima[1]	12.48
Moxee Drain (RM 107.4)	Moxee	.066
RM 89.2	Zillah	.819
East Toppenish Drain (RM 86.0)	Toppenish	.956
Granger Drain (RM 82.8)	Granger	.172
Marion Drain (RM 82.6)	Wapato	.496
RM 82.6	Harrah	.039
RM 60.5	Mabton	.090
RM 59.5	Sunnyside	1.46
RM 46.5	Prosser	.62
RM 28.6	Benton City	.325

[1]Contains industrial wastes.

(7) Moxee Subbasin—in addition to being included in the East side, crop acreage for the Moxee Sub-basin was compiled separately to determine relations between pesticide application and pesticide runoff in an additional subbasin. Crop acreage was determined by using 1992 aerial photography from the former U.S. Soil Conservation Service (now the Natural Resource Conservation Service), driving the subbasin by ground vehicle to identify crops in the photographs, and using planimetry to determine the area of each field.

The largest source of synthetic organic compounds in the Yakima River Basin is the application of pesticides on agricultural crops. Accordingly, pesticide use was estimated to help evaluate the spatial distribution of pesticide concentrations in streams in the Yakima River Basin. Estimates for pesticide use in 1989 in the basin and in major agricultural areas are listed in tables 3 and 59 (at back of report), respectively. In 1989, about 180 pesticides were applied in the Yakima River Basin, of which 54 were analyzed for in this study; 43 of the 54 compounds (80 percent) were detected at trace or quantifiable concentrations.

Table 2. Estimated acreage for agricultural crops, Yakima River Basin, Washington, 1989-92

[East side, the area east of the Yakima River from Moxee Valley downstream to Benton City near Kiona including the Moxee, Granger, Sunnyside, and Whitstran areas; West side, the area west of the Yakima River downstream from Ahtanum Ridge to the city of Mabton including the Wapato, Toppenish, and Mabton areas; Other areas, small areas throughout the basin, for example, crops in the Wenas Creek Subbasin; Total basin, all of the Yakima River Basin downstream to Benton City near Kiona excluding the Kennewick Irrigation District; -, none recorded; see figure 4 for a map delineating agricultural areas; see text for source of information]

Crop	Acres of crop						
	Kittitas area	Tieton area	East side	West side	Other areas	Total basin	Moxee Subbasin[1]
Alfalfa hay	3,478	2,540	14,891	11,193	5,175	37,277	1,238
Alfalfa—timothy and other hay	28,026	-	2,968	2,387	4,510	37,891	-
Apples	1,015	19,162	24,286	10,771	14,274	69,508	5,617
Asparagus	-	-	7,298	5,164	-	12,462	-
Barley	232	590	804	553	315	2,494	-
Cherries	-	406	5,602	410	1,716	8,134	223
Field corn	480	-	16,850	7,422	509	25,261	140
Sweet corn	1,732	-	625	9,496	1,307	13,160	-
Juice grapes	-	-	16,426	3,000	235	19,661	-
Wine grapes	-	-	3,000	474	-	3,474	-
Hops	-	-	18,846	8,998	2,715	30,559	7,515
Mint	-	-	6,068	13,596	-	19,664	-
Oats	520	-	-	-	158	678	-
Pasture	47,200	5,894	22,772	20,947	10,714	107,527	6,600
Peaches, apricots, and nectarines	-	-	1,754	827	819	3,400	45
Pears	345	1,947	5,367	914	2,168	10,741	250
Plums and prunes	-	-	1,257	201	131	1,589	-
Potatoes	838	-	463	-	987	2,288	-
Other vegetables and gardens	250	285	4,019	3,987	704	9,245	100
Wheat	11,876	267	7,251	16,593	3,864	39,851	3,600
Total						454,864	

[1]Acreage included in tabulation of East side, listed in this table.

Pesticide use for each major crop was estimated by compiling the following information:

(1) active ingredients used;

(2) function of the pesticide (for example, insecticide, herbicide, or fungicide);

(3) application rate (in mass per acre per application);

(4) number of applications per year; and

(5) percentage of crop acreage to which the pesticide was applied.

This information was used to estimate an annual rate of pesticide application (mass per acre) for each active ingredient used on a specific crop (multiplication of items 3, 4, and 5 listed above). Annual pesticide use for each of the major agricultural areas was calculated by multiplying the annual rate of application (mass per acre) by the number of acres of a specific crop, and then summing the use for each pesticide applied in the area.

Information for items 1–3, listed above, was obtained from the Pacific Northwest 1992 Insect Control Handbook, Weed Control Handbook, and Plant Disease Control Handbook, prepared by the Oregon State University, Washington State University, and the University of Idaho Extension Services (1992a, 1992b, and 1992c). Compiled data from items 1–3 were sent to County extension agents in the Yakima River Basin; scientists at the Irrigated Agricultural Research and Extension Center at Prosser, Washington; and agencies involved with pesticide use, including personnel with the U.S. Army at the Yakima Firing Range, County and State transportation agencies, U.S. Forest Service, the former U.S. Soil Conservation Service, and others

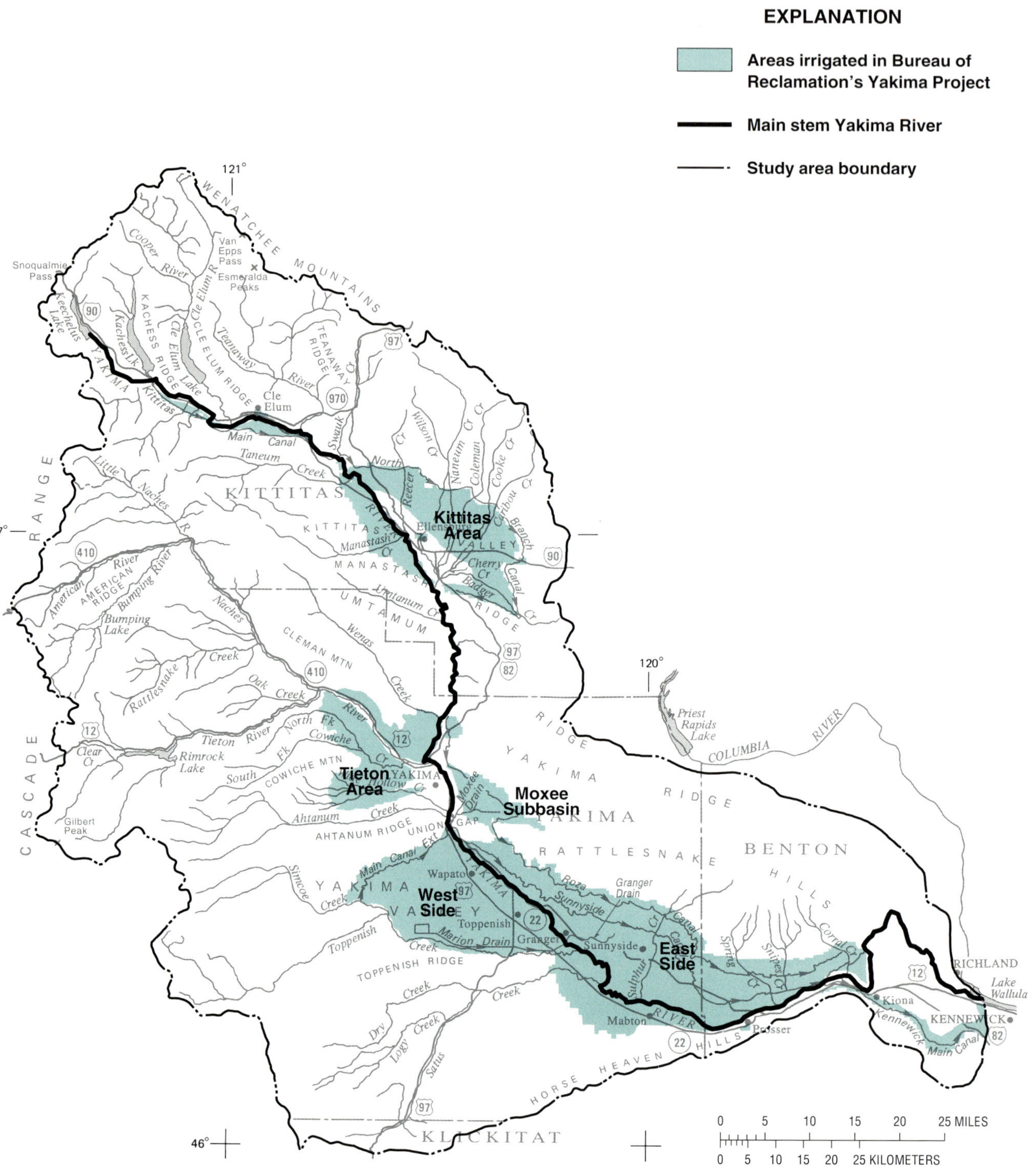

Base modified from U.S. Geological Survey 1:250,000 quadrangles

Figure 4. Location of irrigated areas in the Yakima River Basin, Washington.

Table 3. Estimate of annual pesticide use, Yakima River Basin, Washington, 1989

[Annual use, in kilograms, is for the entire basin downstream to Benton City near Kiona, excluding Kennewick Irrigation District; --, not analyzed; A, analyzed; D, detected in streams at one or more sampling stations in the basin at trace concentrations less than the minimum reporting level or at quantifiable concentrations greater than or equal to the minimum reporting level; N, not detected; 2,4-D, (2,4-dichlorophenoxy) acetic acid; EPTC, *S*-ethyl dipropylthiocarbamate; MCPA, (4-chloro-2-methyl-phenoxy) acetic acid; HCH, hexachlorocyclohexane; MSMA, sodium hydrogen methylarsonate; see text for source of information]

Compound	Annual use	Major applications	Comments
Oil	910,000	Apples, cherries, pears, peaches	--
2,4-D	260,000	Pasture, roads, apples	A, D
Malathion	160,000	Pasture, apples, grapes, cherries	A, D
Glyphosate	150,000	Apples, corn, grapes	--
Azinphos-methyl	130,000	Apples, pears, cherries	A, D
Copper	77,000	Hops, cherries, pears	--
Chlorpyrifos	73,000	Potatoes, urban	A, D
Diazinon	73,000	Hops, alfalfa, urban	A, D
Diuron	58,000	Roads, apples, asparagus	--
Carbaryl	55,000	Asparagus, apples, grapes	A, D
Norflurazon	55,000	Hops, grapes, asparagus, pears	--
Parathion	55,000	Apples, cherries, grapes	A, D
Sulfur	55,000	Cherries, peaches, hops, apples	--
Propargite	47,000	Hops, mint, apples, grapes	A, D
Metam	42,000	Potatoes	--
Monocarbamide dihydrogensulfate	42,000	Potatoes	--
Phosphamidon	36,000	Apples	A, D
Endosulfan	36,000	Apples, pears, peaches	A, D
Dicamba	36,000	Pasture, hay, asparagus	A, D
Disulfoton	30,000	Asparagus, wheat, hops	A, D
Oryzalin	29,000	Grapes, apples, roads	--
Simazine	29,000	Apples, grapes, asparagus, prunes	A, D
Captan	26,000	Corn, apples, grapes	--
4,6-Dinitro-*o*-cresol (DNOC)	26,000	Apples, pears	--
Methidathion	26,000	Hay, apples	A, D
Calcium polysulfide	24,000	Apples, pears	--
Metalaxyl	22,000	Hops	--
Ziram	22,000	Apples, pears	--
Dimethoate	20,000	Grapes, wheat, apples	A, D
Clopyralid	18,000	Pasture, hay	--
Mancozeb	18,000	Asparagus, wheat	--
Alachlor	16,000	Corn	A, D
Butylate	16,000	Corn	A, N
EPTC	16,000	Alfalfa, corn, potatoes	A, D
Terbacil	16,000	Mint, apples, alfalfa	A, N
Paraquat	15,000	Hops, grapes, apples	--
Methomyl	15,000	Grapes, asparagus	A, D
Trifluralin	15,000	Hops, asparagus	A, D
Dodine	13,000	Apples, pears	--
Dichloropropene	11,000	Potatoes	A, N
MCPA	11,000	Pasture, hay	--
Methylbromide	11,000	Urban	A, N
Zinc	11,000	Pears, cherries, grapes	--
Atrazine	9,100	Corn	A, D
Dicofol	9,100	Hops, mint	A, D
Metribuzin	9,100	Asparagus, alfalfa	A, D
Oxamyl	9,100	Apples, pears	A, N
Phorate	9,100	Hops, potatoes, corn	A, D
Phosmet	9,100	Apples, pears	--
Thiram	9,100	Corn, wheat	--
Metiram	8,800	Asparagus	--
Endothal	7,300	Hops	--
Lindane (γ-HCH)	7,300	Wheat, barley	A, D
Napropamide	7,300	Grapes, mint, asparagus	--
Carboxin	5,800	Wheat, corn	A, N
Dalapon	5,800	Field corn	--
Chlorsulfuron	5,500	Roads, wheat	--
Cyanazine	5,500	Field corn, wheat	A, D

Table 3. Estimate of annual pesticide use, Yakima River Basin, Washington, 1989—Continued

Compound	Annual use	Major applications	Comments
Fonofos	5,500	Asparagus, corn	A, D
Hexazinone	5,500	Alfalfa, hay, oats	A, D
Pronamide	5,500	Alfalfa, grapes, apples	--
Sulfuryl fluoride	5,500	Roads, urban	--
Methoxychlor	4,400	Alfalfa, asparagus	A, D
Xylene	4,000	Aquatic application	A, D
Benefluran	3,600	Alfalfa, urban	--
Boron	3,600	Cherries, urban	--
Dichlobenil	3,600	Apples, aquatics, roads, urban	--
Ethephon	3,600	Apples	--
Pendimethalin	3,600	Corn, grapes, apples	--
Triclopyr	3,600	Roads, pasture	--
Amitraz	3,300	Pears	--
Carbofuran	3,300	Corn, grapes	A, D
Iron (micronutrient)	3,300	Grapes, hops	--
Metolachlor	3,300	Corn	A, D
Naled	3,300	Hops, grapes, alfalfa, seed	--
Chinomethionat	2,900	Pears, apples	--
Oxyfluorfen	2,900	Grapes, mint, apples, cherries	--
Propachlor	2,900	Corn	A, N
Methylparathion	2,200	Apples, pears	A, N
Iprodione	1,900	Hops, potatoes	--
Benomyl	1,800	Grapes, apples	--
Bromoxynil	1,800	Wheat, alfalfa	--
Dinocap	1,800	Pears, apples	--
Rotenone	1,800	Apples, cherries	--
Permethrin	1,800	Corn, alfalfa, asparagus	A, D
BEE (bee attractant)	1,600	Pears	--
Bentazone	1,600	Corn, mint	--
Bromacil	1,600	Roads	A, D
Linuron	1,600	Asparagus	A, D
Oxytetracycline	1,600	Pears, apples	--
Acrolein	1,500	Aquatic application, urban	--
Gibberellic acid (GA)	1,500	Apples, cherries	--
Myclobutanil	1,500	Apples, grapes	--
1-Naphthylacetic acid (NAA)	1,500	Apples, pears	--
Picloram	1,500	Firing range, wheat, pasture	A, D
Triadimenol	1,300	Corn, wheat	--
Acephate	1,100	Urban, mint	--
Ethion	1,100	Apples	A, D
Ethoprop	1,100	Potatoes	--
Fenarimol	1,100	Apples, cherries	--
Methamidophos	1,100	Potatoes	A, N
Sulfometuron methyl	1,100	Roads	--
Terbufos	1,100	Corn	A, N
Fenbutatin oxide	910	Pears, apples	--
Thiophanate-methyl	910	Potatoes	--
Tebuthiuron	910	Pasture	--
Chlorothalonil	730	Mint, potatoes	A, D
Formetanate	730	Apples, pears	--
Triadimefon	730	Apples, grapes	A, D
Chloropicrin	550	Potatoes, urban	--
Maleic hydrazide	550	Potatoes	--
Maneb	550	Potatoes	--
Oxydemeton-methyl	550	Apples, mint	--
Sethoxydim	550	Asparagus, grapes	--
Thifensulfuron	550	Wheat	--
Trichlorfon	550	Corn	--
Mevinphos	470	Apples, alfalfa, corn	A, D
Streptomycin	470	Grapes, apples, prunes	--
Byclobutanil	400	Grapes	--
Amitrole	360	Urban	--

Table 3. Estimate of annual pesticide use, Yakima River Basin, Washington, 1989—Continued

Compound	Annual use	Major applications	Comments
Cytokinins	360	Apples	--
Fluazifop-s	360	Asparagus, grapes	--
Mecoprop (MCPP)	360	Urban	--
Metsulfuron	360	Oats, wheat	--
Propyzamide	360	Apples	--
Pyrethrins	360	Apples, urban	--
Ryania	360	Cherries	--
Strychnine	360	Apples, cherries	--
Chlorophacinone	330	Apples	--
Cryolite	330	Apples	--
Diclofop	330	Wheat	--
Difenzoquat	330	Wheat	--
Diphacinone	330	Apples	--
Lactic acid	330	Apples	--
Phosalone	330	Apples	--
Vinclozolin	330	Apples	--
Diatomaceous earth (DE)	290	Pears	--
Imazapyr	260	Firing range, roads	--
1-Naphthylacetamide (NAD)	260	Apples	--
Propoxur	260	Urban	A, N
Triallate	260	Wheat, barley	--
Cyfluthrin	250	Pears	--
Esfenvalerate	220	Corn, pears	--
Prometon	220	Urban, aquatic application	A, D
Fenvalerate	150	Pears	--
Abamectin	130	Pears	--
Avermectin	130	Pears	--
Imazamethabenz	130	Wheat, barley	--
Dichlorvos	110	Urban	--
Dinoseb[1]	110	Urban	--
Trithion (carbophenothion)	110	Apples	A, D
Piperonyl butoxide	100	Urban, prunes	--
Bacillus thuringiensis	90	Apples	--
Ethylan	90	Pears	--
Manganese	90	Oats	--
Propetamphos	90	Urban	--
Primisulfuron	90	Corn	--
MSMA	70	Urban	--
Tribenuron	70	Wheat	--
Triforine	70	Apples, pears	--
Bendiocarb	60	Urban	--
Cacodylic acid	40	Urban	--
Fluridone	40	Aquatic application	--
Oxadiazon	40	Urban	--
Thiabendazole (TBZ)	40	Wheat	--
Dacthal (DCPA)	30	Roads, onions	A, D
Chlorflurenol	30	Urban	--
Fosamine	30	Aquatic application	--
Quintozene (PCNB)	30	Barley, oats	--
Fenoxaprop	20	Wheat	--
Sodium trichloroacetate(TCA)	15	Roads	--
Nicosulfuron	15	Corn	--
Formulated hydrochloride	15	Peaches	--
Demeton	3	Barley	A, D[2]
Imazalil	2	Barley	--

[1]Dinoseb was not sold in 1989.

[2]Demeton-S was detected. Demeton-S and demeton-O were not sold in 1989.

familiar with use in rangelands. These agencies were asked to (1) review and correct, as needed, the compiled data on active ingredients and application rates, (2) provide an estimate for number of applications per year, and (3) provide an estimate of the percentage of crop acreage to which the pesticide was applied in 1989. Ms. Elizabeth H. Beers (Tree Fruit Research and Extension Center, Wenatchee, Washington) provided results of a survey of apple and pear growers for the Yakima River Basin (Elizabeth H. Beers, written commun., February 20, 1992; Beers and Brunner, 1991). Information also was obtained from the Washington Agricultural Statistics Service on chemical usage for apples, sweet cherries, grapes, peaches, pears, prunes, and plums. When pesticide-use information for a crop was available from more than one source, an average value was used for these estimates. An example calculation for pesticide-use estimates for hops is shown in table 4.

Scientists from the Irrigated Agriculture Research and Extension Center provided information on the use of pesticides in ponds and canals. Pesticide use in urban areas was estimated from data collected in the Puget Sound Basin for the EPA in 1988 (Tetra Tech Inc., 1988).

Soils and Erosion Potential

The movement of suspended sediment, including eroded soils, in streams is an important factor in the transport and fate of chemicals in the environment. Many chemicals, especially hydrophobic (low-water solubility) organic compounds, are predominantly associated with soils, suspended sediment, and bed sediment. A major source of these compounds in the Yakima River Basin is the present or historical application of pesticides to agricultural crops. Sediment and associated pesticides erode from the land and may be transported in the water column for long distances or may settle out as streambed sediment for a period of time. Contaminated sediment in the water column or streambed may affect aquatic biota, including benthic fauna (U.S. Environmental Protection Agency, 1988b).

Erosion is caused by flowing water and involves (1) suspension of loose sediment, (2) disaggregation of resistant soil surfaces by suspended-sediment and water impaction, and (3) mutual wear of colliding sediment particles in transport (Guy, 1978). As water falls on the soils, either from irrigation or precipitation, infiltration is reduced, creating a sheet of flowing water. The water then begins to flow in rills (rivulets) and larger channels. In the case of rill irrigation, water is channelled directly into rills between the rows of crop. When compared to sheet flow, rill or channel flow is confined to a small area of resistance and results in increased flow velocities and erosion (Guy, 1978). Rill irrigation is used extensively on row crops throughout the Yakima River Basin.

Factors that affect erosion in the Yakima River Basin are listed in table 5. Soil groups in selected irrigated areas are shown in figure 5, and data on soil characteristics for these areas are listed in table 6. Soils with large erodibility factors on steep slopes have a high probability for water erosion. Soils with a

Table 4. Estimate of pesticide use for hops, Yakima River Basin, Washington, 1989

Active ingredient	Biological activity	Application, in kilograms per hectare per application	Number of applications per year	Portion of crop receiving application, in percent	Pesticide use, in kilograms per hectare per year
Copper	Fungicide	2.2	2	95	4.2
Diazinon	Insecticide	2.2	3	65	4.3
Dicofol	Insecticide	1.1	2	15	.33
Disulfoton	Insecticide	3.5	1	5	.18
Endothal	Herbicide	.74	1	80	.59
Malathion	Insecticide	1.4	1	20	.28
Metalaxyl	Fungicide	2.2	1	80	1.8
Naled	Insecticide	1.1	1	10	.11
Norflurazon	Herbicide	3.4	1	90	3.1
Oxythioquinox	Insecticide	.57	1	< 1	<.006
Paraquat	Herbicide	.52	2	40	.42
Parathion	Insecticide	1.1	2	50	1.1
Phorate	Insecticide	2.7	1	20	.54
Propargite	Insecticide	1.4	2.3	70	2.2
Trifluralin	Herbicide	.89	1	50	.44

Table 5. Factors affecting water erosion and transport of sediment, Yakima River Basin, Washington
[Modified from Guy, 1978]

Major factors	Influences of factors on soil erosion
Precipitation and irrigation water	Nature, amount, and intensity of precipitation or irrigation water affects the rate of runoff (erosive energy) after the infiltration capacity is reached. For example, hops and orchards receive about the same amount of water in the Yakima River Basin; however, hops are irrigated from May through August, and orchards are irrigated from March through October. Consequently, hops receive a higher intensity of irrigation, which increases runoff rate and enhances erosion (Ray Wondercheck, Soil Conservation Service, Yakima, Washington, oral commun., September 1992).
Temperature	Alternate freezing and thawing of soils decreases cohesion and facilitates transport (mean-monthly temperatures at Kennewick, Washington, near the mouth of the Yakima River range from 31 to 77 degrees Fahrenheit).
Topography	Degree and length of slope affect erosion. Increased slope increases the energy of flow as determined by gravity; increased length of slope increases stream velocity, turbulence, and transport capacity of the water (see table 6 for slopes of selected irrigated areas in the Yakima River Basin). A small increase in gradient can greatly affect a stream's ability to transport sediment.
Soil character	Easily erodible soils include unprotected silts, sands, and other loosely consolidated material. The following soil properties affect erosion: permeability rate (infiltration); porosity (water-holding capacity—affects infiltration and runoff rates); moisture content (reduces cohesion); soil swelling; and grain size, shape, and specific gravity (determines force needed for dislodgment of sediment particles). Soil erodibility factors for selected soils in irrigated areas in the Yakima River Basin are listed in table 6.
Land and water use and soil cover	Crop types affect soil cultivation and irrigation practices. For example, hops and other row crops often require frequent tillage, which loosens the soil aggregates, removes the vegetative cover crop (creating barren soils), and makes the soil vulnerable to erosion, especially if rill irrigation is employed. Conversely, many orchards in the Yakima River Basin have a cover crop and are sprinkler irrigated—practices which minimize erosion.
Hydrologic connection between eroding land and streams	The hydrologic connection between the eroding areas and the receiving stream can greatly affect sediment transport. The distance, type of channel conveyance, density of channel system, slope of channel, roughness of channel bottom, channel velocity, and channel turbulence are important factors that affect sediment transport to receiving streams. Channeling agricultural-return flow through a sedimentation pond or wetland can greatly reduce concentrations of suspended sediment in the water column.

large erodibility factor on a relatively flat slope have a slight to moderate probability for erosion.

Concentrations of suspended sediment in surface water throughout the basin during a synoptic sampling conducted June 25–30, 1989 (irrigation season), are shown in figure 6 and listed in table 60 (at back of report). Increased concentrations in agricultural return flows generally reflect steeper subbasin slopes and a higher soil-erodibility factor (table 6). The following characterizations describe soil and stream-sediment conditions in selected irrigated areas (fig. 5) in the Yakima River Basin:

(1) East-side tributaries of the Yakima River downstream from the Naches River (includes Moxee Subbasin and Roza and Sunnyside Irrigation Divisions): Owing to the erosive soil characteristics of the Warden-Esquatzel Association, steeper subbasin slopes (mostly 2–15 percent), rill irrigation practices, and close proximity of the irrigated fields to the Yakima River (mostly less than 5 miles), suspended-sediment loads from the East-side tributaries in June 1989 were the largest in the basin. The instantaneous load was 317,000 kg/d (kilograms per day), and the yield was 490 kg/d/km^2 (kilograms per day per square kilometer) of irrigated crop.

(2) West-side tributaries of the Yakima River (Yakima Indian Reservation): Owing to small subbasin slopes (mostly 0–2 percent, except 0–8 in Warden-Shano Association), the suspended-sediment load and yield from West-side tributaries in June 1989 were smaller than those for the East-side tributaries. The instantaneous load was 59,500 kg/d, and the

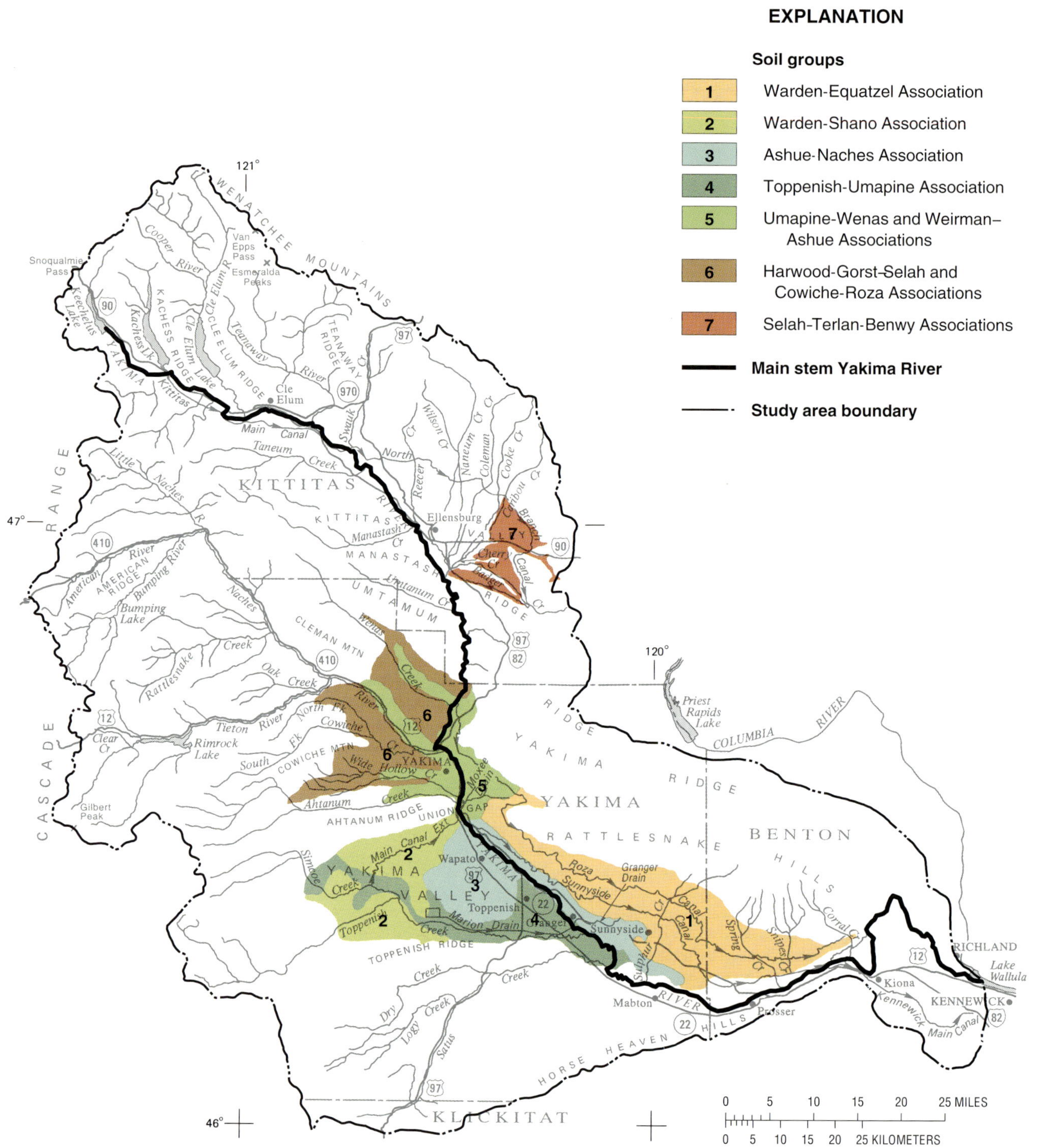

Base modified from U.S. Geological Survey 1:250,000 quadrangles

Figure 5. Soil groups in selected irrigated areas in the Yakima River Basin, Washington.

Table 6. Soil data for selected irrigated areas, Yakima River Basin, Washington

[Data are from Rasmussen (1976), Lenfesty and Reedy (1985), and Bain (1990); soil-erodibility factor is a measure of the susceptibility of a soil to sheet and rill erosion—the larger the factor, the more susceptible the soil is to water erosion; as subbasin slope increases, the hazard to water erosion increases; location of soil groups is shown in figure 5]

Soil group number	Soil group	Slope, in percent	Surface layer	Hazard to water erosion	Soil-erodibility factor
	East side of Yakima River downstream from Naches River (includes Upper Moxee Subbasin and Roza and Sunnyside Irrigation Divisions)				
1	Warden-Esquatzel Association[1]				
	Warden series	0–30 (mostly 2–15 percent)	Silt loam	Slight to high (mostly moderate to high)	0.43–0.49
	Esquatzel series	0–5 (mostly 0–2 percent)	Silt loam	Slight to moderate	.43
	West side of Yakima River downstream from Yakima, Washington (Yakima Indian Nation)				
2	Warden-Shano Association				
	Warden series	0–30 (mostly 0–8 percent)	Silt loam	Slight to high (mostly slight to moderate)	.43–.49
	Shano series	0–30 (mostly 2–8 percent)	Silt loam	Slight to high (mostly slight to moderate)	.49
3	Ashue-Naches Association and Weirman series				
	Ashue series	0–5 (mostly 0–2 percent)	Loam or gravelly loam	Slight to moderate (mostly slight)	.37
	Naches series	0–5 (mostly 0–2 percent)	Loam	Slight to moderate (mostly slight)	.32
	Weirman series	0–5	Fine and gravelly sandy loam	Slight to moderate	.24–.28
4	Toppenish-Umapine Association				
	Toppenish series	0–5 (mostly 0–2 percent)	Silt loam or silt clay loam	None to moderate (mostly none to slight)	.37
	Umapine series	Mostly 1 percent	Silt loam	Slight	.43
	Tieton area (west of Yakima River near Yakima, Washington, includes Naches-Selah Irrigation District and Tieton Irrigation Division)				
5	Umapine-Wenas and Weirman-Ashue Associations				
	Umapine series	0–5 (mostly 0–2 percent)	Silt loam	Slight to moderate (mostly slight)	.43
	Wenas series	0–2	Silt loam	Slight	.37
	Weirman series	0–5	Fine and gravelly sandy loam	Slight	.24–.28
6	Harwood-Gorst-Selah and Cowiche-Roza Associations				
	Harwood series	2–30 (mostly 2–15 percent)	Loam	Moderate	.37
	Gorst series	2–30 (mostly 2–15 percent)	Loam	Moderate to high (mostly moderate)	.32–.37
	Selah series	2–30 (mostly 2–15 percent)	Silt loam	Moderate to high (mostly moderate)	.49
	Cowiche series	2–30 (mostly 8–15 percent)	Loam	Moderate to high (mostly moderate)	.43
	Roza series[2]	5–60 (mostly 15–60 percent)	Clay loam	Moderate to very high (mostly high to very high)	.32
	Kittitas area (southeast corner of Kittitas Valley, includes Cherry Creek and Wilson Creek upstream from Cherry Creek)				
7	Selah-Terlan-Benwy Association				
	Selah series	Row crops on hillside with slopes of 8–15 percent (some in excess of 15 percent)	Silt loam	Moderate to high	.49

[1]Includes some Quincy-Hezel and Umapine-Wenas Associations.
[2]Primarily used for rangeland and wildlife habitat.

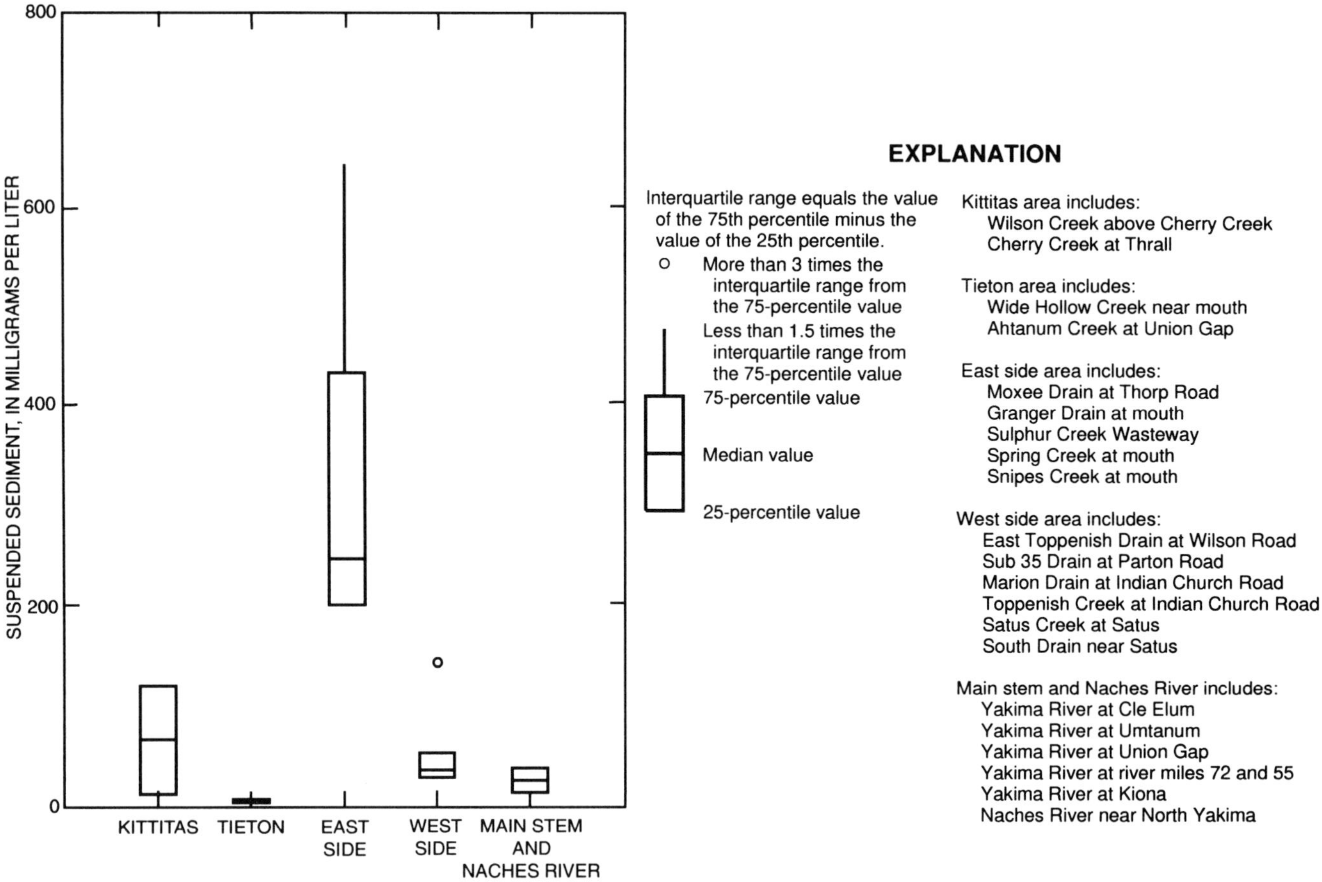

Figure 6. Suspended-sediment concentrations at selected stations, Yakima River Basin, Washington, June 25–30, 1989.

yield was 130 kg/d/km^2 of irrigated crop. Agricultural runoff from the Warden-Shano Association flows 5 or more miles through the irrigation system over relatively flat land (0–2 percent slope) to the Yakima River. These channel-conveyance characteristics tend to increase sediment deposition in the channels and reduce sediment transport to the Yakima River. Furthermore, agricultural return flow in Toppenish Creek flows through wetlands along the western and eastern borders of the Toppenish National Wildlife Refuge and then through a wider slow-moving river reach just southeast of Toppenish near Highway 22. The wetlands and slow-moving reach act as sediment-detention ponds, further reducing sediment loads to the river. In June 1989, the largest instantaneous load of suspended sediment from the West-side tributaries was measured in South Drain at 29,000 kg/d (table 60, at back of report). This load was more than twice the load measured in the other West-side tributaries and can be attributed to rill irrigation and frequent tillage of hop fields located along thc lower reaches of South Drain (Dave Myra, U.S. Soil Conservation Service, Toppenish, Washington, oral commun., September 1992).

(3) Moxee Subbasin: In addition to highly erosive soils on steep slopes in the basin, rill irrigation and frequent tillage of about 7,500 acres of hops accounted for much of the sediment load in June 1989. The instantaneous load was 69,000 kg/d, and the yield was 670 kg/d/km^2 of irrigated crop (Ray Wondercheck, U.S. Soil Conservation Service, Yakima, Washington, oral commun., September 1992). Erosion from orchards in the subbasin is minimal because about 95 percent of the orchards are sprinkler irrigated and have cover crops.

(4) Tieton Area (includes Naches-Selah Irrigation District and Tieton Irrigation Division): The suspended-sediment load and yield were small because agricultural crops primarily consist of

sprinkler-irrigated orchards with cover crops. In June 1989, the instantaneous load was 1,150 kg/d, and the yield was 9 kg/d/km^2 of irrigated crop.

(5) Kittitas Area (includes Cherry Creek and Wilson Creek upstream from Cherry Creek): Irrigated agriculture primarily consists of pasture and hay, which are not conducive to increased erosion. In June 1989, however, the instantaneous suspended-sediment load in Cherry Creek was 37,000 kg/d compared with 4,600 kg/d in Wilson Creek. Increased erosion in Cherry Creek can be attributed to the irrigation of row crops (potatoes and corn) on relatively steep slopes in Badger Pocket, located in the southeast corner of Kittitas Valley. Most of the row crops in Kittitas Valley are located in or near Badger Pocket. The soil in Badger Pocket consists of the Selah-Terlan-Benwy Association, a highly erodible soil (Bain, 1990) on slopes of 8–15 percent. Badger Creek Wasteway flows into Wipple Creek Wasteway and then into Cherry Creek. Wipple Creek Wasteway is located several hundred feet upstream from the water-quality sampling station on Cherry Creek. When samples were collected from Cherry Creek during irrigation season during 1987–90, Wipple Creek Wasteway generally was more turbid than Cherry Creek. In addition, tributaries to Cherry Creek flow through feedlots and corrals that are conducive to erosion (Perry Harvester, Washington Department of Fisheries, written commun., January 1993).

Previous Studies of Pesticides

Prior to 1985, most water-quality data on organic compounds in the Yakima River Basin were collected from the Yakima River at Kiona near the downstream terminus of the basin (fig. 1). More than 80 percent of all determinations for pesticides, PAHs, and volatile organic compounds were reported as being below the minimum reporting levels. For several compounds, however, historical minimum reporting levels were 10 or more times larger than water-quality criteria for freshwater aquatic life (U.S. Environmental Protection Agency, 1986) and more recent reporting levels (Rinella, McKenzie, Crawford, and others, 1992; Johnson and others, 1986; Hopkins and others, 1985). There also is a lack in the historical record of data on pesticides that are commonly used today.

Compounds that consistently were detected in the water column, bed sediment, and (or) tissues of resident fish during 1968–85 were aldrin, 2,4-D, DDT (4,4'-dichlorodiphenyltrichloroethane), DDE (4,4'-dichlorodiphenyldichloroethylene), DDD (4,4'-dichlorodiphenyl-dichloroethane), diazinon, dieldrin, endrin, heptachlor, heptachlor epoxide, lindane, PCB (polychlorinated biphenyls), and (2,4,5-trichlorophenoxy) acetic acid [2,4,5-T] (Rinella, McKenzie, and Fuhrer, 1992; Johnson and others, 1986). The highest concentrations of organic compounds were detected in the lower Yakima River Basin downstream from the city of Yakima, where stream quality is controlled by agricultural return flow during irrigation season. Concentrations of several compounds (aldrin, dieldrin, endosulfan I, DDT+DDE+DDD, endrin, parathion, and PCB) in water exceeded chronic-toxicity water-quality criteria for freshwater aquatic life; however, concentrations of all compounds were below acute-toxicity criteria (U.S. Environmental Protection Agency, 1986).

Data collected in 1979 and 1984 revealed that fish from the lower Yakima River Basin downstream from the city of Yakima had high concentrations of PCBs and DDT+DDE+DDD (Hopkins and others, 1985). A liver sample from a northern squawfish from the Yakima River at Kiona had a PCB concentration of 2 μg/g (micrograms per gram), wet weight; another liver sample from a northern squawfish from the Yakima River at Birchfield Drain (also called Moxee Drain) had a concentration of 26 μg/g, wet weight (Hopkins and others, 1985). In 1985, concentrations of DDT+DDE+DDD in resident whole fish (Johnson and others, 1986) exceeded the guideline of 1 μg/g, wet weight, for the protection of fish predators recommended by the National Academy of Science-National Academy of Engineering (NAS-NAE) Committee on Water Quality Criteria (1973). In whole fish collected from the Yakima River Basin from 1980 to 1983, concentrations of DDT+DDE+DDD ranged from 0.58 to 5.1 μg/g, wet weight. Comparison of these concentrations in fish from the Yakima River Basin to data collected by U.S. Fish and Wildlife Service (USFWS) in their National Contaminant Biomonitoring Program (NCBP) indicates that resident fish from the lower Yakima River Basin have among the highest concentrations of DDT+DDE+DDD in the Nation (Schmitt and others, 1990).

ASSESSMENT METHODS

The focus of the NAWQA study in the Yakima River Basin was to assess surface-water-quality conditions at a regional scale, which generally entailed characterizing persistent water-quality conditions in stream reaches that extend 25 or more river miles (RM). In contrast, a local-scale study, which was beyond the scope of this study, would attempt to describe water-quality patterns within shorter reaches of perhaps several hundreds of yards or less. Although the local scale was not the focus of this assessment, data from this study may reveal several, previously unknown, local-scale concerns about human health and ecosystems.

The occurrence, magnitude, and spatial distribution of concentrations of hydrophobic and hydrophilic pesticides and other organic compounds in water and suspended sediment were examined initially by collecting data from eight stations during different seasons and hydrologic periods. Results from this temporal sampling were used to select a time period when pesticide concentrations and loads were large for collecting synoptic data from a larger number of stations throughout the basin. Synoptic data were used to determine the occurrence, magnitude, spatial distribution, and major sources of pesticides along the main stem of the Yakima River, upstream and downstream from agricultural activities. Also sampled were large tributaries (including the Naches River), background stations in forested areas, and the mouths of major agricultural return flows (tributaries that receive agricultural return drainage).

In bed sediment and aquatic biota, concentrations of organic compounds were determined by collecting data in a variety of areas associated with different land- and water-use activities throughout the basin (for example, subbasins influenced by forest activities, urbanization, and several types or combinations of irrigated crops). Bed-sediment samples were collected during relatively uniform streamflow conditions during late irrigation season and (or) fall months to represent influences from recently transported pesticides (sorbed and dissolved phases). Aquatic-biota samples generally were collected in the fall about 2 weeks after irrigation season during a period of low streamflows. Low flows facilitated fish collection because the fish congregated and were easier to collect in the smaller stream-channel areas. In addition, streams were accessible for wading with electrofishing gear during the low flows. In this study, the seasonal variability of concentrations of pesticides and other organic compounds in bed sediment and aquatic biota was not determined.

The specific sampling plan (sampling purpose and scope) and types of analyses for water, bed-sediment, soil, and aquatic-biota samples from the Yakima River Basin are outlined in table 7. Locations of the water, bed-sediment, soil, and aquatic-biota stations are given in table 8 and shown in figure 7. A listing of the biological activity of pesticides that were analyzed is shown in table 9.

The occurrence, transport, and fate of organic compounds in the environment is determined by physical, chemical, and (or) biological characteristics of the compounds, soils, stream sediment, stream water, and aquatic biota. Additional factors that influence transport and fate include irrigation and pesticide-application practices. The movement of these compounds to streams may occur by direct inflow (point-source discharges), direct application of herbicides to irrigation canals, direct application and drift from aerial applications on crops and forests, and erosion of contaminated soils. Physical, chemical, and biological properties that influence transport and fate of organic compounds are listed in table 61 (at back of report). Important properties include water solubility, sediment or soil sorption, and persistence in soil (half-life). Actual solubility, sorption coefficients, and half-lives can vary widely depending on station-specific soil and water characteristics; therefore, values listed in table 61 should be used only as relative indicators for predicting the occurrence and fate of these compounds in the environment.

Water solubility is a measure of the tendency for a compound to dissolve in pure water. Compounds with solubilities larger than 30 mg/L (milligrams per liter) are likely to wash off crop foliage and soils and leach into surface-water runoff or ground water (U.S. Environmental Protection Agency, 1988a). Compounds with solubilities between 1 and 30 mg/L are transitional and may dissolve into ground water or surface-water runoff, depending on other controlling factors—for example, irrigation practices; amount and frequency of pesticide application; intensity of storm events; and physical, chemical, and biological properties of the soil and water (Becker and others, 1989). Compounds with solubilities less than 1 mg/L are likely to sorb to the soils.

Table 7. Sampling plan for collecting organic-compound samples, Yakima River Basin, Washington, 1987–91

[See table 8 for a listing of the chemical families analyzed at each station and table 9 for a listing of the compounds in each chemical family; DDT+DDE+DDD, 4,4'-dichlorodiphenyltrichloroethane (DDT) plus 4,4'-dichlorodiphenyldichloroethylene (DDE) plus 4,4'-dichlorodiphenyldichloroethane (DDD)]

Sampling medium	Sampling period	Number of stations	Chemical family analyzed	Purpose of sampling
Bed sediment	August-November 1987	36	Organochlorines Semivolatiles	Two main-stem samples, one urban-runoff sample, and seven composited samples (four to five stations per composite) representing seven different crop types were collected to examine land-use effects on bed-sediment chemistry.
Bed sediment	August-September 1988	8	Organochlorines Semivolatiles	One main-stem station, one pristine station, one urban station, and five agricultural-return flows were sampled to compare organic-compound concentrations in bed sediment and in the water column.
Bed sediment	September 1988	4	Organochlorines Semivolatiles	Stations were resampled and individually analyzed to examine variability in agricultural-return flows that drain apple orchards. Preliminary results from the 1987 sampling showed that drains from apple orchards (one composited sample of five orchards) had higher concentrations of organic compounds in bed sediment than those concentrations in drains from six other crop types.
Bed sediment	1989–90	16	Organochlorines Semivolatiles	Stations were sampled to compare organic-compound concentrations in bed sediment and aquatic biota.
Soil	July 1989	3	Organochlorines Organophosphorus	Soil samples were collected from the A and B horizons at several locations in Moxee Subbasin. Moxee Subbasin was selected because relatively high concentrations of DDT+DDE+DDD were observed in Moxee Drain in 1988. Hop fields were selected because surface-water drainage from these fields is laden with sediment and may be transporting organic compounds that are sorbed to the sediment. An apple orchard was selected because preliminary results from the 1987 sampling showed that drains from apple orchards (one composited sample of five orchards) had higher concentrations of organic compounds in bed sediment than those concentrations in drains from six other crop types.
Unfiltered water	May, June, July, August, November 1988, and March 1989	8	Organochlorines Organophosphorus Triazines Carbamates Chlorophenoxy acids Volatiles Phenols and cresols	One main-stem station, one pristine station, one urban station, and five agricultural-return flows were sampled to examine seasonal variability in concentrations of organic compounds in unfiltered-water samples. Three storm samples also were collected from the main-stem station, Yakima River at Kiona, in December 1989.

Table 7. Sampling plan for collecting organic-compound samples, Yakima River Basin, Washington, 1987–91—Continued

Sampling medium	Sampling period	Number of stations	Chemical family analyzed	Purpose of sampling
Unfiltered water	July 1988	18	Organochlorines Organophosphorus Triazines Carbamates Chlorophenoxy acids Phenols and cresols	Five main-stem stations, one pristine station, one urban station, the Naches River, and ten major agricultural-return flows were sampled synoptically (within a 4-day period) during peak irrigation to determine the spatial distribution of concentrations and major sources of organic compounds in unfiltered-water samples.
Filtered and suspended phases (separate analyses of each phase)	June 1989	29	Organochlorines Organophosphorus Triazines Carbamates Chlorophenoxy acids	The main stem, major tributaries, and major agricultural-return flows were sampled synoptically (within a 6-day period) to determine transport characteristics of organic compounds in the Basin during peak irrigation. In addition, two drinking-water wells were sampled in the vicinity of intense agricultural activity to explore surface- and ground-water relations. Large sample volumes were collected, processed, and analyzed to reduce minimum reporting levels.
Filtered and suspended phases (separate analyses of each phase)	May, June, July, September 1991	4	Organochlorines Organophosphorus	One main-stem station and three major agricultural-return flows were sampled to examine the variability of filtered- and suspended-phase transport of organic compounds during irrigation season.
Ground water	January 1989	3	Organochlorines Organophosphorus	Three shallow wells were sampled near the mouth of Moxee Drain to explore surface- and ground-water relations.
Aquatic biota	May 1989	4	Organochlorines Polycyclic aromatic hydrocarbons	A preliminary field test was conducted to provide estimates of time and resources required for collection and analysis of contaminants in tissues.
Aquatic biota	October-November 1989	19	Organochlorines Polycyclic aromatic hydrocarbons	Biota from reference stations and potentially-contaminated stations were sampled to determine the occurrence of contaminants in tissues.
Aquatic biota	October-November 1990	25	Organochlorines Polycyclic aromatic hydrocarbons	Biota were sampled to provide a spatial distribution of contaminants in tissues.

Table 8. Sampling locations for pesticides and other organic compounds in water, bed sediment, soil, and aquatic biota, Yakima River Basin, Washington, 1987–91

[Latitude and longitude are reported in degrees, minutes, and seconds; Carb, carbamates and other compounds; Chlor, chlorophenoxy-acid herbicides; OC, organochlorine compounds; OP, organophosphorus compounds; TR, triazine herbicides; V, volatile organic compounds; PC, phenols and cresols; SV, semivolatile organic compounds; PAH, polycyclic aromatic hydrocarbons; map reference numbers correspond to figure 7]

Map reference number	Station number	Station name	Latitude	Longitude	Chemical families analyzed at each station
		Water samples			
2	12478200	Cooper River at Salmon LaSac near Roslyn	47 24 29	121 06 11	Carb, Chlor, OC, OP, TR
3	12478300	Cle Elum River above Cle Elum Lake near Roslyn	47 21 19	121 06 22	Carb, Chlor, OC, OP, TR
4	12479500	Yakima River at Cle Elum	47 11 35	120 56 55	Carb, Chlor, OC, OP, TR
11	12484100	Wilson Creek above Cherry Creek at Thrall	46 55 35	120 30 01	Carb, Chlor, OC, OP, TR
13	12484480	Cherry Creek at Thrall	46 55 34	120 29 51	Carb, Chlor, OC, OP, TR
14	12484500	Yakima River at Umtanum	46 51 46	120 28 44	Carb, Chlor, OC, OP, TR
15	12484550	Umtanum Creek near mouth at Umtanum	46 51 27	120 29 46	Carb, Chlor, OC, OP, TR
19	12496510	Pacific Power & Light Company Wasteway	46 41 44	120 39 11	Carb, Chlor, OC, OP, TR, V
20	12496511	City of Yakima-Finish Water	46 41 43	120 39 10	Carb, Chlor, OC, OP, TR, V
21	12496550	Buckskin Slough below Gleed Ditch near Gleed	46 38 01	120 34 50	Carb, Chlor, OC, OP
22	12499000	Naches River near North Yakima	46 37 42	120 31 10	Carb, Chlor, OC, OP, TR
23	12500420	Moxee Drain at Birchfield road near Union Gap	46 32 46	120 26 13	OC, OP
24	12500430	Moxee Drain at Thorp Road near Union Gap	46 32 18	120 27 19	Carb, Chlor, OC, OP, PC, TR, V
27	12500445	Wide Hollow Creek near mouth at Union Gap	46 32 19	120 28 17	Carb, Chlor, OC, OP, PC, TR, V
28	12500450	Yakima River above Ahtanum Creek at Union Gap	46 32 04	120 27 58	Carb, Chlor, OC, OP, PC, TR
30	12502500	Ahtanum Creek at Union Gap	46 32 10	120 28 20	Carb, OC, OP, TR
33	12505350	East Toppenish Drain at Wilson Road near Toppenish	46 22 04	120 15 00	Carb, Chlor, OC, OP, TR
34	12505410	Sub 35 Drain at Parton Road near Granger	46 20 11	120 13 48	Carb, Chlor, OC, OP, TR
35	12505450	Granger Drain at Granger	46 20 37	120 11 09	OC, OP
36	12505460	Granger Drain at mouth near Granger	46 20 10	120 11 38	Carb, Chlor, OC, OP, PC, TR, V
37	12505510	Marion Drain at Indian Church Road at Granger	46 19 52	120 11 54	Carb, Chlor, OC, OP, TR
38	12507508	Toppenish Creek at Indian Church Road near Granger	46 18 52	120 11 53	Carb, Chlor, OC, OP, TR
40	12507585	Yakima River at river mile 72 above Satus Creek near Sunnyside	46 16 11	120 05 30	Carb, Chlor, OC, OP, TR
41	12507594	Satus Creek above Wilson-Charley Canyon near Toppenish	46 01 00	120 40 54	Carb, Chlor, OC, OP, TR
43	12508620	Satus Creek at gage at Satus	46 16 26	120 08 32	Carb, Chlor, OC, OP, TR
45	12508630	South Drain near Satus	46 15 35	120 07 57	Carb, Chlor, OC, OP, TR
46	12508850	Sulphur Creek Wasteway near Sunnyside	46 15 03	120 01 07	Carb, Chlor, OC, OP, PC, TR, V
47	12509050	Yakima River at Euclid Bridge at river mile 55 near Grandview	46 13 01	119 55 00	Carb, Chlor, OC, OP, TR
48	12509499	Chandler Canal at Bunn Road at Prosser	46 13 27	119 44 08	Carb, Chlor, OC, OP
49	12509710	Spring Creek at mouth at Whitstran	46 14 00	119 40 38	Carb, Chlor, OC, OP, TR
50	12509829	Snipes Creek at mouth at Whitstran	46 14 02	119 40 37	Carb, Chlor, OC, OP, TR

Table 8. Sampling locations for pesticides and other organic compounds in water, bed sediment, soil, and aquatic biota, Yakima River Basin, Washington, 1987–91—Continued

Map reference number	Station number	Station name	Latitude	Longitude	Chemical families analyzed at each station
		Water samples—Continued			
51	12510500	Yakima River at Kiona	46 15 13	119 28 37	Carb, Chlor, OC, OP, PC, TR, V
53	461720120043201	Well 1, Sunnyside, 09N/22E-04P01	46 17 20	120 04 32	Carb, Chlor, OC, OP, TR
55	462510120323901	Well 2, Harrah, 11N/18E-22R02	46 25 10	120 32 39	Carb, Chlor, OC, OP, TR
58	463233120262101	Well 12N/19E-09H02	46 32 33	120 26 21	OC, OP
59	463234120261601	Well 12N/19E-09H01	46 32 34	120 26 16	OC, OP
61	463257120260801	Well 12N/19E-10D02	46 32 57	120 26 08	OC, OP
		Bed-sediment samples			
1	12478100	Waptus River at mouth near Roslyn	47 25 13	121 05 15	SV
3	12478300	Cle Elum River above Cle Elum Lake near Roslyn	47 21 19	121 06 22	OC, SV
5	12479720	Jungle Creek near mouth near Cle Elum	47 20 30	120 51 59	OC, SV
6	12479750	North Fork Teanaway River below bridge at Dickey Creek Campground	47 17 21	120 51 30	OC
7	12480000	Teanaway River below forks near Cle Elum	47 14 48	120 51 36	OC, SV
8	12481900	Taneum Creek at Taneum Meadow near Thorp	47 06 47	120 52 01	OC
9	12483190	South Fork Manastash Creek near Ellensburg	46 58 18	120 48 32	OC
10	12483750	Naneum Creek below High Creek near Ellensburg	47 10 55	120 26 44	OC
13	12484480	Cherry Creek at Thrall	46 55 34	120 29 51	OC, SV
14	12484500	Yakima River at Umtanum	46 51 46	120 28 44	OC, SV
15	12484550	Umtanum Creek near mouth at Umtanum	46 51 27	120 29 46	OC, SV
16	12487200	Little Naches River at mouth near Cliffdell	46 59 20	121 05 55	OC
17	12488250	American River at Hells Crossing near Nile	46 58 04	121 15 45	OC, SV
18	12489100	Rattlesnake Creek above North Fork Rattlesnake Creek near Nile	46 48 34	121 04 08	OC, SV
24	12500430	Moxee Drain at Thorp Road near Union Gap	46 32 18	120 27 19	OC, SV
25	12500437	Wide Hollow Creek at West Valley Middle School near Ahtanum	46 34 56	120 36 34	OC, SV
27	12500445	Wide Hollow Creek near mouth at Union Gap	46 32 19	120 28 17	OC, SV
28	12500450	Yakima River above Ahtanum Creek at Union Gap	46 32 04	120 27 58	OC, SV
29	12500900	South Fork Ahtanum Creek above Conrad Ranch near Tampico	46 29 32	120 57 23	OC
31	12503640	Unnamed drain at Lateral & Riggs Roads near Wapato	46 28 40	120 31 59	OC, SV
36	12505460	Granger Drain at mouth near Granger	46 20 10	120 11 38	OC, SV
38	12507508	Toppenish Creek at Indian Church Road near Granger	46 18 52	120 11 53	OC, SV
42	12508500	Satus Creek below Dry Creek near Toppenish	46 15 00	120 22 40	OC
44	12508625	Yakima River below Satus Creek at river mile 68 near Satus	46 15 06	120 05 45	OC, SV
46	12508850	Sulphur Creek Wasteway near Sunnyside	46 15 03	120 01 07	OC, SV
49	12509710	Spring Creek at mouth at Whitstran	46 14 00	119 40 38	SV
51	12510500	Yakima River at Kiona	46 15 13	119 28 37	OC, SV
52	461652119522000	Unnamed drain at County Line Road near Grandview	46 16 52	119 52 20	OC, SV

Table 8. Sampling locations for pesticides and other organic compounds in water, bed sediment, soil, and aquatic biota, Yakima River Basin, Washington, 1987–91—Continued

Map reference number	Station number	Station name	Latitude	Longitude	Chemical families analyzed at each station
		Bed-sediment samples—Continued			
54	461744119522400	Unnamed drain to East Turbine Lateral near Grandview	46 17 44	119 52 24	OC, SV
62	463359120281400	Unnamed urban runoff at Union Gap	46 33 59	120 28 14	OC, SV
63	463501120353300	Wide Hollow Creek at Ahtanum Boulevard near Ahtanum	46 35 01	120 35 33	OC, SV
		Composited bed-sediment samples by crop type (stations not shown in figure 7)			
Pear Orchards					OC, SV
		Wide Hollow Creek at Ahtanum Boulevard near Ahtanum	46 35 01	120 35 33	
		Waste ditch south of Old Naches Road near Naches	46 43 02	120 39 49	
		Road ditch at Houghton Road above Roza Canal	46 25 34	120 11 33	
		Waste ditch, 0.2 miles south of East Zillah Drive	46 24 06	120 13 08	
Apple orchards					OC, SV
		Wide Hollow Creek at Ahtanum Boulevard near Ahtanum	46 35 01	120 35 33	
		Road ditch north of Old Naches Road near Gleed	46 40 56	120 38 01	
		Waste ditch at intersection of Lateral Road and Riggs Road	46 28 40	120 32 00	
		Unnamed drain to East Turbine Lateral near Grandview	46 17 44	119 52 24	
		Unnamed drain at County Line Road near Grandview	46 16 52	119 52 20	
Hops fields					OC, SV
		Waste ditch near Faucher Road near Moxee Drain	46 32 47	120 22 24	
		Waste ditch near intersection of Fort Road and Bench Road	46 22 37	120 36 19	
		Waste ditch at McDonald Road, 0.6 miles north of Hanks Road	46 16 52	119 43 24	
		Waste ditch at intersection of Hanks Road and Hinzerling Road	46 16 23	119 46 03	
		Waste ditch northeast of intersection of Bonnieview Road and Olmstead Road	46 15 41	119 52 53	
Corn fields					OC, SV
		Waste ditch near intersection of Fort Road and Island Road	46 22 32	120 35 00	
		Waste ditch at Fort Road, 0.4 miles east of Becker Road	46 22 32	120 20 49	
		Waste ditch at Cemetary Road, 0.3 miles north of Gap Road	46 19 28	120 04 53	
		Waste ditch at Indian Church Road near Granger	46 20 01	120 11 58	
		Waste ditch at Sorensen Road, 0.5 miles east of Denmark Hall	46 56 30	120 25 22	
Asparagus fields					OC, SV
		Waste ditch at Yost Road, 0.5 west of South Wapato Road	46 20 48	120 25 45	
		Waste ditch at Cemetary Road, 0.3 miles north of Gap Road	46 19 28	120 04 53	
		Waste ditch at intersection of Factory Road and Bethany Road	46 19 02	119 55 28	
		Waste ditch at intersection of Lemley Road and Wilgus Road	46 15 06	119 49 46	
		Waste ditch near intersection of Colwash Road and Winnier Road	46 13 02	120 05 51	

Table 8. Sampling locations for pesticides and other organic compounds in water, bed sediment, soil, and aquatic biota, Yakima River Basin, Washington, 1987–91—Continued

Map reference number	Station number	Station name	Latitude	Longitude	Chemical families analyzed at each station
	Potato fields				OC, SV
		Waste ditch at Orchard Road, 0.1 miles south of Thrall Road	46 55 44	120 28 28	
		Composited bed-sediment samples by crop type (stations not shown in figure 7)—Continued			
		Waste ditch at Dodge Road, 0.4 miles south of Orchard Road	46 55 03	120 27 15	
		Waste ditch at Sorensen Road, 0.4 miles west of Denmark Hall	46 56 30	120 26 31	
		Waste ditch near intersection of Hanks Road and Missimer Road	46 16 26	119 48 55	
		Waste ditch at intersection of Olden Way Road and Jensen Road	46 19 57	120 22 28	
	Grape orchards				OC, SV
		Waste ditch, intersection—McCreadie and Griffin Roads	46 15 33	119 51 02	
		Waste ditch, intersection—McCreadie and Missimer Roads	46 15 32	119 48 31	
		Waste ditch, intersection—Snipes and Hinzerling Roads	46 18 11	119 46 15	
		Waste ditch, intersection—Phipps and Scoon Roads	46 22 38	120 00 51	
		Soil samples			
56	463202120223600	Apple orchard near Moxee City	46 32 02	120 22 36	OC, OP
57	463232120234900	Hop field near Moxee City	46 32 32	120 23 49	OC, OP
60	463247120222300	Hop field near Moxee City	46 32 47	120 22 23	OC, OP
		Aquatic-biota samples			
1	12478100	Waptus River at mouth near Roslyn	47 25 13	121 05 15	OC
4	12479500	Yakima River at Cle Elum	47 11 35	120 56 55	OC
5	12479720	Jungle Creek near mouth near Cle Elum	47 20 30	120 51 59	OC
6	12479750	North Fork Teanaway River below bridge at Dickey Creek Campground	47 17 21	120 51 30	OC
7	12480000	Teanaway River below forks near Cle Elum	47 14 48	120 51 36	OC
8	12481900	Taneum Creek at Taneum Meadow near Thorp	47 06 47	120 52 01	OC
9	12483190	South Fork Manastash Creek near Ellensburg	46 58 18	120 48 32	OC
10	12483750	Naneum Creek below High Creek near Ellensburg	47 10 55	120 26 44	OC
12	12484440	Cherry Creek above Wipple Wasteway at Thrall	46 55 44	120 29 48	OC
13	12484480	Cherry Creek at Thrall	46 55 34	120 29 51	OC, PAH
14	12484500	Yakima River at Umtanum	46 51 46	120 28 44	OC, PAH
15	12484550	Umtanum Creek near mouth at Umtanum	46 51 27	120 29 46	OC
16	12487200	Little Naches River at mouth near Cliffdell	46 59 20	121 05 55	OC
17	12488250	American River at Hells Crossing near Nile	46 58 04	121 15 45	OC
18	12489100	Rattlesnake Creek above North Fork Rattlesnake Creek near Nile	46 48 34	121 04 08	OC
22	12499000	Naches River near North Yakima	46 37 42	120 31 10	OC
24	12500430	Moxee Drain at Thorp Road near Union Gap	46 32 18	120 27 19	OC, PAH
25	12500437	Wide Hollow Creek at West Valley Middle School near Ahtanum	46 34 56	120 36 34	OC
26	12500442	Wide Hollow Creek at old sewage treatment plant at Union Gap	46 32 35	120 28 26	OC, PAH

Table 8. Sampling locations for pesticides and other organic compounds in water, bed sediment, soil, and aquatic biota, Yakima River Basin, Washington, 1987–91—Continued

Map reference number	Station number	Station name	Latitude	Longitude	Chemical families analyzed at each station
		Aquatic-biota samples—Continued			
29	12500900	South Fork Ahtanum Creek above Conrad Ranch near Tampico	46 29 32	120 57 23	OC
30	12502500	Ahtanum Creek at Union Gap	46 32 10	120 28 20	OC
32	12503950	Yakima River at Parker	46 30 22	120 27 07	OC, PAH
36	12505460	Granger Drain at mouth near Granger	46 20 10	120 11 38	OC
38	12507508	Toppenish Creek at Indian Church Road near Granger	46 18 52	120 11 53	OC, PAH
39	12507525	Yakima River below Toppenish Creek at river mile 79.6 near Granger	46 18 58	120 09 13	OC
40	12507585	Yakima River at river mile 72 above Satus Creek near Sunnyside	46 16 11	120 05 30	OC, PAH
41	12507594	Satus Creek above Wilson-Charley Canyon near Toppenish	46 01 00	120 40 54	OC
43	12508620	Satus Creek at gage at Satus	46 16 26	120 08 32	OC, PAH
46	12508850	Sulphur Creek Wasteway near Sunnyside	46 15 03	120 01 07	OC
47	12509050	Yakima River at Euclid Bridge at river mile 55 near Grandview	46 13 01	119 55 00	OC, PAH
49	12509710	Spring Creek at mouth at Whitstran	46 14 00	119 40 38	OC, PAH
51	12510500	Yakima River at Kiona	46 15 13	119 28 37	OC, PAH

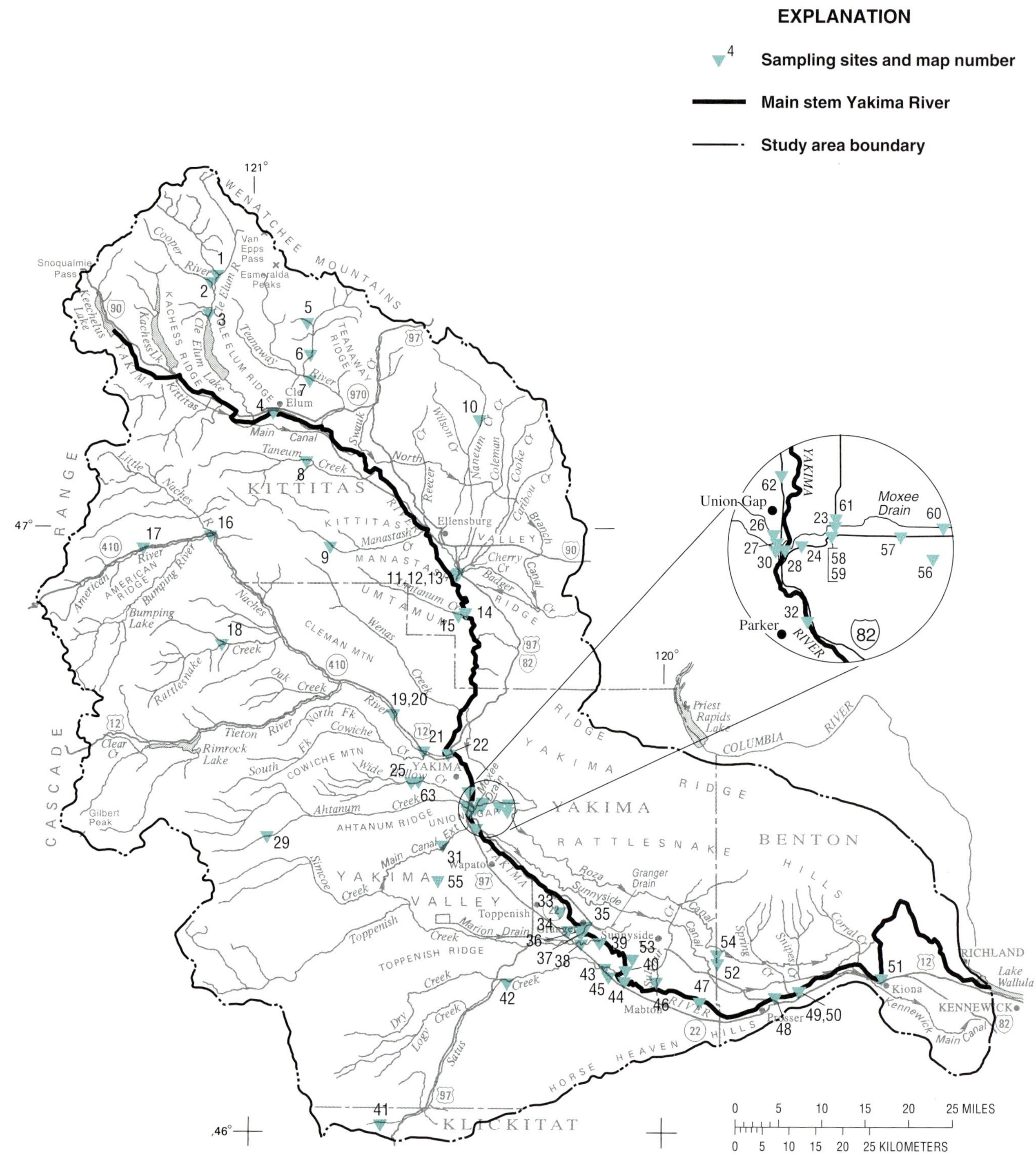

Base modified from U.S. Geological Survey 1:250,000 quadrangles

Figure 7. Sampling stations for organic compounds in water, suspended sediment, bed sediment, soil, and aquatic biota, Yakima River Basin, Washington, 1987–91. (Map reference numbers are shown on map and are listed in table 8).

Table 9. Common technical name and biological activity of pesticides analyzed in water, sediment, soil, and aquatic biota, Yakima River Basin, Washington, 1987–91

[DDD, dichlorodiphenyldichloroethane; DDE, dichlorodiphenyldichloroethylene; DDT, dichlorodiphenyltrichloroethane; HCB, hexachlorobenzene; HCH, hexachlorocyclohexane; DEF, *S,S,S*-tributyl phosphorotrithioate; 2,4-D, (2,4-dichlorophenoxy) acetic acid; 2,4-DP, 2-(2,4-dichlorophenoxy) propionic acid; 2,4,5-T, (2,4,5-trichlorophenoxy) acetic acid; EPTC, *S*-ethyl dipropylthiocarbamate]

Common technical name	Biological activity
Organochlorine compounds	
Aldrin	Insecticide
cis-Chlordane	Insecticide—major component of technical chlordane
trans-Chlordane	Insecticide—major component of technical chlordane
Chlorothalonil	Fungicide
4,4'-DDD and 2,4'-DDD	Insecticide and degradation product of DDT
4,4'-DDE and 2,4'-DDE	Degradation product of DDT
4,4'-DDT and 2,4'-DDT	Insecticide
Dacthal	Herbicide
1,3-Dichloropropene	Nematicide
Dieldrin	Insecticide and degradation product of aldrin
Dicofol	Acaricide
Endosulfan I	Insecticide and acaricide
Endosulfan II	Insecticide and acaricide
Endosulfan sulfate	Degradation product of endosulfan I and II
Endrin	Insecticide
Endrin aldehyde	Transformation product of endrin
Endrin ketone	Metabolite of endrin
HCB	Fungicide and industrial intermediate
α-HCH	Insecticide
β-HCH	Insecticide
δ-HCH	Insecticide
Heptachlor	Insecticide—component of technical chlordane
Heptachlor epoxide	Degradation product of heptachlor
Kepone	Insecticide
Lindane (γ-HCH)	Insecticide and rodenticide
2,4'-Methoxychlor and 4,4'-Methoxychlor	Insecticides
Mirex	Insecticide and fire retardant
cis-Nonachlor	Insecticide—component of technical chlordane
trans-Nonachlor	Insecticide—component of technical chlordane
Oxychlordane	Degradation product of cis- and trans-chlordane
Perthane	Insecticide
Toxaphene	Insecticide, nematicide, herbicide, and fish poison
Organophosphorus compounds	
Azinphos-methyl	Insecticide and acaricide
Chlorpyrifos	Insecticide
DEF	Plant growth regulator
Demeton-S	Insecticide and acaricide

Table 9. Common technical name and biological activity of pesticides analyzed in water, sediment, soil, and aquatic biota, Yakima River Basin, Washington, 1987–91—Continued

Common technical name	Biological activity
Organophosphorus compounds—Continued	
Diazinon	Insecticide and acaricide
Dimethoate	Insecticide and acaricide
Disulfoton	Insecticide and acaricide
Ethion	Acaricide and insecticide
Fonofos	Insecticide
Isofenphos	Insecticide
Malathion	Insecticide and acaricide
Methidathion	Insecticide and acaricide
Methyl parathion	Insecticide and acaricide
Methyl trithion (methyl-carbophenothion)	Insecticide and acaricide
Mevinphos	Insecticide and acaricide
Parathion	Insecticide and acaricide
Phorate	Insecticide, acaricide, and nematicide
Phosphamidon	Insecticide and acaricide
Terbufos	Insecticide and nematicide
Trithion (carbophenothion)	Insecticide and acaricide
Chlorophenoxy-acid herbicides	
2,4-D	Herbicide
2,4-DP (dichlorprop)	Herbicide and plant growth regulator
Fenoprop (Silvex)	Herbicide
2,4,5-T	Herbicide
Triazine compounds	
Ametryn	Herbicide
Atrazine	Herbicide
Cyanazine	Herbicide
Desethyl-atrazine	Degradation product of atrazine and propazine
Desisopropyl-atrazine	Degradation product of atrazine and simazine
Hexazinone	Herbicide
Metribuzin	Herbicide
Prometon	Herbicide
Prometryn	Herbicide
Propazine	Herbicide
Simazine	Herbicide
Simetryn	Herbicide
Carbamate compounds	
Aldicarb	Insecticide, acaricide, and nematicide
Aldicarb sulfone	Nematicide and insecticide; degradation product of aldicarb
Aldicarb sulfoxide	Degradation product of aldicarb
Carbaryl	Insecticide and plant growth regulator

Table 9. Common technical name and biological activity of pesticides analyzed in water, sediment, soil, and aquatic biota, Yakima River Basin, Washington, 1987–91—Continued

Common technical name	Biological activity
Carbamate compounds—Continued	
Carbofuran	Insecticide, acaricide, and nematicide
Methiocarb	Molluscicide, insecticide, acaricide, and repellent
Methomyl	Insecticide and acaricide
1-Naphthol	Insecticide; hydrolysis product of carbaryl
Oxamyl	Insecticide, acaricide, and nematicide
Propham	Herbicide and plant growth regulator
Propoxur	Insecticide
Thiocarbamate compounds	
Butylate	Herbicide
Cycloate	Herbicide
EPTC (also in carbamate family)	Herbicide
Vernolate	Herbicide
Acetamide compounds	
Alachlor (also in chloroacetanilide family)	Herbicide
Diphenamid	Herbicide
Metolachlor	Herbicide
Anilide compound	
Carboxin	Fungicide
Benzoic compound	
Dicamba	Herbicide
Chloroacetanilide compounds	
Butachlor	Herbicide
Propachlor	Herbicide
Conazole compound	
Triadimefon	Fungicide
Pyrethoid compounds	
Flucythrinate	Insecticide
cis-Permethrin	Insecticide
trans-Permethrin	Insecticide
Pyridine compound	
Picloram	Herbicide
Sulfite compound	
Propargite	Acaricide
Trifluoromethyl compound	
Trifluralin	Herbicide
Uracil compounds	
Bromacil	Herbicide
Terbacil	Herbicide
Urea compound	
Linuron	Herbicide

K_d is the sediment-water partition coefficient and is a measure of the tendency of a compound to sorb to sediment: $K_d = C_s / C_w$, where C_s is the concentration of the organic compound sorbed to a specific mass of sediment, and C_w is the concentration of the compound dissolved in an equal mass of water (Smith and others, 1988). Research data suggest that dissolved nonionic and acidic organic compounds partition to organic matter associated with the sediment as a result of molecular forces in the solubilization process. Conversely, sorption of organic bases occurs primarily by adsorption (physical and chemical bonds to the particle's surface) (Smith and others, 1988). The partition coefficient commonly is expressed on an organic-carbon basis: $K_{oc} = K_d / F_{oc}$, where K_{oc} is the sorption coefficient and is a measure of an organic compound's tendency to partition to the organic-carbon component of sediment, and F_{oc} is the fractional mass of organic carbon in the sediment.

Assuming equilibrium conditions, theoretical partitioning of a nonionic organic compound in water and sediment may be estimated using the following equation derived from equations in Smith and others (1988):

$$\textit{Fraction of compound's mass partitioned to sediment} \quad (1)$$

$$= 1 - \frac{1}{1 + (ss \times 10^{-6} \times F_{oc} \times K_{oc})}$$

where,

ss = suspended-sediment concentration in mg/L,
F_{oc} = fraction of sediment that is organic carbon, and
K_{oc} = sediment-water partition coefficient in mL/g (milliliters per gram).

For example, if the suspended-sediment concentration is 250 mg/L, F_{oc} is 0.01 (1 percent is typical of sediment in the Yakima River Basin), and the K_{oc} for DDT is 243,000 mL/g, then:

$$\textit{Fraction of DDT partitioned to sediment} \quad (2)$$

$$= 1 - \frac{1}{1 + (250 \times 10^{-6} \times 0.01 \times 243,000)} = 0.38$$

Therefore, theoretically, 38 percent of the DDT is partitioned to suspended sediment and 62 percent is dissolved in water, assuming that dissolved organic carbon is not enhancing the compound's solubility.

In the Yakima River Basin, compounds that are most likely to leach from soils and dissolve into ground water and surface water have K_{oc} values less than about 300 to 500 mL/g (fig. 8). For compounds with K_{oc} values less than 50 mL/g (for example, see 2,4-D, dimethoate, and phosphamidon in table 61, at back of report), more than 50 percent of the compound is associated with the dissolved phase in the soil-pore water. When K_{oc} values exceed 1,000 mL/g, most of the compound is highly partitioned to soil at equilibrium and is less likely to leach into the ground water. The compound, however, may be transported to streams by way of soil erosion. For example, a soil sample from an apple orchard in the Yakima River Basin had the following measured concentrations of organic carbon and DDT in the A horizon (0 to 6 inches) and B horizon (6 to 9 inches) of the soil: 1.4 and 0.52 percent organic carbon, respectively, and 0.65 and 0.058 µg/g of DDT, respectively. Assuming a soil density of 2.65 grams per cubic centimeter and a moisture content of 10 percent in both the A and B horizons, more than 99.9 percent of the DDT would be partitioned to soil at equilibrium. Therefore, the major transport of DDT to streams should occur through soil erosion. Theoretical calculations also indicate that concentrations in soil-pore water would be about 800 ηg/L (nanograms per liter) and 200 ηg/L in the A and B horizons, respectively. Consequently, soils may provide a long-term reservoir of DDT as the pore water

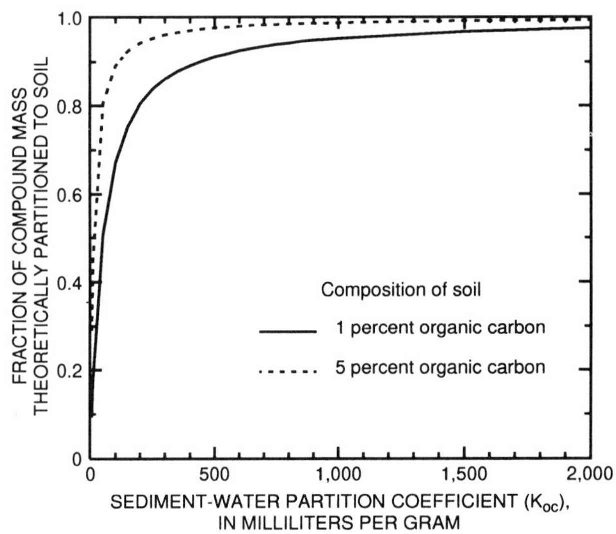

Figure 8. Theoretical relation between sediment-water partition coefficients and the fraction of organic compounds sorbed to the soil. (Equilibrium partitioning is assumed. The remaining compound fraction is dissolved in the soil-pore water.)

flows through the soil profile to the ground water. Depending on the persistence of compounds and on sorption characteristics of deeper soils, concentrations of DDT and other hydrophobic compounds may be measurable in ground water.

Once soils are eroded, DDT and other hydrophobic compounds that are partitioned to soils will equilibrate with stream water, resulting in the partial desorption of compounds into the dissolved phase. Partitioning of these compounds among bed-sediment, suspended-sediment, and dissolved phases typically is controlled by the concentration of suspended sediment, F_{oc}, and the compound's water solubility. However, at dissolved-organic-carbon (DOC) concentrations higher than those typically observed in the Yakima River Basin, DOC also may enhance solubility by enabling partitioning to colloidal particles or high-molecular-weight dissolved-organic matter (Smith and others, 1988).

In the Yakima River at Kiona and three major agricultural return flows during 1988–91, suspended-sediment concentrations generally ranged from 10 to 1,000 mg/L, and F_{oc} values generally ranged from 0.4 to 5 percent (fig. 9). Theoretical partitioning for compounds at these concentrations over a range of K_{oc} values is shown in figure 10. The K_{oc} value for DDT is about 243,000 mL/g (Mercer and others, 1990). In Yakima River Basin soils, more than 99 percent of the DDT should be partitioned to soil at equilibrium (fig. 8); conversely, in the Yakima River at Kiona and in major agricultural return flows, about 45 percent or more of the DDT in the water column should be in the dissolved phase when suspended-sediment concentrations are less than 600 mg/L (fig. 9).

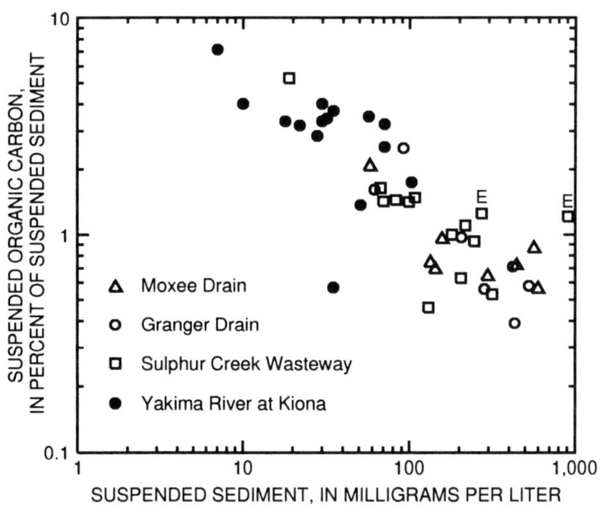

Figure 9. Concentrations of suspended sediment and suspended organic carbon in major agricultural-return flows and the Yakima River at Kiona, Yakima River Basin, Washington, 1988–91. ("Greater-than values are excluded. E, the value for suspended organic carbon is an estimate.")

The half-life (a measure of persistence) of a compound in soils is the length of time required for the compound to be reduced to one-half of its original concentration. Persistence depends on many physical, chemical, and biological processes, including volatilization, hydrolysis (breakdown reaction with water), photolysis (sunlight-induced breakdown reactions), and microbial transformations. Many of these processes vary greatly depending on the compound's physical/chemical properties and on soil characteristics that include moisture, temperature, oxygen, pH, and nutrients available for microbial populations (Becker and others, 1989). As a result, estimates of compound half-lives may vary by a factor of three or more from station to station and should be used only as relative indicators of persistence. Sometimes, breakdown products also are of environmental concern. Compounds with half-lives longer than 3 weeks tend to have a higher potential to leach into ground water or runoff into the surface water. Compounds with half-lives shorter than 3 weeks also may be detected in surface water if (1) applications are large, (2) sampling stations in streams are located close to application areas, or (3) sampling occurs soon after application and irrigation (or a precipitation runoff event).

In summary, compounds with water solubilities higher than 30 mg/L, K_{oc} values lower than 300–500 mL/g and soil half-lives longer than 3 weeks are considered to have a high potential to leach (dissolve) into ground and surface water (Becker and others, 1989; U.S. Environmental Protection Agency, 1988a). Compounds with water solubilities lower than 1 mg/L, K_{oc} values higher than 500 mL/g, and soil half-lives longer than 3 weeks tend to partition to soils and be transported to streams during periods of erosion. Once these soils enter the water, the sorbed compounds will tend to equilibrate with the water column, and some of the compound will partition into the dissolved phase. As shown in figure 10 for stream-water samples collected from the Yakima River Basin, compounds with K_{oc} values lower than 50,000 mL/g will be associated mostly with the dissolved phase.

Water solubility and sediment-water-partition coefficients were the primary characteristics used to select which environmental compartments (bed sediment, suspended sediment, soil, dissolved phase, and

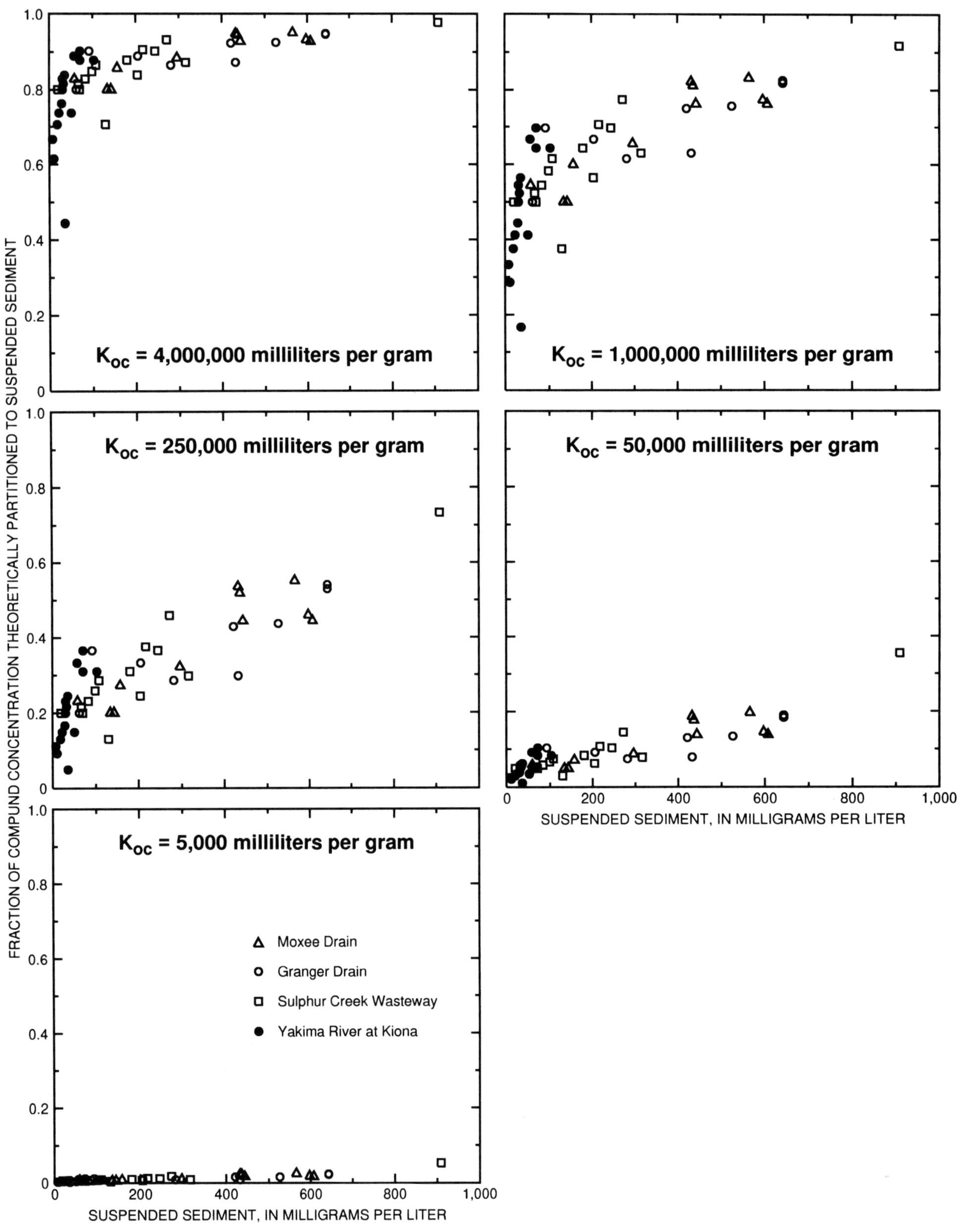

Figure 10. Theoretical relation, in unfiltered water, between concentrations of suspended sediment and fractions of organic compounds sorbed to suspended sediment for selected stations, Yakima River Basin, Washington, 1988–91. (Equilibrium partitioning is assumed. K_{oc}, sediment-water partition coefficient.)

aquatic biota) were sampled for different classes of compounds. Organic compounds with high aqueous solubilities or high sediment-water-partition coefficients are associated with high bioconcentration factors (Chiou and others, 1977). General comments concerning the selection follow:

(1) Carbamates—Solubilities generally are higher than 30 mg/L, and K_{oc} values are lower than 500 mL/g. Filtered- and unfiltered-water samples were analyzed for these compounds because carbamates primarily occur in the water column.

(2) Chlorophenoxy-acid herbicides—Solubilities are higher than 30 mg/L, and K_{oc} values are lower than 5,000 mL/g (K_{oc} values for 2,4-D and 2,4,5-T are both lower than 100 mL/g). Filtered- and unfiltered-water samples were analyzed for these compounds because these herbicides primarily occur in the water column.

(3) Organochlorine compounds—Solubilities generally are lower than 1 mg/L and K_{oc} values vary widely from about 1,000 to 24,000,000 mL/g. These compounds are relatively insoluble and partition to soils, bed sediment, suspended sediment, and lipids in aquatic biota. In streams, several of these compounds may be transported in the dissolved phase, especially in streams where suspended-sediment concentrations are lower than 1,000 mg/L and F_{oc} values are lower than 0.04. Filtered-water, unfiltered-water, suspended-sediment, bed-sediment, soil, and aquatic-biota samples were analyzed because these compounds are likely to partition into all of these environmental compartments.

(4) Organophosphorus compounds—Solubilities generally are higher than 20 mg/L, and K_{oc} values generally are higher than 300 and lower than 15,000 mL/g. At equilibrium in streams, these compounds should occur primarily in the dissolved phase (fig. 10). In soils, however, many of these compounds should be partitioned primarily to the soil rather than in the soil-pore water. Bioaccumulation into aquatic biota generally is not significant (Smith and others, 1988). In addition to analyzing filtered- and unfiltered-water samples, suspended-sediment and soil samples were analyzed to determine how quickly these compounds undergo dissolution from soils and suspended sediment.

(5) Triazine, acetamide, and chloroacetanilide compounds—Solubilities generally are higher than 30 mg/L, and K_{oc} values generally lower than 500 mL/g. Filtered- and unfiltered-water samples were analyzed for these compounds because they occur primarily in the water column.

(6) Semivolatile organic compounds (ethers, phenols, cresols, phthalate esters, and monocyclic and polycyclic aromatic hydrocarbons)—Because of low solubilities and high K_{oc} values, many of these compounds strongly partition into soils, sediment, and to certain aquatic biota (Smith and others, 1988). Polycyclic aromatic hydrocarbons, however, do not highly accumulate in fish because they readily oxidize (Malins and Hodgins, 1981; Eisler, 1987). Conversely, invertebrates are generally less efficient in oxidizing polycyclic aromatic hydrocarbons (Pearson and others, 1980; Hale, 1988; Pereira and others, 1988). Therefore, bed-sediment and invertebrate samples (specifically mollusks) were collected for analysis of these compounds in a reconnaissance sampling of the basin. In addition, some aquatic-plant samples also were analyzed to explore relations among concentrations in bed sediment, invertebrates, and aquatic plants.

Data Collection, Laboratory Methods, and Data Compilation

Collection, processing, preservation, and laboratory procedures for analyses of unfiltered water, filtered water, suspended sediment, bed sediment, soil, and aquatic biota (fish, mollusks, crayfish, and plants) from the Yakima River Basin are outlined or referenced in Rinella, McKenzie, Crawford, and others (1992). Previous studies (Johnson and others, 1986; Rinella, McKenzie, and Fuhrer, 1992) found that many organic compounds had concentrations in the water column that were lower than conventional minimum reporting levels. To reduce the reporting levels, large volumes of water were collected and processed in a synoptic sampling during June 25–30, 1989 (Goulden and Anthony, 1985; Foster and Rogerson, 1990; Foster and others, 1991). The water was filtered, and the filtrate and suspended sediment were analyzed separately to determine the importance of dissolved- and suspended-transport phases in the Yakima River Basin. A conventional organic analysis typically requires solvent extraction from a 1-L water sample.

By analyzing samples as large as 100 liters or more, minimum reporting levels were reduced by one to two orders of magnitude. Foster and others (1993) have shown that the number of pesticides detected in water samples usually increases in direct proportion to increases in sample volume. Data on organic compounds in the Yakima River Basin during 1987–91 are published in a report by Rinella, McKenzie, Crawford, and others (1992).

Quality Assurance

Quality-assurance data on blanks, replicates, and (or) spiked samples were collected for unfiltered-water, filtered-water, suspended-sediment, bed-sediment, and aquatic-biota samples. The data are summarized in this report; sampling protocol and analytical data for individual quality-assurance samples are listed in the data report by Rinella, McKenzie, Crawford, and others (1992).

The accuracy of water, sediment, and aquatic-biota analyses was assessed by analyzing sample blanks and samples spiked with dissolved analytes. Analytical results from spiked samples provide an estimate of method performance for extracting dissolved compounds from the water matrix; however, the analytical recoveries do not necessarily reflect extraction efficiencies of compounds from mixtures of sediment and water (unfiltered-water samples), bed-sediment, and tissue samples.

For unfiltered-water samples collected during 1988–91, extraction recoveries from spiked samples ranged from 54 to 110 (median 80) percent for carbamate compounds (except one sample where spiked compounds were not detected); from 49 to 82 (median 54) percent for chlorophenoxy-acid herbicides; from less than 17 to 160 (median 93) percent for phenols and cresols (median 93 percent); from 58 to 120 (median 100) percent for triazine herbicides and other miscellaneous pesticides; and from 22 to 180 (median 54) percent for volatile-organic compounds. For filtered-water samples collected in June 1989, mean extraction recoveries for surrogate spikes ranged from 28 percent for isodrin to 84 percent for diazinon d-10. For a spiked bed-sediment sample collected in August 1988, extraction recoveries ranged from less than 0.4 to 143 (median 76) percent for organochlorine compounds and from less than 12 to 250 (median 83) percent for semivolatile organic compounds, including phthalate esters, ethers, phenols, cresols, and monocyclic and polycyclic aromatic hydrocarbons.

The majority of spiked compounds in this study had extraction recoveries less than 100 percent. Because the quality-control data are inadequate in number for estimating a reliable correction for each of the sediment and (or) water matrices, measured concentrations in this study are not corrected for this low bias. As typically observed for reported analyses of organic compounds, the concentrations summarized in this report may underestimate actual concentrations.

One unfiltered-water sample was spiked with organochlorine compounds that included a complex mixture of chlordane, PCBs, and toxaphene. Minimum reporting levels for toxaphene are 1,000 ηg/L, so the sample was spiked with toxaphene to obtain a concentration of 4,600 ηg/L. The other organochlorine compounds were spiked to obtain concentrations less than 100 ηg/L (typical minimum reporting levels are 1 ηg/L). The laboratory was able to quantify 4,000 ηg/L of toxaphene; however, this high concentration of toxaphene resulted in analytical interference, so the other organochlorine compounds could not be detected. Another problem occurred in two bed-sediment samples that were spiked with the organochlorine compounds and semivolatile compounds. The semivolatile compounds and most of the organochlorine compounds were successfully detected with spike recoveries generally larger than 60 percent (Rinella, McKenzie, Crawford, and others, 1992); however, PCB and toxaphene were not detected. These few analyses are insufficient to be conclusive, but the spiking of water and sediment samples with a large number of organic compounds may result in laboratory interferences or erroneous nondetections using conventional laboratory procedures as outlined by Wershaw and others (1987). Analytical precision for these conventional analyses of spiked organic compounds in distilled water also is listed in Wershaw and others (1987).

Precision associated with sample processing and analysis was assessed by analyzing duplicate samples that were thought to be essentially identical in composition; however, the results also may reflect short-term changes in stream quality because duplicate unfiltered-water samples were collected within 60 minutes of one another in this study. In this study, precision is expressed as a relative difference, as follows:

Relative difference, in percent (3)

$$= \frac{(\textit{concentration in duplicate 1} - \textit{concentration in duplicate 2}) \times 100}{\dfrac{(\textit{concentration in duplicate 1} + \textit{concentration in duplicate 2})}{2}}$$

Unfiltered-water samples were collected during 1988–91 for duplicate analyses of carbamates, chlorophenoxy-acid herbicides, organochlorine compounds, organophosphorus compounds, phenols and cresols, triazine pesticides, and other organic compounds. Relative differences ranged from 0 to 140 percent. Most of the differences were reported as 0 percent because most of the duplicate concentrations were lower than analytical reporting levels. For filtered-water samples collected in June 1989, the relative difference ranged from 0 to 47 percent, with most of the differences less than 16 percent. For suspended-sediment samples collected from May to September 1991, the relative difference for organochlorine and organophosphorus compounds ranged from 0 to 86 percent, with most of the differences equaling 0 percent.

Blank solutions consisted of distilled water that presumably was free of target analytes. These solutions were analyzed concurrently with filtered-water samples that were extracted using the Goulden large-sample extractor (Rinella, McKenzie, Crawford, and others, 1992). Blank filters (clean, oven-baked filters without suspended sediment) also were analyzed. Except for β-hexachlorocyclohexane (HCH), which had quantifiable concentrations in most of the blank filters, target-compound concentrations for the blank solutions and blank filters generally were less than minimum reporting levels. As a result of contamination from the filter, concentrations of β-HCH were not reported for the suspended-sediment phase. Blanks for soil and bed-sediment samples were not analyzed in this study.

For aquatic organisms, quality-assurance measures were applied in the field, laboratory, and data management. In the field, each composite sample was limited to a single species, and all composites contained at least 5 individuals of approximately the same size, both within and among composites. This protocol, however, could not be achieved in many cases (Rinella, McKenzie, Crawford, and others, 1992). Sample containers and dissection instruments were precleaned according to guidelines in Crawford and Luoma (1993). All samples were field processed within 8 hours of collection, most within 4 hours of collection and many within 2 hours of collection. Replicate field samples of biota were taken for 4 of the 56 samples (7 percent) collected for analysis of organic compounds.

For organochlorine compounds in biological samples, laboratory quality-assurance measures included analyses of reagent blanks, blank (uncontaminated) fish and plant tissue, and spiked and split samples. Eighteen reagent blanks were analyzed, and all analyte concentrations were less than minimum reporting levels. Matrix blanks for fish samples consisted of ocean fish that were previously determined to have low or nondetectable concentrations of target compounds. Three fish-tissue matrix blanks were analyzed for organochlorine compounds, resulting in two samples with detections of cis-chlordane, dieldrin, 4,4'-DDE and 4,4'-DDD at concentrations slightly higher than the minimum reporting levels. The minimum reporting level was 0.01 μg/g for most analytes, except toxaphene and PCBs for which it was 0.05 μg/g. Blank plant samples consisted of plant samples that were collected in nonagricultural areas. One matrix blank for vegetation was analyzed, and no target analytes were detected. Eight samples of fish and two samples of vegetation spiked with organochlorine compounds were analyzed. The rate of recovery for most analytes was 85 percent or better. Recoveries for octachlorostyrene, however, ranged from 71 to 85 percent, for hexachlorobenzene ranged from 57 to 88 percent, and for kepone ranged from 62 to 82 percent. In the lab, one sample was split into five subsamples and analyzed for organochlorine compounds. Differences in concentrations were not detected among the split samples for most target analytes. The largest difference occurred in split samples of largescale suckers, which were analyzed for trans-nonachlor. One split measured 0.03 μg/g and the second split measured 0.04 μg/g for a relative difference of 28 percent.

For PAHs, six reagent blanks were analyzed, and no target compounds were detected. Although PAHs were not analyzed in fish because fish quickly metabolize PAHs, a matrix blank of fish was used in the quality-assurance program of the contract laboratory. A matrix blank for mollusk tissue, however, was not available. Three matrix blanks for fish and one for vegetation were analyzed for PAHs, and target PAHs were not detected at a minimum reporting level of 0.01 μg/g. Three fish samples and one vegetation sample were spiked with PAHs, and percent recoveries ranged from 49 to 100 percent. Recoveries were usually greater than 70 percent; however, recoveries for naphthalene, fluoranthene, 1,2-benzanthracene, and chrysene in one of the samples (not necessarily the same sample) and 1,2,5,6-dibenzanthracene, benzo(k)fluoranthene, benzo(a)pyrene, and benzo(g,h,i)perylene in two of the samples were less than 70 percent. Two biota samples analyzed for PAHs were split in the lab. Results for target compounds were identical except for naphthalene (a relatively volatile compound) in one split sample of the western pearlshell clam. For this sample, one split had a concentration of 0.01 μg/g whereas the other split had a concentration of less than 0.01 μg/g.

Overall, quality-assurance measures indicated that the quality of the results was acceptable for analyses of unfiltered water, filtered water, suspended sediment, and aquatic biota. Future studies should include analyses of standard-reference samples for bed sediment and aquatic biota to better evaluate laboratory accuracy. Additionally, an increased number of quality-assurance samples could be collected to estimate error bounds in concentrations for the collection, preservation, and analysis of organic compounds.

DISTRIBUTION OF PESTICIDES IN THE AQUATIC ENVIRONMENT

During 1987–91, pesticides and other organic compounds were detected in surface water throughout the Yakima River Basin. The highest concentrations were detected in heavily irrigated agricultural areas downstream from the city of Yakima (Rinella, McKenzie, Crawford, and others, 1992). These compounds were detected in agricultural soils, water, suspended sediment, bed sediment, and aquatic biota (fish, mollusks, crayfish, and plants).

Annually, concentrations of hydrophilic and hydrophobic pesticides began to increase in streams in June in response to increased irrigation following springtime pesticide applications to cropland. This general pattern of seasonal occurrence is similar to the pattern reported by Thurman and others (1991), Thurman and others (1992), and Goolsby and others (1991) for herbicides in the Mississippi River and other streams in the Midwestern United States. They reported that the highest concentrations occurred in response to flushing by late spring and early summer rainfall. Unlike Midwestern streams, the amount of rainfall in the Yakima River Basin usually is small during the spring and summer; however, peak water use during irrigation season flushes pesticides to the streams. Although not monitored in this study, occasional thunderstorms during the summer also would flush pesticides from the fields.

Results from this study show that concentrations of several organic compounds exceeded water-quality criteria and interim bed-sediment-quality criteria for the protection of freshwater aquatic life (U.S. Environmental Protection Agency, 1986, 1988b). In addition, concentrations of several organochlorine compounds in fish tissue exceeded guidelines recommended for the protection of freshwater aquatic life and fish predators (National Academy of Sciences-National Academy of Engineering Committee on Water Quality Criteria, 1973).

Water Column

The first objective for assessing pesticides and other organic compounds in the basin was to determine the temporal distribution of organic-compound concentrations in water in the Yakima River and its major tributaries. This information led to the design of a synoptic-sampling plan that was to be implemented when organic-compound concentrations and loads were expected to be large. Objectives of the synoptic study in June 1989 were to (1) determine the spatial distribution of organic compounds, (2) identify and quantify major sources, and (3) develop a conceptual transport model for dissolved and suspended organic compounds. Results from the study were designed to aid water managers in evaluating and prioritizing management alternatives.

Temporal Variation

To determine the temporal distribution of organic-compound concentrations, unfiltered-water samples were collected once a month at eight stations in May, June, July, August/September, and November 1988, and March 1989 (Rinella, McKenzie, Crawford, and others, 1992). The stations included a background station in the Wenatchee National Forest (Cle Elum River above Cle Elum Lake), an urban/agricultural runoff station (Wide Hollow Creek near mouth), five agricultural return flows (Cherry Creek at Thrall, Moxee Drain at Thorp Road, Granger Drain at mouth, Toppenish Creek at Indian Church Road, and Sulphur Creek Wasteway near Sunnyside), and the main stem (Yakima River at Kiona at RM 29.9). The main stem also was sampled during storm runoff in December 1989. Moxee Drain, Granger Drain, and Sulphur

Creek Wasteway were selected for sampling because organochlorine compounds had been detected consistently in the water column in 1985 (Johnson and others, 1986). Cherry and Toppenish Creeks were selected to gather information on runoff from Kittitas Valley and the Yakama Indian Nation Reservation, respectively.

Temporal patterns in the concentrations of hydrophilic and hydrophobic compounds were similar (table 10 and fig. 11). Compounds that consistently were detected using conventional laboratory techniques (Wershaw and others, 1987) were DDT, DDE, DDD, and dieldrin (organochlorine compounds); diazinon and parathion (organophosphorus compounds); atrazine and simazine (triazines); 2,4-D (a chlorophenoxy-acid herbicide); and dicamba (a benzoic compound).

Highest concentrations of hydrophilic and hydrophobic organic compounds generally occurred near or during peak irrigation in June and July and during storm runoff from agricultural land. Storm runoff from croplands was observed in March 1989 at Cherry Creek, Wide Hollow Creek, and Yakima River at Kiona. The temporal pattern for atrazine was slightly offset, with highest concentrations occurring in March, May, and June. Similar to other compounds, however, the lowest atrazine concentrations occurred during low flows in November. Variations in this pattern probably were caused by differences in the time and frequency of applications, hydrologic connection to streams, and other physical, chemical, and biological characteristics of the compounds and the soils. For example, the primary application for atrazine and simazine in the basin occurred in the spring, and samples from several stations contained the highest concentrations of these compounds at that time (Dr. Rick Boydston, Weed Scientist, U.S. Department of Agriculture, Irrigated Agriculture Research and Extension Center, Prosser, Washington, oral commun., October 1992). Maximum use of 2,4-D and dicamba, however, typically occurs from February to April and to a lesser extent during the summer and fall. This seasonal use may account for some of the variability in the pattern of 2,4-D concentrations detected in Sulphur Creek Wasteway.

The highest concentrations of suspended sediment also occurred in June and July and during storm runoff in March, suggesting that eroding soils are transporting sorbed pesticides from fields to streams during periods of overland flow. Theoretical equilibrium partitioning for selected compounds is shown in table 11. These calculations, along with relations shown in figure 12, indicate that compounds with K_{oc} values higher than about 85 mL/g (organochlorine compounds and many organophosphorus, triazine, carbamate, and other modern-day pesticides) are sorbed mostly to soils prior to soil erosion. As soil begins eroding, compounds with K_{oc} values lower than about 85 mL/g initially are transported with the suspended soil (commonly called suspended sediment) even though several of the compounds have solubilities larger than 30 mg/L. Once soils are suspended in water in the fields and the suspended-sediment concentration begins to decrease from the fields to the streams as a result of sediment deposition, partitioning favors desorption of hydrophilic compounds into the dissolved phase (table 11) (Squillace and Thurman, 1992).

The temporal pattern for pesticide concentrations in the basin suggests an annual cycle of increasing pesticide concentrations in streams and agricultural return flows immediately following pesticide applications (Goolsby and others, 1991). Concentrations probably increased because of increased erosion of sediment and sorbed compounds (K_{oc} values higher than 85 mL/g), or increased flushing of more hydrophilic compounds (K_{oc} values lower than 85 mL/g) from the soil by irrigation water and (or) storm runoff. Compound concentrations decreased by the end of irrigation season because of chemical and biological degradation, volatilization, decreases in pesticide and irrigation-water applications, overland runoff, and erosion. Nonetheless, compounds that are persistent in soils continued to be transported throughout the year, especially during storm runoff or snowmelt from the agricultural areas.

The importance of pesticide transport during storm runoff from agricultural fields can be observed in storm-runoff data collected in March 1989 from Cherry Creek, Wide Hollow Creek, and Yakima River at Kiona (table 10). Concentrations of suspended sediment during the March 1989 storm were the highest concentrations observed at these stations in 1988–89. Increased suspended-sediment concentrations were accompanied by increased concentrations of DDT+ DDE+DDD (sum of 4,4'-DDT, 4,4'-DDE, and 4,4'DDD in water samples), dieldrin, atrazine, and simazine at Cherry Creek; increased DDT+DDE+ DDD, dicamba, atrazine, and simazine at Wide Hollow Creek; and increased DDT+DDE+DDD at Yakima River at Kiona. The more hydrophobic compounds, including DDT+

Table 10. Concentrations of selected pesticides, suspended sediment, and specific conductance in unfiltered-water samples, Yakima River Basin, Washington, 1988–89

[Concentrations reported in nanograms per liter, except for suspended sediment, which is in milligrams per liter, and specific conductance, which is in microsiemens per centimeter; range of values is shown when multiple determinations were made in a month, otherwise values are for one sample; DDD+DDE+DDD, 4,4'-dichlorodiphenyltrichloroethane (DDT) plus 4,4'-dichlorodiphenyldichloroethylene (DDE) plus 4,4'-dichlorodiphenyl-dichloroethane (DDD); <, less than; *, highest concentration at station; **, second highest concentration at station; --, no data available]

Sampling stations	May 1988	June 1988	July 1988	August/ September 1988	November 1988	March 1989	December 1989
DDT + DDE + DDD							
Cle Elum River above Cle Elum Lake	<3	<3	<3	<3	<3	<3	--
Cherry Creek at Thrall	2	1	4**	4**	2	39*	--
Moxee Drain at Thorp Road	8–9	21–42**	31–76*	14	3	5	--
Wide Hollow Creek near mouth	4	10**	4	2	1	91*	--
Granger Drain at mouth	41	40–110**	96–122*	36	19	28	--
Toppenish Creek at Indian Church Road	1	1	2**	1	5*	<3	--
Sulphur Creek Wasteway near Sunnyside	12	27–51*	35**	13–14	10	6	--
Yakima River at Kiona	5**	<3–3	1	2	1	14*	1
Dieldrin							
Cle Elum River above Cle Elum Lake	<1	<1	<1	<1	<1	<1	--
Cherry Creek at Thrall	3	3	12**	5	1	41*	--
Moxee Drain at Thorp Road	1	<1–4**	6–8*	2	<1	<1	--
Wide Hollow Creek near mouth	2**	2**	3*	<1	2**	<1	--
Granger Drain at mouth	6	<1–9**	17–30*	8	5	<1	--
Toppenish Creek at Indian Church Road	1	7*	3**	1	<1	<1	--
Sulphur Creek Wasteway near Sunnyside	2	<1–5	14–15*	6**	4	3	--
Yakima River at Kiona	1	<1–2**	3*	2**	<1	<1	<1
Diazinon							
Cle Elum River above Cle Elum Lake	<10	<10	<10	<10	<10	<10	--
Cherry Creek at Thrall	<10	<10	<10	<10	<10	<10	--
Moxee Drain at Thorp Road	<10	10–370**	130–630*	30	<10	<10	--
Wide Hollow Creek near mouth	<10	120*	10**	<10	<10	<10	--
Granger Drain at mouth	<10	<10–10	10–30*	10**	<10	<10	--
Toppenish Creek at Indian Church Road	<10	<10	150*	10**	<10	<10	--
Sulphur Creek Wasteway near Sunnyside	<10	<10–20**	10–20*	<10–10	<10	<10	--
Yakima River at Kiona	<10	<10–30**	250*	10	<10	<10	<10
Parathion							
Cle Elum River above Cle Elum Lake	<10	<10	<10	<10	<10	<10	--
Cherry Creek at Thrall	<10	<10	<10	<10	<10	<10	--
Moxee Drain at Thorp Road	<10	<10	10–100*	<10	<10	<10	--
Wide Hollow Creek near mouth	<10	10*	<10	<10	<10	<10	--
Granger Drain at mouth	<10	<10	<10	<10	<10	<10	--
Toppenish Creek at Indian Church Road	<10	<10	20*	<10	<10	<10	--
Sulphur Creek Wasteway near Sunnyside	<10	<10–10*	<10	<10	<10	<10	--
Yakima River at Kiona	<10	<10–10**	60*	<10	<10	<10	<10
(2,4-dichlorophenoxy) acetic acid [2,4-D]							
Cle Elum River above Cle Elum Lake	<10	<10	<10	<10	220*	<10	--
Cherry Creek at Thrall	7,500*	140–150	480**	60	<10	200	--
Moxee Drain at Thorp Road	20–120	<10–150	200–1,900*	50	<10	270**	--
Wide Hollow Creek near mouth	<10	250*	20**	<10	<10	<10	--
Granger Drain at mouth	<10	<10–430*	330–410**	<10	<10	<10	--
Toppenish Creek at Indian Church Road	<10	160*	30	40**	40**	<10	--
Sulphur Creek Wasteway near Sunnyside	10	<10–100*	<10–90**	50–60	100*	<10	--
Yakima River at Kiona	10	70–110*	70**	40	<10	<10	--

Table 10. Concentrations of selected pesticides, suspended sediment, and specific conductance in unfiltered-water samples, Yakima River Basin, Washington, 1988–89—Continued

Sampling stations	May 1988	June 1988	July 1988	August/ September 1988	November 1988	March 1989	December 1989
Atrazine							
Cle Elum River above Cle Elum Lake	<100	<100	<100	<100	<100	<100	--
Cherry Creek at Thrall	70	20–60	78**	61	48	600*	--
Moxee Drain at Thorp Road	30*	10	<100–17	16**	<100	<100	--
Wide Hollow Creek near mouth	<100	40*	<100	<100	<100	11**	--
Granger Drain at mouth	90	90	40–47	120**	<100	280*	--
Toppenish Creek at Indian Church Road	50	60*	54**	43	23	<100	--
Sulphur Creek Wasteway near Sunnyside	60*	20–22	25–31	34	<100	44**	--
Yakima River at Kiona	50*	<100–30	49**	30	<100	<100	--
Simazine							
Cle Elum River above Cle Elum Lake	<100	<100	<100	<100	<100	<100	--
Cherry Creek at Thrall	90	<100–40	220**	23	<100	6,600*	--
Moxee Drain at Thorp Road	20**	<100–10	<100–24*	<100	<100	18	--
Wide Hollow Creek near mouth	26	40	33**	22	<100	43*	--
Granger Drain at mouth	80	170–460*	91–100**	24	<100	100**	--
Toppenish Creek at Indian Church Road	<100	40**	45*	16	<100	11	--
Sulphur Creek Wasteway near Sunnyside	120*	60–77**	68–74	24–29	<100	34	--
Yakima River at Kiona	40	20–44**	49*	17	<100	28	--
Dicamba							
Cle Elum River above Cle Elum Lake	<10	<10	<10	<10	<10	<10	--
Cherry Creek at Thrall	2,600*	140	520**	50	<10	240	--
Moxee Drain at Thorp Road	30*	10	10–20**	10	<10	<10	--
Wide Hollow Creek near mouth	<10	20*	<10	<10	<10	10**	--
Granger Drain at mouth	<10	70–130*	40**	<10	<10	20	--
Toppenish Creek at Indian Church Road	10	70*	10	10	<10	<10	--
Sulphur Creek Wasteway near Sunnyside	20*	10–20*	<10–20**	10	<10	<10	--
Yakima River at Kiona	10	10–20*	10	20*	<10	<10	--
Suspended sediment							
Cle Elum River above Cle Elum Lake	<1	3*	<1	1	2*	1	--
Cherry Creek at Thrall	91	64–121**	82	45	25	1,020*	--
Moxee Drain at Thorp Road	134–143	296–436**	443–607*	157	58	47	--
Wide Hollow Creek near mouth	17	8–28**	8	5	5	211*	--
Granger Drain at mouth	205	526–643*	421–432**	282	62	92	--
Toppenish Creek at Indian Church Road	32**	32	13	11	337*	30	--
Sulphur Creek Wasteway near Sunnyside	70	204–245*	99–128**	67–83	19	108	--
Yakima River at Kiona	28	30–35	22	35**	10	103*	57–71
Specific conductance							
Cle Elum River above Cle Elum Lake	47	37	31	45	50**	118*	--
Cherry Creek at Thrall	262	245–363	412**	334	460*	359	--
Moxee Drain at Thorp Road	279	275–295	266–320	280	649*	606**	--
Wide Hollow Creek near mouth	247	202–260	322	324**	446*	253	--
Granger Drain at mouth	418	357–360	328–350	304	652**	689*	--
Toppenish Creek at Indian Church Road	217	207–262	262	267	451*	402**	--
Sulphur Creek Wasteway near Sunnyside	264	239–259	333–381	290	669**	693*	--
Yakima River at Kiona	242	222–267	306**	320*	251	277	152–251

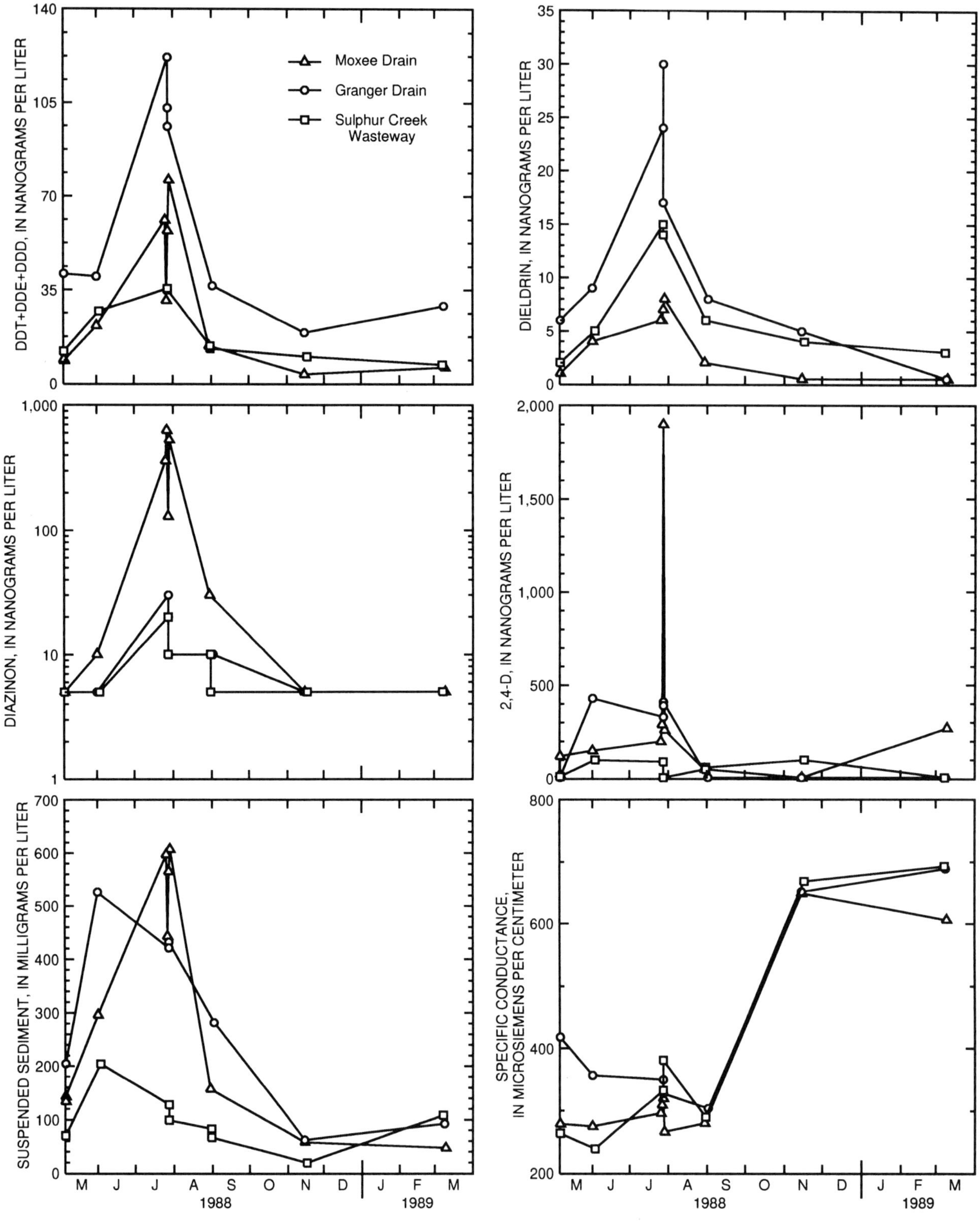

Figure 11. Concentrations of selected pesticides, suspended sediment, and specific conductance in unfiltered-water samples from major agricultural-return flows for selected months, Yakima River Basin, Washington, 1988–89. (Values less than the minimum reporting level are graphically represented as one-half the minimum reporting level. DDT+DDE+ DDD, 4,4'-dichlorodiphenyltrichloroethane [DDT] plus 4,4'-dichlorodiphenyldichloroethylene [DDE] plus 4,4'-dichlorodiphenyl-dichloroethane [DDD]; 2,4-D, (2,4-dichlorophenoxy) acetic acid. Lines between symbols are not intended to be used for interpolation.)

Table 11. Theoretical equilibrium partitioning of selected organic compounds sorbed to soil and dissolved in water

[K_{oc}, sediment-water partition coefficient normalized to organic carbon, in milliliters per gram; <, less than; >, greater than; 2,4-D, (2,4-dichlorophenoxy) acetic acid; DDT, 4,4'-dichlorodiphenyltrichloroethane; DDE, 4,4'-dichlorodiphenyldichloroethylene; DDD, 4,4'-dichlorodiphenyldichloroethane]

		Partitioning of compounds in soils prior to erosion		Partitioning of compounds to soil suspended in water	
Compound	K_{oc}	Percent of compound mass sorbed to soils prior to erosion[1]	Percent of compound mass dissolved in soil-pore water[1]	Percent of compound mass sorbed to suspended soils[2]	Percent of compound mass dissolved in water[2]
Atrazine	163	77	23	< 1	> 99
2,4-D	20	29	71	< 1	> 99
DDT	243,000	> 99	< 1	55	45
DDE	4,400,000	> 99	< 1	96	4
DDD	770,000	> 99	< 1	79	21
Diazinon	85	64	36	< 1	> 99
Dicamba	2	4	96	< 1	> 99
Dieldrin	1,700	97	3	< 1	> 99
Parathion	10,700	> 99	< 1	5	95
Simazine	138	74	26	< 1	> 99

[1]Calculation assumes sorption is controlled by partitioning to organic carbon, fraction organic carbon equals 1 percent, and soil moisture equals 10 percent.

[2]Calculation assumes sorption is controlled by partitioning to organic carbon, fraction organic carbon in soil equals 1 percent, and concentration of soil suspended in water is 500 milligrams per liter.

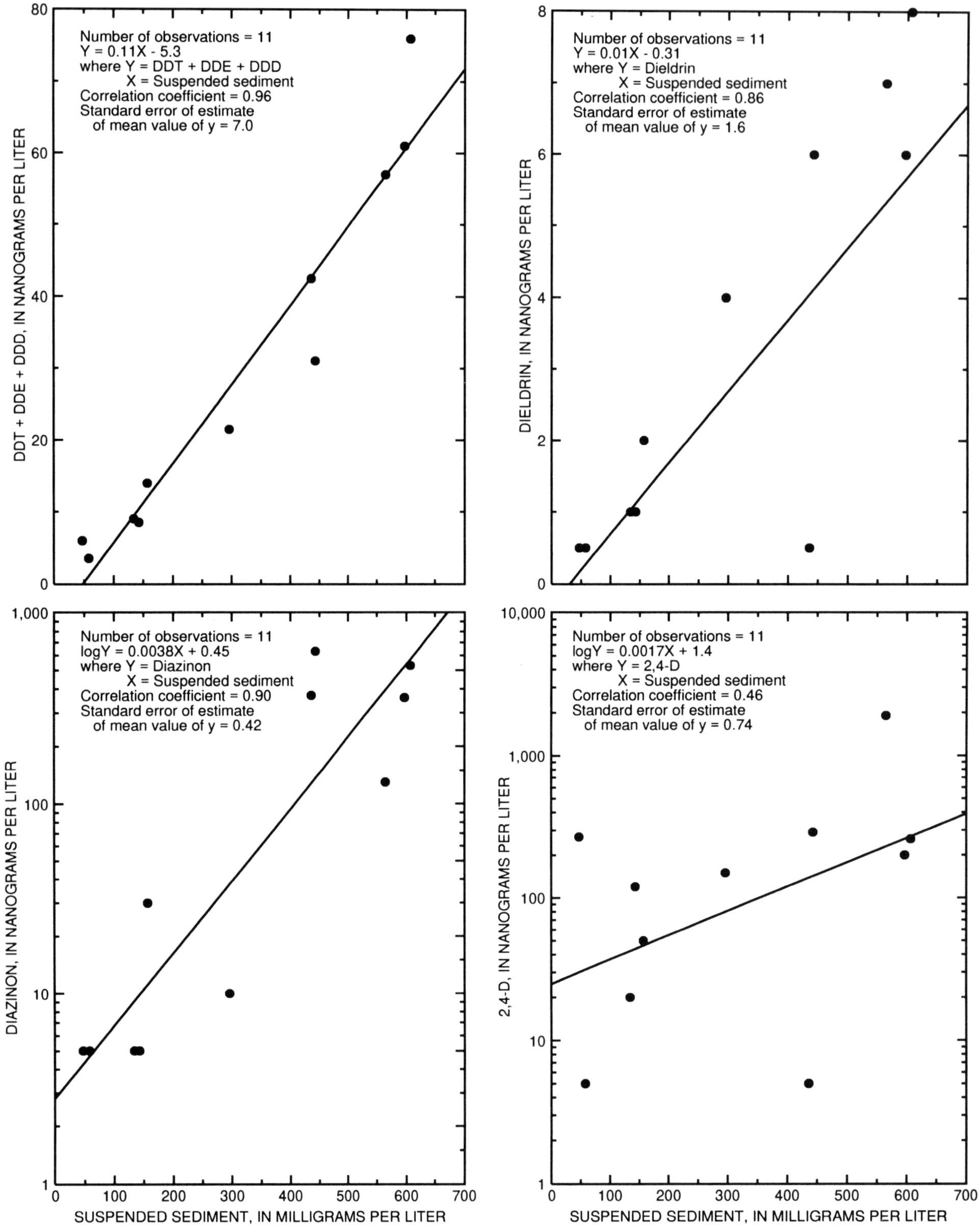

Figure 12. Relation of concentrations of suspended sediment and selected organic compounds in unfiltered-water samples from Moxee Drain at Thorp Road near Union Gap, Yakima River Basin, Washington, 1988–89. Values less than the minimum reporting level are graphically represented as one-half the minimum reporting level. DDT+DDE+DDD, 4,4'-dichlorodiphenyltrichloroethane [DDT] plus 4,4'-dichlorodiphenyldichloroethylene [DDE] plus 4,4'-dichlorodiphenyldichloroethane [DDD]; 2,4-D,(2,4-dichlorophenoxy) acetic acid.)

DDE+DDD and dieldrin, were sorbed to the soils and were transported to the streams by erosion. The hydrophilic compounds, including atrazine, simazine, and dicamba, probably were sorbed partially to soil and were dissolved partially in the soil-pore water. Consequently, they probably were transported to streams by erosion and by the flushing of compounds from the pore water. Hurle and Freed (1972) showed that the adsorption of hydrophilic nonionic pesticides to soils was increased in the presence of various salts (for example, fertilizers), so that the transport of these pesticides to the surface water would be enhanced by soil erosion of the sorbed pesticides or the flushing of the salts from the soil.

In December 1989, additional storm-runoff data were collected from the Yakima River at Kiona. Unlike the storm in March 1989, the December 1989 runoff originated from a rain-on-snow storm that occurred primarily in the Wenatchee and Snoqualamie National Forests upstream from agricultural activities. Even though suspended-sediment concentrations doubled from those levels typically observed during irrigation season in the Yakima River at Kiona, concentrations of DDT+DDE+DDD (hydrophobic compounds) did not increase, suggesting that (1) eroding soils from the forested headwater streams were minimally contaminated, and (2) resuspension of contaminated bed sediment from the Yakima River did not drastically increase concentrations of DDT+DDE+ DDD in the water column. Available information indicates that DDT was not applied historically in the Wenatchee and Snoqualamie National Forests (Bill Garrigues, U.S. Forest Service, Naches, Washington, oral commun., October 1992). Atmospheric deposition, however, is a possible source of the low-level concentrations of DDT in the National forests.

In addition to collecting data for describing monthly patterns, a few samples were collected during peak irrigation season in July and August 1988 to examine hourly and daily variability. Variability in pesticide concentrations over a 4-day period at Moxee Drain was notably higher than the hourly variability at Granger Drain and Sulphur Creek Wasteway (table 12). Samples collected from Granger Drain or Sulphur Creek Wasteway within a 5-hour period had minimum and maximum pesticide concentrations that generally ranged within a factor of two; the median factor for all pesticides listed in table 12 is about 1, indicating nearly identical concentrations. Conversely, samples collected daily over a 4-day period at Moxee Drain had concentrations that ranged within a factor of 12; the median factor was about 2.5. The larger variability in compound concentrations over a 4-day period suggests the river system was not at steady-state conditions relative to source loads and concentrations of organic compounds in the streams. For modeling pesticide concentrations and loads, this variability indicates the necessity for collecting data in carefully timed Lagrangian synoptic samplings.

Occurrence and Spatial Distribution

Six general groups of synthetic organic compounds are discussed in this section of the report: (1) carbamates and other miscellaneous compounds; (2) chlorophenoxy-acid herbicides, dicamba, and picloram; (3) organophosphorus compounds; (4) triazine herbicides and other nitrogen-containing compounds; (5) organochlorine compounds; and (6) volatile organic compounds. Information concerning properties, occurrence, and exceedances of water-quality criteria are presented and discussed for each group. Long-term trends in compound concentrations also are discussed for several pesticides that were monitored in the 1970s and in 1985. Sources and loads of organic compounds primarily are discussed in the next section of the report. In this section of the report, two different data sets are used to evaluate occurrence, spatial distribution, and exceedances of water-quality criteria: (1) conventional laboratory analyses of unfiltered-water samples (generally, 1-L samples) collected primarily in 1988–89 (analyses by U.S. Geological Survey National Water-Quality Laboratory, Arvada, Colorado), and (2) research analyses of large sample volumes of filtered water and large masses of suspended sediment collected in June 1989 (U.S. Geological Survey's Methods Research and Development Program, Arvada, Colorado).

The first data set consists of conventional analyses of unfiltered-water samples collected primarily at 8 stations in May, June, July, August/September, November, 1988 and March 1989, and at an additional 10 stations in July 1988 (Rinella, McKenzie, Crawford, and others, 1992). A few unfiltered-water samples also were collected in 1991 for carbamate and chlorophenoxy-acid compounds. Most of these stations were located at agricultural return flows or at reaches in the Yakima River downstream from agricultural activities. Samplings from May 1988 through

Table 12. Maximum and minimum concentrations of selected pesticides in unfiltered-water samples collected daily and (or) hourly from major agricultural-return flows, Yakima River Basin, Washington, July-August 1988

[Range of minimum to maximum concentrations reported in nanograms per liter; 2,4-D, (2,4-dichlorophenoxy) acetic acid; DDD, 4,4'-dichlorodiphenyldichloroethane; DDE, 4,4'-dichlorodiphenyldichloroethylene; DDT, 4,4'-dichlorodiphenyltrichloroethane; <, less than; --, not analyzed]

	Moxee Drain at Thorp Road	Granger Drain at mouth	Sulphur Creek Wasteway	
	Sampling dates	Sampling date and times	Sampling dates and times	
Compounds	July 26, 27, 28, 29	July 28 at 0730 0930, 1015 hours	July 28 at 1200 and 1700 hours	August 31 at 1530 and 1630 hours
Benzoic compound and chlorophenoxy-acid herbicides				
Dicamba	10–20	40–40	<10–20	10–10
2,4-D	200–1,900	330–410	<10–90	50–60
Organochlorine compounds				
DDD	1–2	7–11	<1–3	1–1
DDE	16–40	47–61	15–19	7–7
DDT	14–34	39–50	16–17	5–6
Dieldrin	6–8	17–30	6–6	6–6
Endosulfan I	24–190	76–100	5–11	5–11
Endrin	<1–1	<1–<1	<1–<1	<1–<1
Organophosphorus compounds				
Diazinon	130–630	10–30	<10–10	<10–10
Ethion	40–410	<10–<10	<10–<10	<10–<10
Parathion	10–100	<10–<10	<10–<10	<10–<10
Phorate	20–240	--	10–10	--
Triazine compounds				
Atrazine	9–17	40–47	25–31	34–34
Simazine	15–24	91–100	68–74	24–29

March 1989 provide data on temporal variability, and the sampling in July 1988 provides synoptic data on pesticide sources and transport in the Yakima River Basin during peak irrigation. These samples were analyzed using conventional laboratory analyses, so minimum reporting levels were about 10 to 100 times larger than those in the June 1989 synoptic sampling. The second data set consists of analyses of filtered-water and suspended-sediment samples collected in a synoptic sampling of 29 stations in June 1989. Research techniques were used in this sampling to separately analyze the dissolved and suspended phases for pesticides (Rinella, McKenzie, Crawford, and others, 1992). Sampling stations in June 1989 included 4 stations upstream from agricultural activities, 3 drinking-water supplies, 15 agricultural return flows, 6 stations on the Yakima River, and the mouth of the Naches River. The primary objective for this research was to reduce minimum reporting levels in order to better describe the occurrence, major sources, and transport of dissolved and suspended pesticides in the Yakima River Basin. The synoptic sampling occurred near or during peak irrigation, when pesticide concentrations were expected to be relatively high on the basis of temporal data collected in 1988.

The volume of streamflow, which is extensively regulated by storage reservoirs and irrigation diversions in the Yakima River Basin, directly affects dilution and the transport of organic compounds in streams. Instantaneous streamflow in the Yakima River during the June 1989 synoptic sampling is shown in figure 13. This streamflow pattern is typical of flow conditions in the basin from late June through August. Upstream from Roza Canal, the flow in the Yakima River is relatively large as a result of flow augmentation from Keechelus, Kachess, and Cle Elum Lakes. Major agricultural return flows to the Yakima River from Kittitas Valley are Wilson and Cherry Creeks at RM 147. The large flow in the Yakima River in this reach (about 3,600 ft^3/s [cubic feet per second]) dilutes concentrations of organic compounds from these return flows by as much as 10 times. At Yakima RM 127.9, Roza Canal diverts about 2,000 ft^3/s, leaving 1,600 ft^3/s in the main stem. Streamflows from the Naches River (RM 116.3) and the Roza Power Plant return (RM 113.3) contain relatively low concentrations of organic compounds and increase the main-stem flow to about 3,500 ft^3/s. Wide Hollow Creek, Moxee Drain, and Ahtanum Creek then contribute agricultural return flows at RMs 107.4, 107.3, and 106.9, respectively. Concentrations of organic compounds from these agricultural return flows also are

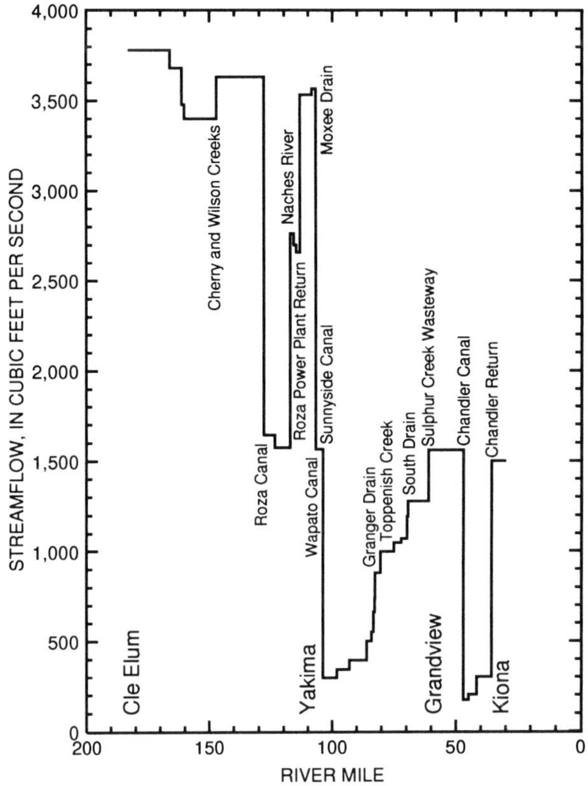

Figure 13. Instantaneous streamflow in the Yakima River resulting from tributary contributions and canal diversions, Yakima River Basin, Washington, June 25–30, 1989.

diluted by the relatively large flow in the Yakima River at Union Gap (RM 107.3). Even though Moxee Drain contributes major loads of several organic compounds to the Yakima River during irrigation season, the compound loads in the Yakima River are reduced quickly because a large portion of the Yakima River flow and accompanying pesticide loads are diverted into Wapato Canal (RM 106.7) and Sunnyside Canal (RM 103.8) a few miles downstream. Much of the canal water is then distributed as irrigation water onto agricultural fields. The flow remaining in the Yakima River near Parker (RM 103.7) is only about 200 ft^3/s, which provides minimal flow for dilution, making the lower Yakima River vulnerable to inflowing organic compounds from agricultural return flows and point-source discharges. Downstream from Parker, numerous agricultural return flows feed into the Yakima River, typically accounting for as much as 80 percent of the main-stem flow. As a result, return flows in the lower basin control water-quality conditions in the lower Yakima River during most of the irrigation season. Owing to extensive irrigation-water diversions that reduce main-stem flow, and increasing contributions of agricultural runoff from the lower Basin, concentrations of many organic compounds increase down the main stem.

Carbamates and Other Compounds

The use of carbamate pesticides has been increasing since the 1970s. These compounds have relatively short half-lives in soils, especially when compared to organochlorine compounds, (table 61, at back of report) (Smith and others, 1988). Carbamates are not stable in the environment and undergo rapid chemical and biological degradation. Under alkaline conditions, they quickly hydrolyze. For example, the half-life of carbaryl is 10 days in water at pH 7 and 15 minutes at pH 10 (Biggar and Seiber, 1987). Carbamates have relatively high solubilities and low K_{oc} values (table 61, at back of report), so that sorption and bioaccumulation generally are not important fate processes. As a result of their high solubility and short half-lives, carbamate concentrations may vary widely in streams, and consequently, their occurrence is difficult to document (Liz Block, U.S. Fish and Wildlife Service, Olympia, Washington, written commun., January 1993).

Analyses of unfiltered-water samples collected from 19 stations in the Yakima River Basin during 1988–91 (table 13) show that carbaryl and methomyl were the only carbamates quantified using conventional analyses. Pesticide-use estimates indicate that carbaryl and methomyl also are among the most abundantly used carbamates in the basin (table 14). The small number of carbamate quantifications in the 1988–91 data is consistent with a previous study by Johnson and others (1986), who did not detect carbamates in the basin during a synoptic sampling in the summer of 1985 (minimum reporting levels of 5,000 ηg/L). In contrast, research data (minimum reporting levels of about 1 ηg/L for large sample volumes) collected in June 1989 showed that carbaryl and *S*-ethyl dipropylthiocarbamate (EPTC) were quantified at 2 and 17 of the 21 and 29 stations that were sampled, respectively (table 14) (Rinella, McKenzie, Crawford, and others, 1992).

In the June 1989 synoptic sampling, pesticides that were used in the basin in greater abundance generally were detected at a higher frequency in streams (table 14). Two notable exceptions were bromacil and butylate. Bromacil is more persistent in soils than the other compounds, which may account for its higher frequency of detections. The reason butylate was not detected is unknown. Carbamate- and thiocarbamate-compound concentrations were less than 40 ηg/L, which probably reflects their short half-lives in soils.

EPTC and propargite were detected frequently in the June 1989 synoptic sampling (table 60, at back of report). EPTC is an herbicide used extensively on alfalfa, corn, and potatoes, and propargite is an acaricide used extensively on hops, mint, apples, and grapes. These compounds were detected at most of the agricultural return flows and main-stem stations located downstream from the city of Yakima.

Water-quality criteria have not been established for carbamates and other compounds listed in table 13 and 14. The NAS-NAE (1973), however, has recommended that carbaryl not exceed 20 ηg/L for the protection of freshwater aquatic life (table 62, at back of report). The maximum concentration detected in the Yakima River Basin was 14 ηg/L at Cherry Creek at Thrall (tables 13 and 14).

Chlorophenoxy-acid Herbicides, Dicamba, and Picloram

Chlorophenoxy-acid herbicides investigated in this study include 2,4-D, 2,4-DP (dichlorprop), fenoprop (silvex), and 2,4,5-T. Dicamba is a benzoic-acid herbicide, and picloram is a pyridine herbicide. Compounds in these groups that were applied in the Yakima River Basin in 1989 include 2,4-D (260,000 kg), dicamba (36,000 kg), and picloram (1,500 kg). Production and sale of 2,4,5-T and fenoprop were discontinued in the United States in 1978 and 1984, respectively. 2,4-D is the most heavily applied herbicide in the basin, and dicamba ranks within the top 10 herbicides used (table 3). These herbicides (2,4-D, dicamba, and picloram) are highly soluble in water and have low sediment-water partition coefficients (table 61, at back of report), so they do not significantly bioaccumulate or sorb to sediment (Smith and others, 1988). Dicamba and 2,4-D have half-lives of less than 2 weeks in soils (table 61, at back of report). Picloram is more persistent, with a half-life of about 3 months.

Analyses of 1-L unfiltered-water samples collected from 20 stations in the Yakima River Basin during 1988–91 (table 15) show that 2,4-D and dicamba were the most frequently detected compounds of this group (minimum reporting level of 10 ηg/L). These herbicides were detected in more than 50 percent of the samples (Rinella, McKenzie, Crawford, and others, 1992). Similar to the occurrence of carbamates, the number of detections coincides with the degree of

Table 13. Summary of carbamate concentrations in 1-liter unfiltered-water samples, Yakima River Basin, Washington, 1988–91

[Concentrations and minimum reporting levels reported in nanograms per liter; --, insufficient data to calculate statistic; some compound concentrations were quantified below the minimum reporting level]

Compound	Number of samples	Approximate minimum reporting level	Number of quantifiable concentrations	Quantifiable concentrations		
				Minimum	Median	Maximum
Aldicarb	77	500	0	--	--	--
Aldicarb sulfone	79	500	0	--	--	--
Carbaryl	79	500	4	2	5	14
Carbofuran	79	500	0	--	--	--
3-Hydroxycarbofuran	79	500	0	--	--	--
Methiocarb	70	500	0	--	--	--
Methomyl	79	500	4	30	80	140
1-Naphthol	79	500	0	--	--	--
Oxamyl	79	500	0	--	--	--
Propham	77	500	0	--	--	--
Propoxur	70	500	0	--	--	--

pesticide use in the basin. To a lesser extent, picloram and 2,4-DP also were detected (table 15). Previous sampling of the Yakima River at Kiona and several major agricultural return flows by Johnson and others (1986) in August 1985 showed that 2,4-D, dicamba, fenoprop, and 2,4,5-T were not detected (minimum reporting level of 100 ηg/L). Apparent inconsistencies in detections between the 1985 data and the 1988–91 data probably resulted from lower reporting levels and increased sampling frequency in the latter study.

Analyses of filtered-water samples collected from 28 stations in June 1989 show that 2,4-D and dicamba were detected at 12 and 7 stations, respectively (table 60, at back of report). Both of these compounds were applied throughout the basin in the Kittitas, Tieton, East-side, and West-side areas (see fig. 4 for location of areas and table 59, at back of report, for pesticide applications). As expected, according to areas of application, 2,4-D was detected at most of the agricultural return flows and main-stem stations. Even though dicamba was used throughout the basin, it was not detected in the West-side tributaries or in the main stem. One hypothesis is that travel time from the fields, where dicamba is applied, to the stream sampling stations is longer for the West-side area because of longer travel distances, smaller subbasin slopes and extensive reuse of irrigation water. As a result, dicamba has more time to degrade to concentrations below minimum reporting levels. This hypothesis also is consistent with the lack of detections in the lower Yakima River. Even though East-side tributaries are contributing dicamba to the main stem, dicamba has time to degrade to concentrations below minimum reporting levels in the lower main stem.

Water-quality criteria have not been established for chlorophenoxy-acid herbicides, dicamba, or picloram. The NAS-NAE (1973), however, recommends that 2,4-D, fenoprop, and dicamba not exceed 3,000, 1,400, and 200,000 ηg/L, respectively, for the protection of freshwater aquatic life (table 62, at back of report). These recommendations were not exceeded except for one sample for 2,4-D at Cherry Creek at Thrall (May 5, 1988), which had a concentration of 7,500 ηg/L (Rinella, McKenzie, Crawford, and others, 1992). Concentrations of 2,4-D at Cherry Creek in 6 other samples collected from June 1988 to March 1989 were more than an order of magnitude lower. The higher concentration at Cherry Creek is consistent with the greater usage of 2,4-D in the Kittitas area (table 59, at back of report).

Long-term historical data are available for 2,4-D concentrations in the Yakima River at Kiona (fig. 14). The limited amount of data suggest that concentrations of 2,4-D have decreased by as much as a factor of 10 since 1968. Historical data on the annual use of 2,4-D are not available to explain the decrease. Monthly data collected from the Yakima River at Kiona in 1970 and 1971 show that 2,4,5-T was detected four times at 10 ηg/L (minimum reporting level of 10 ηg/L). In subsequent samplings during 1972–91, including the NAWQA sampling during 1988–91, 2,4,5-T was not detected (minimum reporting level also 10 ηg/L) even though its use in the United States was not discontinued until 1978. The detection of 2,4-D and 2,4,5-T indicates that polychlorinated dibenzo-*p*-dioxins may be present. These compounds are extremely toxic and are produced inadvertently during manufacture of 2,4,5-T and 2,4-D (Smith and others, 1988).

Table 14. Estimated annual use of carbamates and other compounds, half-life in soil, number of detections, and summary of concentrations in large-volume filtered-water samples, Yakima River Basin, Washington, June 25–30, 1989

[Concentrations reported in nanograms per liter; Number of detections, the number of quantifiable determinations plus the number of determinations when the compound was detected but could not be quantified because its concentration was less than the minimum reporting level; EPTC, *S*-ethyl dipropylthiocarbamate; --, insufficient data to calculate statistic; na, not analyzed; data and minimum reporting levels reported in Rinella, McKenzie, Crawford, and others, 1992]

Compound	Compound group	Annual use, in kilograms	Half-life in soil, in days	Number of samples	Number of detections	Number of quantifiable concentrations	Quantifiable concentrations		
							Minimum	Median	Maximum
Carbaryl	Carbamate	55,000	7–28	21	11	2	1.5	--	13
Propargite	Sulfite	47,000	14–56	29	19	16	.57	23	260
EPTC	Carbamate	16,000	6–30	29	19	17	1.0	3.1	37
Butylate	Thiocarbamate	16,000	12–13	29	0	0	--	--	--
Methomyl	Carbamate	15,000	30–33	na	na	na	na	na	na
Oxamyl	Carbamate	9,100	4–7	na	na	na	na	na	na
Carbofuran	Carbamate	3,300	30–60	29	7	3	.96	1.5	3.7
cis- and trans-Permethrin	Pyrethoid	1,800	30–38	29	2 (cis) 3 (trans)	0 (cis) 0 (trans)	--	--	--
Bromacil	Uracil	1,600	60–350	29	12	3	1.7	13	27
Linuron	Urea	1,600	60–150	3	1	1	--	1.9	--
Triadimefon	Conazole	730	6–26	28	5	0	--	--	--
Propoxur	Carbamate	260	30	na	na	na	na	na	na
Flucythrinate	Pyrethroid	0	21	29	2	0	--	--	--
Vernolate	Thiocarbamate	0	8–>64	29	8	1	--	.94	--

Table 15. Summary of concentrations of chlorophenoxy-acid herbicides, dicamba, and picloram in 1-liter unfiltered- and filtered-water samples, Yakima River Basin, Washington, 1988–91

[Concentrations reported in nanograms per liter; 2,4-D, (2,4-dichlorophenoxy) acetic acid; 2,4-DP, 2-(2,4-dichlorophenoxy) propionic acid; 2,4,5-T represents (2,4,5-trichlorophenoxy) acetic acid; <, less than]

Compound	Number of samples	Minimum concentration	Concentration at indicated percentile					Maximum concentration
			10	25	50	75	90	
Unfiltered-water samples collected from 1988–91								
2,4-D	78	<10	<10	<10	40	110	290	7,500
2,4-DP (dichlorprop)	78	<10	<10	<10	<10	<10	<10	200
Dicamba	78	<10	<10	<10	10	20	70	2,600
Fenoprop (silvex)	78	<10	<10	<10	<10	<10	<10	<10
Picloram	78	<10	<10	<10	<10	<10	10	70
2,4,5-T	78	<10	<10	<10	<10	<10	<10	<10
Filtered-water samples collected from June 25–30, 1989								
2,4-D	28	<10	<10	<10	<10	70	140	290
2,4-DP (dichlorprop)	28	<10	<10	<10	<10	<10	<10	<10
Dicamba	28	<10	<10	<10	<10	10	50	100
Fenoprop	28	<10	<10	<10	<10	<10	<10	<10
Picloram	28	<10	<10	<10	<10	<10	<10	30
2,4,5-T	28	<10	<10	<10	<10	<10	<10	<10

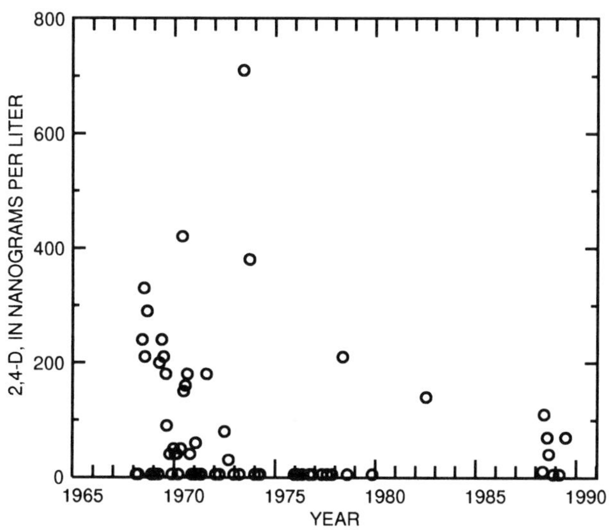

Figure 14. Concentrations of (2,4-dichlorophenoxy) acetic acid [2,4-D] in unfiltered-water samples from the Yakima River at Kiona, Yakima River Basin, Washington, 1968–89. (Values less than the minimum reporting level are graphically represented as one-half the minimum reporting level.)

Organophosphorus Compounds

Organophosphorus compounds became important as insecticides after World War II. These compounds were developed in the course of chemical-warfare research in Germany, and several of the compounds were found to be highly toxic to humans as well as insects (Sawyer and McCarty, 1967). The insecticide parathion was introduced into the United States in 1946 and has been used cautiously because it is quite toxic to humans. A more widely used organophosphorus insecticide is malathion, because it is highly toxic to insects yet has low toxicity to mammals (Sawyer and McCarty, 1967). In response to environmental concern over the persistence of organochlorine compounds, the use of organophosphorus pesticides has increased since the 1970s. Organophosphorus compounds have relatively short half-lives in soils (table 16), and most are highly soluble, with water solubilities higher than 30 mg/L. Similar to carbamate compounds, bioaccumulation is not an important fate process for organophosphorus insecticides (Smith and others, 1988), and K_{oc} values generally are less than 5,000 mL/g (table 61, at back of report).

Analyses of 1-L unfiltered-water samples collected from 20 stations, mostly agricultural return flows, in the Yakima River Basin during 1988–89 show that diazinon was the most frequently quantified organophosphorus compound (table 17 of this report; Rinella, McKenzie, Crawford, and others, 1992). Other compounds that were quantified include chlorpyrifos, disulfoton, ethion, malathion, parathion, and phorate. A previous sampling of the Yakima River at Kiona and several major agricultural return flows by Johnson and others (1986) in August 1985 showed that diazinon was the only organophosphorus compound quantified, with a concentration of 1,700 ηg/L in Moxee Drain (minimum reporting level of 500 ηg/L). The 1985 and

Table 16. Estimated annual use of organophosphorus compounds, number of detections in large-volume filtered-water samples from the June-1989 synoptic sampling, and half-life in soil, Yakima River Basin, Washington

[Number of detections, the number of quantifiable determinations plus the number of determinations when the compound was detected but could not be quantified because it was less than the minimum reporting level; DEF, *S,S,S*-tributyl phosphorotrithioate; --, no data available]

Compound	Annual use, in kilograms	Number of detections[1]	Half-life in soil, in days
Malathion	160,000	27	1
Azinphos-methyl	130,000	[2]9	10–40
Chlorpyrifos	73,000	14	7–141
Diazinon	73,000	22	32–40
Parathion	55,000	14	14–18
Phosphamidon	36,000	20	17
Disulfoton	30,000	9	5–30
Methidathion	26,000	3	7–21
Dimethoate	20,000	11	7
Phorate	9,100	9	2–90
Fonofos	5,500	2	21–112
Methyl parathion	2,200	6	5–15
Ethion	1,100	9	150
Terbufos	1,100	0	5–27
Mevinphos	470	1	3
Demeton-S	3	3	1–10
DEF	0	0	--
Isofenphos	0	2	150
Methyl trithion	0	2	--
Trithion	0	1	--

[1]Twenty-nine samples from 29 stations were analyzed in June 1989.
[2]Seventeen samples were analyzed.

1988–89 data are consistent, considering that most quantified concentrations in the 1988–89 data were below the minimum reporting levels applicable or achievable in 1985.

In June 1989, large quantities of filtered water and suspended sediment, collected synoptically from 29 stations throughout the basin, were analyzed for organophosphorus compounds. Equilibrium-partitioning calculations indicate that organophosphorus insecticides should be sorbed mostly to soils prior to transport into streams. After the suspension of soil particles in the runoff, however, partitioning calculations indicate that these compounds should desorb and partition mostly into the dissolved phase. This transport mechanism helps to explain why increasing suspended-sediment concentrations correlate with increasing concentrations of hydrophilic organophosphorus compounds (rank-correlation coefficients for diazinon, malathion, and phosphamidon are significant—ρ [probability level] less than or equal to 0.02) (fig. 15). If dissolution from suspended soils is rapid in streams in the Yakima River Basin, then the primary mode of stream transport is the dissolved phase. Owing to uncertainty concerning rates of dissolution for organophosphorus compounds, both the dissolved and suspended phases were analyzed in the June 1989 synoptic sampling.

Synoptic data show that malathion, diazinon, phosphamidon, azinphos-methyl, parathion, and chlorpyrifos were detected in about 50 percent or more of the samples (table 18). Because of instrumental calibration problems, azinphos-methyl was analyzed quantitatively in only three samples; it was, however, qualitatively detected in 9 of 17 samples. As observed for carbamates and chlorophenoxy-acid herbicides, detections of organophosphorus compounds generally increased with increases in annual use (table 16). Several inconsistencies in this relation include fewer than expected detections for methidathion and an increased number of detections for ethion, considering the small annual use of ethion. The unexpected increase in detections for ethion may be explained by its considerably longer half-life.

Data from the synoptic sampling in June 1989 (table 18) and a monthly sampling of selected agricultural return flows during May-September 1991 (table 19) show that the highest concentrations of organophosphorus compounds occur in the dissolved phase (filtered water). Observed partitioning and calculated equilibrium partitioning of diazinon at selected stations indicate that diazinon exists primarily in the filtered water (table 20). Data from several agricultural return flows (table 20) suggest that dissolution of most of the diazinon from agricultural soils is rapid once the soils are suspended in the runoff. The bias between the observed and calculated partitioning using a K_{oc} value of 85 mL/g indicates that diazinon may not be in equilibrium or that the K_{oc} for sediment from the Yakima River Basin might actually be 1,000 mL/g, which is the maximum K_{oc} value listed for diazinon in table 61 (at back of report).

In 1989, organophosphorus compounds were applied throughout agricultural areas in the basin (table 59, at back of report). Accordingly, the compounds were detected in most of the agricultural return flows in June 1989, except in Kittitas Valley (table 60, at back of report). Field applications of those organophosphorus compounds listed in table 60 (at back of report) were relatively small in Kittitas Valley, which likely accounts for the lack of analytical detections in Cherry and Wilson Creeks in June 1989. As a result of

Table 17. Summary of concentrations of organophosphorus compounds in 1-liter unfiltered-water samples, Yakima River Basin, Washington, 1988–89

[Concentrations reported in nanograms per liter; <, less than; --, insufficient data to calculate statistic]

Compound	Number of samples	Minimum concentration	Concentration at indicated percentile					Maximum concentration
			10	25	50	75	90	
Chlorpyrifos	3	<10	--	--	10	--	--	10
Diazinon	77	<10	<10	<10	<10	20	130	630
Disulfoton	5	<10	--	--	<10	--	--	70
Ethion	77	<10	<10	<10	<10	<10	<10	410
Malathion	77	<10	<10	<10	<10	<10	<10	60
Methyl parathion	77	<10	<10	<10	<10	<10	<10	<10
Methyl trithion	77	<10	<10	<10	<10	<10	<10	<10
Parathion	77	<10	<10	<10	<10	<10	10	100
Phorate	13	<10	<10	10	10	100	200	240
Trithion	77	<10	<10	<10	<10	<10	<10	<10

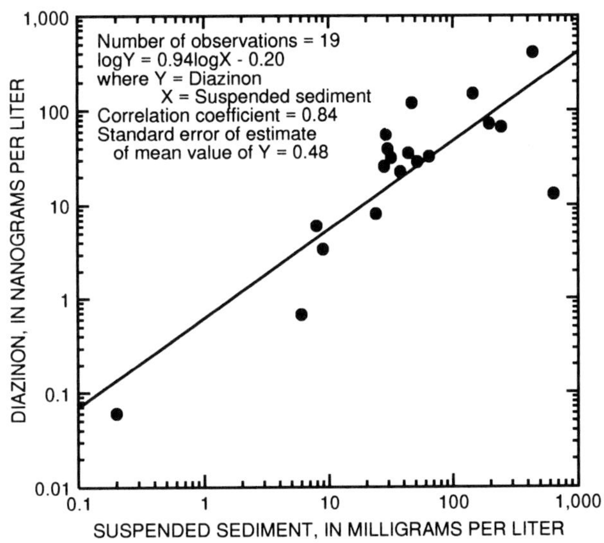

Figure 15. Relation of concentrations of suspended sediment and diazinon in unfiltered-water samples from selected stations, Yakima River Basin, Washington, June 25–30, 1989. (Values less than the minimum reporting level are excluded.)

pesticide loads from agricultural return flows in the lower basin and reduced main-stem streamflows caused by diversions through Wapato and Sunnyside Canals, concentrations of organophosphorus compounds generally increased in the lower Yakima River downstream from the city of Yakima (fig. 16 and table 60, at back of report).

Chronic-toxicity water-quality criteria (U.S. Environmental Protection Agency, 1986) for the protection of freshwater aquatic life have been established for azinphos-methyl, chlorpyrifos, demeton, malathion, and parathion (table 62, at back of report). Water-quality guidelines have been recommended by the NAS-NAE (1973) for diazinon, disulfoton, ethion, mevinphos, and phosphamidon. During 1988–91, measured concentrations of azinphos-methyl, chlorpyrifos, demeton, and mevinphos did not exceed water-quality criteria or guidelines. Azinphos-methyl, however, was analyzed quantitatively at only three stations, and its mean concentration of 7.4 ηg/L (table 18) in the Yakima River at Kiona in June 1989 was close to the chronic-toxicity criterion of 10 ηg/L. These few data suggest that azinphos-methyl concentrations may be exceeding the criterion in agricultural return flows. Two samples collected by the North Yakima Conservation District (1991) in the Moxee Subbasin had azinphos-methyl concentrations of 64,000 ηg/L in a ditch at the downstream end of an orchard during spraying, and 3,000 ηg/L in the return flow about one mile downstream from the orchard. These concentrations substantially exceed the criterion for azinphos-methyl by 6,400 and 300 times, respectively. Other organophosphorus compounds that exceeded EPA chronic-toxicity criteria or NAS-NAE guidelines during 1988–91 are listed in table 21. Parathion and malathion exceeded EPA criteria at 11 and 1 station(s), respectively; diazinon, disulfoton, ethion, and phosphamidon exceeded NAS-NAE guideline at 16, 1, 1, and 4 station(s), respectively. In 1989, diazinon and parathion were used extensively in the lower Yakima River Basin (table 59, at back of report); as a result of their extensive use and low water-quality guideline/criterion values for the protection of freshwater aquatic life (9 ηg/L for diazinon and 13 ηg/L for parathion), these compounds exceeded guideline/criterion values downstream from the city of Yakima (fig. 16). These compounds were used minimally in Kittitas Valley; therefore, exceedances were not

Table 18. Summary of concentrations of organophosphorus compounds in large-volume water samples, Yakima River Basin, Washington, June 25–30, 1989

[Concentrations reported in nanograms per liter; Number of detections, the number of quantifiable determinations plus the number of determinations when the compound was detected but could not be quantified because it was less than the minimum reporting level; DEF, *S,S,S*-tributyl phosphorotrithioate; --, insufficient data to calculate statistic; 29 stations were sampled in June 1989; data and minimum reporting levels reported in Rinella, McKenzie, Crawford, and others, 1992]

Compound	Summary for filtered-water samples						Summary for suspended-sediment samples					
	Number of samples	Number of detections	Number of quantifiable concentrations	Quantifiable concentration			Number of samples	Number of detections	Number of quantifiable concentrations	Quantifiable concentrations		
				Minimum	Median	Maximum				Minimum	Median	Maximum
Azinphos-methyl	[1]17	[1]9	[1]1	--	[1]7.4	--	0	--	--	--	--	--
Chlorpyrifos	29	14	4	1.0	3.0	9.6	29	12	8	0.1	0.2	0.48
DEF	29	0	0	--	--	--	16	0	0	--	--	--
Demeton-S	29	3	0	--	--	--	0	--	--	--	--	--
Diazinon	29	22	18	.67	32	410	29	16	14	.06	.3	3.3
Dimethoate	29	11	5	4.6	7.3	330	0	--	--	--	--	--
Disulfoton	29	9	0	--	--	--	29	1	1	--	.6	--
Ethion	29	9	0	--	--	--	29	6	2	.5	--	.52
Fonofos	29	2	0	--	--	--	16	0	0	--	--	--
Isofenphos	29	2	0	--	--	--	16	1	0	--	--	--
Malathion	29	27	17	1.5	13	70	29	0	0	--	--	--
Methidathion	29	3	2	4.9	13	140	0	--	--	--	--	--
Methyl parathion	29	6	3	5.6	13	16	29	2	2	.3	--	.6
Methyl trithion	29	2	0	--	--	--	29	0	0	--	--	--
Mevinphos	29	1	0	--	--	--	0	--	--	--	--	--
Parathion	28	14	8	7.4	14	180	29	10	9	.1	.17	.5
Phorate	29	9	5	3.6	10	130	29	3	2	.3	--	1.4
Phosphamidon	29	20	11	5.3	16	56	0	--	--	--	--	--
Terbufos	29	0	0	--	--	--	0	--	--	--	--	--
Trithion	29	1	0	--	--	--	29	0	0	--	--	--

[1]Azinphos-methyl was detected in three replicate filtered-water samples from the Yakima River at Kiona at concentrations of 9.0, 6.5, and 6.7 nanograms per liter, which equals an average concentration of 7.4 nanograms per liter.

Table 19. Summary of concentrations of organophosphorus compounds in water samples, Yakima River Basin, Washington, May-September 1991

[Concentrations reported in nanograms per liter; DEF, *S,S,S*-tributyl phosphorotrithioate; --, insufficient data to calculate statistic; 23 samples from 6 stations were analyzed in May-September 1991; data and minimum reporting levels reported in Rinella, McKenzie, Crawford, and others, 1992]

Compound	Summary for filtered-water samples: Number of quantifiable concentrations	Quantifiable concentration: Minimum	Median	Maximum	Summary for suspended--sediment samples: Number of quantifiable concentrations	Quantifiable concentrations: Minimum	Median	Maximum
Chlorpyrifos	4	10	10	10	1	--	0.8	--
DEF	0	--	--	--	0	--	--	--
Diazinon	11	10	20	1,300	5	1	1	30
Disulfoton	0	--	--	--	0	--	--	--
Ethion	2	20	--	20	2	1	--	1
Fonofos	1	--	10	--	0	--	--	--
Malathion	9	10	20	290	0	--	--	--
Methyl parathion	6	10	25	60	11	.4	1	7
Methyl trithion	0	--	--	--	0	--	--	--
Parathion	8	10	10	20	0	--	--	--
Phorate	4	10	10	10	1	--	.6	--
Trithion	0	--	--	--	0	--	--	--

Table 20. Partitioning of diazinon in filtered water and suspended sediment for selected stations, Yakima River Basin, Washington, June 25–30, 1989

[For theoretical equilibrium-partitioning calculations, the sediment-water partition coefficient = 85 milliliters per gram; concentrations reported in nanograms per liter]

Station number	Station name	Diazinon concentration: Filtered water	Diazinon concentration: Suspended sediment	Percent of diazinon mass associated with suspended sediment: Observed	Percent of diazinon mass associated with suspended sediment: Theoretical equilibrium partitioning (calculated)
12507585	Yakima River at river mile 72	35	0.16	0.4	0.01
12508630	South Drain near Satus	150	2.9	2	.02
12508850	Sulphur Creek Wasteway	66	.18	.3	.02
12509829	Snipes Creek at mouth	32	.3	.9	.01
12510500	Yakima River at Kiona	39	.07	.2	.01

observed in the basin upstream from Moxee Drain. All stations, listed in table 21, are located at or downstream from Moxee Drain.

Long-term historical data were available for diazinon concentrations in the Yakima River at Kiona (fig. 17). The data show that diazinon concentrations at that location have exceeded the NAS-NAE (1973) guideline value of 9 ηg/L for the protection of freshwater aquatic life since 1972.

Triazine Herbicides and Other Organonitrogen Compounds

Triazine herbicides are used for selective preemergence and early postemergence control of broadleaf weeds and some grasses (Biggar and Seiber, 1987). Most triazine compounds analyzed in this study have water solubilities higher than 30 mg/L and K_{oc} values between 100 and 500 mL/g (table 61, at back of report). With K_{oc} values higher than 100 mL/g, these compounds readily partition to the soils in the agricultural fields (Biggar and Seiber, 1987). In a study of the Cedar River in Minnesota and Iowa, Squillace and Thurman (1992) showed that triazine herbicides undergo rapid dissolution from soils once the soils are suspended in overland flow and are transported to streams. Accordingly, triazine and other organonitrogen compounds were expected to occur primarily in the dissolved phase in Yakima River Basin streams. Similar to carbamates and organophosphorus compounds, these compounds with relatively high water solubilities and low K_{oc} values do not readily partition in lipids of aquatic organisms (Smith and others, 1988).

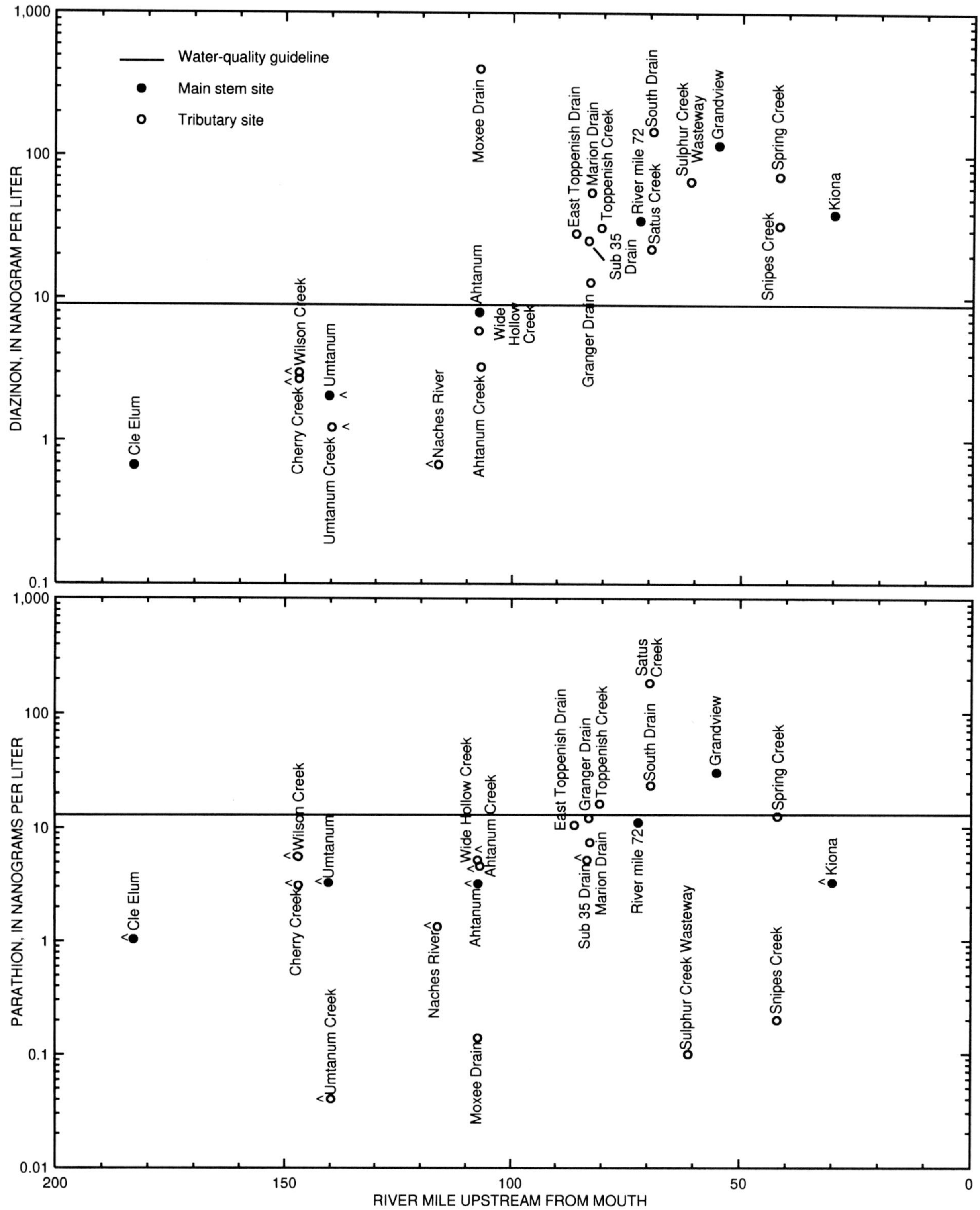

Figure 16. Concentrations of diazinon and parathion in large-volume water samples from selected stations, Yakima River Basin, Washington, June 25–30, 1989. (Concentrations represent the sum of filtered-water and suspended-sediment analyses. Water-quality guideline, the guideline for the protection of freshwater aquatic life recommended by the National Academy of Sciences-National Academy of Engineering Committee on Water Quality Criteria [1973]; <, data point represents a value less than the minimum reporting level.)

Table 21. Exceedances of water-quality criteria for the protection of freshwater aquatic life for organophosphorus compounds in water samples, Yakima River Basin, Washington, 1988–91

[Data are shown only when criteria or guidelines are equaled or exceeded and are based on analyses of either concentrations in unfiltered water or concentrations in both filtered water and suspended sediment; only one determination is included in listing for replicate samples collected within a 1-hour period; see table 62 for criteria and guidelines; --, no exceedance detected]

Station number	Station name	Number of samples equaling or exceeding criteria or guidelines and number of samples collected					
		Diazinon	Disulfoton	Ethion	Malathion	Parathion	Phosphamidon
12500430	Moxee Drain at Thorp Road[1]	12 of 16	--	7 of 16	--	5 of 16	1 of 1
12500445	Wide Hollow Creek near mouth	2 of 7	--	--	--	1 of 7	
12500450	Yakima River above Ahtanum Creek	1 of 2	--	--	--	--	--
12505350	East Toppenish Drain at Wilson Road	1 of 1	--	--	--	--	1 of 1
12505410	Sub 35 Drain at Parton Road	1 of 1	--	--	--	--	--
12505460	Granger Drain at mouth[2]	8 of 15	--	--	--	2 of 15	1 of 1
12505510	Marion Drain at Indian Church Road	2 of 2	--	--	--	1 of 2	--
12507508	Toppenish Creek at Indian Church Road	3 of 7	--	--	--	2 of 7	--
12507585	Yakima River at River Mile 72	1 of 1	--	--	--	--	--
12508620	Satus Creek at gage at Satus	2 of 2	--	--	--	2 of 2	--
12508630	South Drain near Satus	1 of 1	--	--	--	1 of 1	--
12508850	Sulphur Creek Wasteway	8 of 14	2 of 14	--	--	4 of 14	--
12509050	Yakima River at Euclid Bridge at River Mile 55	2 of 2	--	--	--	2 of 2	--
12509710	Spring Creek at mouth	2 of 2	--	--	--	2 of 2	--
12509829	Snipes Creek at mouth	2 of 2	--	--	--	--	1 of 1
12510500	Yakima River at Kiona[3]	6 of 15	--	--	1 of 4	5 of 15	--

[1]Includes data collected from Moxee Drain at Birchfield Road.
[2]Includes data collected upstream at Granger Drain at Granger.
[3]Exceedance for malathion occurred during summer 1991.

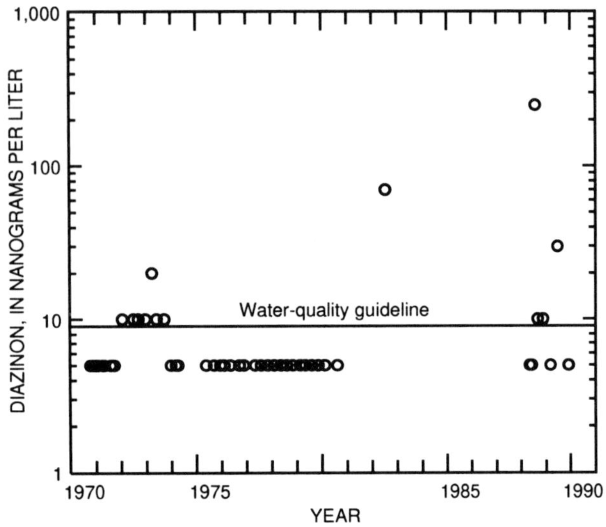

Figure 17. Concentrations of diazinon in unfiltered-water samples from the Yakima River at Kiona, Washington, 1970–89. (Values less than the minimum reporting level are graphically represented as one-half the minimum reporting level (10 ηg/L). Water-quality guideline, the guideline for the protection of freshwater aquatic life recommended by the National Academy of Sciences-National Academy of Engineering Committee on Water Quality Criteria [1973].)

Conventional analyses of 1-L unfiltered-water samples from 17 stations (mostly agricultural return flows) in the Yakima River Basin collected during 1988–89 show that atrazine and simazine were the most frequently detected organonitrogen compounds (table 22). These compounds were quantified in nearly 70 percent of the samples. The typical minimum reporting level was 100 ηg/L; the laboratory, however, was able to quantify several compounds at concentrations lower than 100 ηg/L. A previous sampling of the Yakima River at Kiona and several of its major agricultural return flows by Johnson and others (1986) in August 1985 showed that all measured concentrations of simazine and cyanazine in the basin were below the minimum reporting levels of 3,000 and 2,000 ηg/L, respectively. In a national study of pesticides in 144 rivers from 1975 to 1980, atrazine was the most frequently detected pesticide in water (24 percent of the stations and 4.8 percent of the samples [Gilliom and others, 1985]). The minimum reporting level in the national study was relatively high at 500 ng/L, so the number of detections could have been higher if a minimum reporting level of 100 ng/L had been obtainable in the 1970s.

In June 1989, large volumes of filtered-water samples (as large as 112 L) from 29 stations throughout the basin were collected synoptically and analyzed for triazine herbicides and other organonitrogen compounds (table 23). Atrazine, desethyl-atrazine (degradation product of atrazine), simazine, prometon, trifluralin, alachlor, and metolachlor were detected at 45 percent or more of the stations, with atrazine, desethyl-atrazine, and simazine detected at more than 75 percent of the stations. The numbers of detections of these compounds in streams generally increased with their increased annual use (table 23). Assuming that pesticide-use estimates for 1989 are reliable, several discrepancies between use and the number of detections are apparent. The numbers of detections for metribuzin, cyanazine, and propachlor were less than expected, and the numbers for metolachlor and prometon were greater than expected. The few detections for cyanazine and propachlor may be explained by their relatively short half-lives of 2–14 days and 6–7 days, respectively. The greater number of detections for prometon could be explained by its relatively long half-life (120–500 days). Metribuzin has one of the highest solubilities and lowest K_{oc} values in this group of organonitrogen compounds (table 61, at back of report), which likely results in its episodic leaching and transport early in the irrigation season immediately following application. In contrast, compounds with K_{oc} values higher than about 100 tend to partition to soils in the fields and are transported during periods of increased erosion. In the Yakima River Basin, increased erosion in agricultural areas typically occurs during peak irrigation and storm runoff.

In 1989, several organonitrogen compounds were applied throughout much of the basin (table 59, at back of report). Alachlor, atrazine, metolachlor, prometon, and simazine were detected in many agricultural return flows and in the lower Yakima River downstream from the city of Yakima (table 60, at back of report; fig. 18). These herbicides were applied primarily to corn, apples, asparagus, prunes, and grapes, and also in urban areas. The reason for the unexpected increased number of detections for metolachlor is unknown. Metolachlor was applied primarily in the West-side and East-side areas of the basin (table 59, at back of report) and was detected in the West-side tributaries and in the lower Yakima River (table 60, at back of report).

Water-quality criteria for freshwater aquatic life have not been established for organonitrogen compounds. A water-quality guideline has been recommended by the NAS-NAE (1973) for simazine not to exceed 10,000 ηg/L for the protection of freshwater aquatic life (table 62, at back or report). The highest

Table 22. Summary of concentrations of triazine herbicides and other organonitrogen compounds in 1-liter unfiltered-water samples, Yakima River Basin, Washington, 1988–89

[Minimum reporting levels and concentrations reported in nanograms per liter; some concentrations were quantified below the minimum reporting level; --, insufficient data to calculate statistic]

Compound	Number of samples	Approximate minimum reporting level	Number of quantifiable concentrations	Summary of quantifiable concentrations		
				Minimum	Median	Maximum
Alachlor	72	100	1	--	30	--
Ametryn	72	100	0	--	--	--
Atrazine	72	100	48	9	42	600
Bromacil	72	100	3	67	110	200
Butachlor	72	100	0	--	--	--
Butylate	72	100	0	--	--	--
Carboxin	72	100	0	--	--	--
Cyanazine	72	100	10	12	64	145
Cycloate	72	100	0	--	--	--
Diphenamid	72	100	0	--	--	--
Hexazinone	72	100	3	15	30	90
Metolachlor	72	100	0	--	--	--
Metribuzin	72	100	1	--	42	--
Prometon	72	100	9	20	50	210
Prometryn	72	100	1	--	100	--
Propachlor	72	100	0	--	--	--
Propazine	72	100	0	--	--	--
Simazine	72	100	50	10	40	6,600
Simetryn	72	100	0	--	--	--
Terbacil	72	100	0	--	--	--
Trifluralin	72	100	5	19	55	100
Vernolate	72	100	0	--	--	--

concentration detected in the basin was 6,600 ηg/L at Cherry Creek at Thrall during storm runoff from fields in Kittitas Valley on March 9, 1989. Most concentrations of simazine during irrigation season were lower than 100 ηg/L.

Organochlorine Compounds

Organochlorine compounds are among the most persistent in the environment when compared to other groups of synthetic organic compounds. These compounds have low water solubility (generally lower than 0.5 mg/L) and a strong tendency to sorb to soils, suspended sediment, and bed sediment (K_{oc} values generally higher than 1,000 mL/g) (table 61, at back of report; Gilliom and others, 1985). Unlike carbamate, chlorophenoxy-acid, organophosphorus, and organonitrogen compounds, organochlorine compounds tend to strongly partition into lipids of aquatic organisms. Although production and use of many of these compounds have been banned or restricted for more than a decade (table 24), these compounds and (or) their metabolites are still commonly detected in many of our Nation's surface waters (Gilliom and others, 1985; Smith and others, 1988; Schmitt and others, 1990). For example, the half-life of technical chlordane in the environment is about 10 to 20 years. Assuming that the production of chlordane peaked in the 1970s, about 25 to 50 percent of the technical chlordane applied still exists unaltered in the environment (Dearth and Hites, 1991).

Organochlorine compounds used in the Yakima River Basin in 1989 include endosulfan (36,000 kg), lindane (7,300 kg), methoxychlor (4,400 kg), permethrin (1,800 kg), and chlorothalonil (730 kg) (table 59, at back of report). As late as the mid-1980s, pesticide-use estimates made in this study indicate that chlordane (1,800 kg), heptachlor (360 kg), and aldrin (60 kg) also were applied. Even though the compounds listed above were applied most recently, they were not the most frequently detected. Analyses of 1-L unfiltered-water samples in 1988–89 from 18 stations in the Yakima River Basin show that 4,4'-DDE, 4,4'-DDT, 4,4'-DDD, dieldrin, endosulfan I, endrin, and lindane were the most frequently detected organochlorine compounds (table 25). DDT+DDE+DDD was detected in more than 75 percent of the samples, whereas dieldrin and endosulfan I were detected in more than 50 and 25

Table 23. Estimated annual use of triazine herbicides and other organonitrogen compounds, half-life in soil, number of detections, and summary of concentrations in large-volume filtered-water samples, Yakima River Basin, Washington, June 25–30, 1989

[Concentrations reported in nanograms per liter; Number of detections, the number of quantifiable detections plus the number of determinations when the compound was detected but could not be quantified because it was less than the minimum reporting level; na, not analyzed; >, greater than; --, insufficient data to calculate statistic; data and minimum reporting levels reported in Rinella, McKenzie, Crawford, and others, 1992]

Compound	Annual use, in kilograms	Half-life in soil, in days	Number of detections in 29 samples	Number of quantifiable concentrations	Quantifiable concentrations		
					Minimum	Median	Maximum
Simazine	29,000	60–75	22	22	0.24	10	134
Alachlor	16,000	15–18	13	11	4.5	17	98
Terbacil[1]	16,000	50–120	na	na	na	na	na
Trifluralin	15,000	21–>190	17	4	3.1	7.2	28
Atrazine	9,100	40–70	26	23	.26	32	130
Metribuzin	9,100	30–60	3	0	--	--	--
Carboxin[1]	5,800	3–7	na	na	na	na	na
Cyanazine	5,500	2–14	3	1	--	8.6	--
Hexazinone[1]	5,500	30–180	na	na	na	na	na
Metolachlor	3,300	20–90	13	7	.93	2.2	6.0
Propachlor	2,900	6–7	0	0	--	--	--
Prometon	220	120–500	20	13	1.1	3.0	32
Ametryn	0	60–120	0	0	--	--	--
Butachlor	0	--	3	0	--	--	--
Cycloate[1]	0	30–56	na	na	na	na	na
Diphenamid[1]	0	30	na	na	na	na	na
Prometryn	0	40–70	3	0	--	--	--
Propazine	0	135	4	1	--	.47	--
Simetryn	0	--	0	0	--	--	--
Desethyl-atrazine[2]	0	--	22	2	1.6	--	4.0
Desisopropyl-atrazine[3]	0	--	2	0	--	--	--

[1]Compounds were not analyzed in June 1989. Other samples (1-liter conventional analyses) were collected from 1988–89, and only hexazinone was detected (Rinella, McKenzie, Crawford, and others, 1992).

[2]Degradation product of atrazine and propazine.

[3]Degradation product of atrazine and simazine.

percent, respectively (minimum reporting levels of 1 ηg/L). Several samplings during irrigation season by Johnson and others (1986) in 1985 at main-stem stations and major agricultural return flows also showed that DDT+DDE+DDD, dieldrin, and endosulfans I and II were the predominant organochlorine compounds detected. In the 1985 study, minimum reporting levels were about 10 to 20 ηg/L, which are about 10–20 times higher than those levels in the 1988–89 study. As a result, DDT+DDE+DDD, dieldrin, and endosulfan were detected less frequently: 24, 26, and 2 percent of the samples, respectively (Johnson and others, 1986). In addition, many of the detections in the 1985 study were reported as estimated concentrations because they were lower than the minimum reporting level.

In June 1989, large quantities of filtered water and suspended sediment from 29 stations throughout the basin were collected synoptically and analyzed for organochlorine compounds (table 26). Minimum reporting levels typically were lower than 1 ηg/L (Rinella, McKenzie, Crawford, and others, 1992). Results show that DDT, DDE, and/or DDD were ubiquitous in the basin and were quantified at 97 percent of the stations (table 60, at back or report). Dieldrin (69 percent of the stations), endosulfan I (50 percent), and chlordane (39 percent) also were quantified frequently. Concentrations of hexachlorobenzene (HCB) and α-HCH generally were lower than minimum reporting levels, they were detected qualitatively, however, at 62 and 34 percent of the stations, respectively (Rinella, McKenzie, Crawford, and others, 1992).

Results from the June 1989 synoptic sampling show that concentrations of suspended DDT+DDE+DDD and suspended dieldrin increased as concentrations of suspended sediment and suspended organic carbon increased (fig. 19). These relations were expected because these nonionic organic compounds readily sorb to suspended-organic carbon. Sampling stations with the highest concentrations of DDT+DDE+

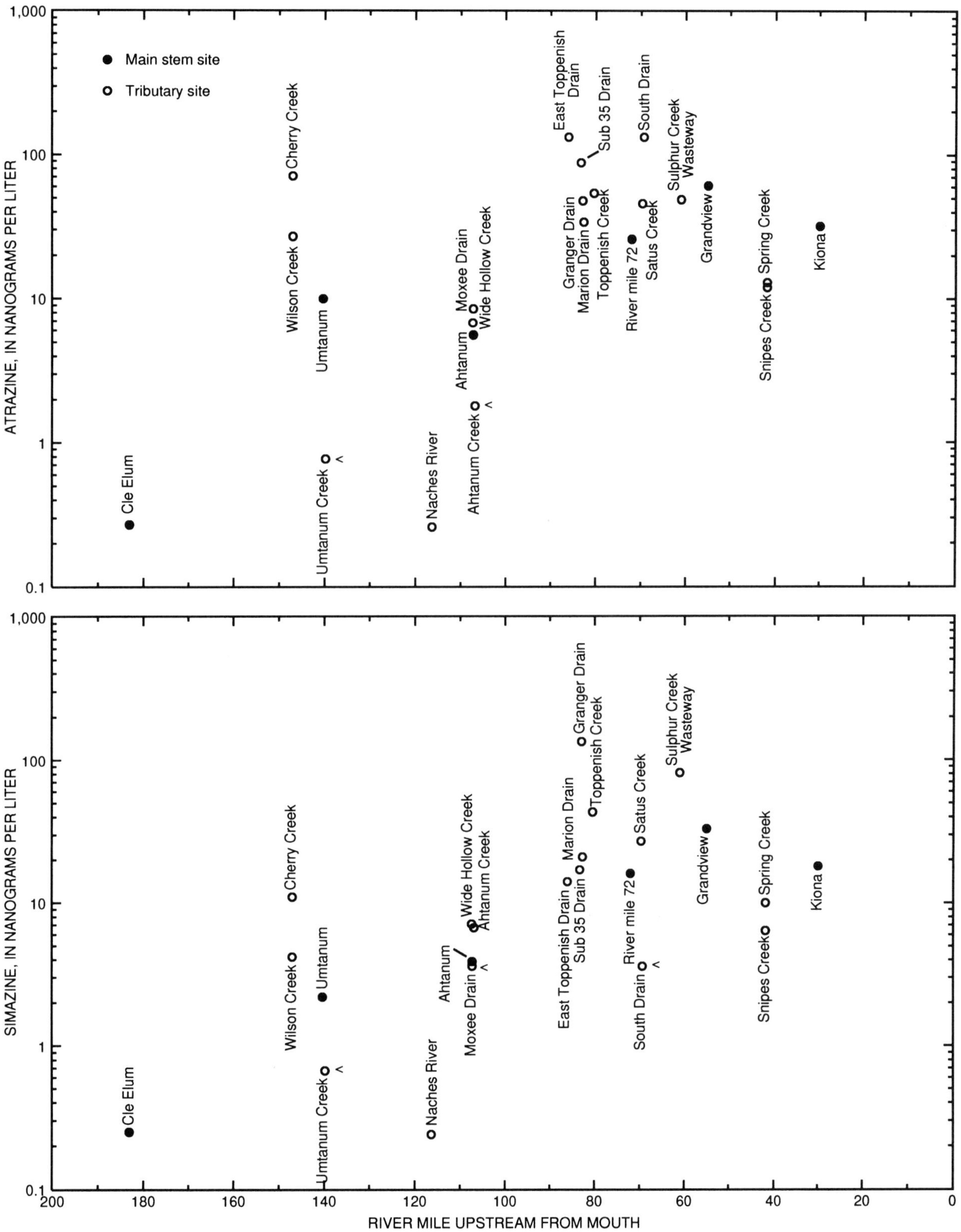

Figure 18. Concentrations of atrazine and simazine in filtered-water samples from selected stations, Yakima River Basin, Washington, June 25–30, 1989. (<, data point represents a value less than the minimum reporting level.)

Table 24. Status of use for selected organochlorine compounds detected in the Yakima River Basin, Washington, 1988–91

[DDT, dichlorodiphenyltrichloroethane; DDE, dichlorodiphenyldichloroethylene; DDD, dichlorodiphenyldichloroethane; PCB, polychlorinated biphenyls]

Compound	Comments
Aldrin	U.S. Department of Agriculture cancelled registration in 1970.[1] In 1974, production was prohibited for all products containing aldrin and dieldrin in the United States.[2] All uses have been cancelled in United States. Aldrin is converted to dieldrin via oxidation.[3] Production and sale discontinued in 1985.[4]
DDT	Imported in quantity into United States in 1944.[3] Use banned in United States in December 1972.[2]
DDE	Breakdown product of DDT in aerobic environments. Impurity in dicofol.
DDD	Breakdown product of DDT in anaerobic environments. Insecticide was discontinued by Rohm and Hass Company. Impurity in dicofol.[3]
Chlordane (technical grade)	Registered as a pesticide in 1948.[1] All uses cancelled in 1987, except for subsurface insertion for termite control.[1] Production was discontinued voluntarily in July 1987, and use stopped in 1988.[4]
Dieldrin	Suspended use on food products in 1974.[1] All uses banned in 1985 except for subsurface termite control, dipping of nonfood roots and tops, and moth proofing.[1] All uses have been voluntarily cancelled by industry.[1] In 1974, production was prohibited for all products containing aldrin and dieldrin in the United States.[2]
Endrin	Registered as pesticide in 1951.[1] Restricted uses on apple orchards and ornamentals in 1979.[1] Voluntary cancellation requested by producer in 1984.[1]
Endosulfan	No restrictions found.
Heptachlor	Registered as insecticide in 1952.[1] Restrictions instituted in 1978.[1] All uses banned in 1983 except for subsurface termite control, dipping of nonfood roots and tops.[1] No longer sold in United States as of August 1987.[1]
Heptachlor epoxide	Breakdown product of heptachlor.[1]
Hexachlorobenzene	Registration as pesticide voluntarily cancelled in 1984.[1]
Lindane	Application permitted under direct supervision of a certified applicator.[1]
Methoxychlor	Introduced in 1945.[1] No restrictions found.[1]
Mirex	All registered uses cancelled in 1977.[1]
Nonachlor	No restrictions or bans.[1] First produced in 1948.[3]
Perthane	All uses cancelled in 1980.[1]
PCB	Production began in 1929.[1] Manufacturing, processing, and distribution banned in 1979.[1]
Toxaphene	Introduced as an insecticide in 1948.[2] Discontinued in 1989.[3]

[1]U.S. Environmental Protection Agency, 1991b.
[2]Sittig, 1981.
[3]Poplyk, 1989.
[4]Dr. Mark Lacey, Wilmington, Delaware, oral commun., November 1992.

Table 25. Summary of concentrations of organochlorine compounds in 1-liter unfiltered-water samples, Yakima River Basin, Washington, 1988–89

[Concentrations reported in nanograms per liter; summary is based on 77 samples from 18 stations; DDD, 4,4'-dichlorodiphenyldichloroethane; DDE, 4,4'-dichlorodiphenyldichloroethylene; DDT, 4,4'-dichlorodiphenyltrichloroethane; HCH, hexachlorocyclohexane; PCB, polychlorinated biphenyls; PCN, polychlorinated naphthalenes; <, less than]

Compound	Minimum concentration	Concentration at indicated percentile					Maximum concentration
		10	25	50	75	90	
Aldrin	<1	<1	<1	<1	<1	<1	<1
Gross chlordane	<100	<100	<100	<100	<100	<100	<100
DDD	<1	<1	<1	<1	1	2	11
DDE	<1	<1	1	3	12	29	61
DDT	<1	<1	<1	<1	10	28	60
DDT + DDE + DDD	<3[1]	<3[1]	2	4	24	58	122
Dieldrin	<1	<1	<1	2	5	10	41
Endosulfan I	<1	<1	<1	<1	13	67	190
Endrin	<1	<1	<1	<1	<1	<1	2
Heptachlor	<1	<1	<1	<1	<1	<1	<1
Heptachlor epoxide	<1	<1	<1	<1	<1	<1	<1
Lindane (γ-HCH)	<1	<1	<1	<1	<1	<1	1
4,4'-Methoxychlor	<10	<10	<10	<10	<10	<10	<10
Mirex	<10	<10	<10	<10	<10	<10	<10
Gross PCB	<100	<100	<100	<100	<100	<100	<100
Gross PCN	<100	<100	<100	<100	<100	<100	<100
Perthane	<100	<100	<100	<100	<100	<100	<100
Toxaphene	<1,000	<1,000	<1,000	<1,000	<1,000	<1,000	<1,000

[1]DDD, DDE, and DDT are each less than 1 nanogram per liter.

Table 26. Summary of concentrations of organochlorine compounds in large-volume water samples, Yakima River Basin, Washington, June 25–30, 1989

[Concentrations reported in nanograms per liter; Number of detections, the number of quantifiable determinations plus the number of determinations when the compound was detected but could not be quantified because it was less than the minimum reporting level; DDD, dichlorodiphenyldichloroethane; DDE, dichlorodiphenyldichloroethylene; DDT, dichlorodiphenyltrichlorethane; HCB, hexachlorobenzene; HCH, hexachlorocyclohexane; PCB, polychlorinated biphenyl; PCN, polychlorinated naphthalenes; --, insufficient data to calculate statistic; ≤, less than or equal to; E, estimated; data and minimum reporting levels reported in Rinella, McKenzie, Crawford, and others, 1992]

	Summary for filtered-water samples						Summary for suspended-sediment samples					
				Quantifiable concentrations						Quantifiable concentrations		
Compound	Number of samples	Number of detections	Number of quantifiable concentrations	Minimum	Median	Maximum	Number of samples	Number of detections	Number of quantifiable concentrations	Minimum	Median	Maximum
Aldrin	29	7	0	--	--	--	29	1	1	--	0.09	--
Gross chlordane	0	--	--	--	--	--	13	10	10	0.15	.43	4
cis-Chlordane	29	3	0	--	--	--	16	3	1	--	.05	--
trans-Chlordane	29	3	1	--	0.31	--	16	3	1	--	.09	--
Chlorothalonil	2	1	0	--	--	--	16	1	1	--	.08	--
Dacthal	0	--	--	--	--	--	16	4	3	.15	.23	.31
2,4'-DDD	0	--	--	--	--	--	16	5	4	.03	.28	1.9
4,4'-DDD	0	--	--	--	--	--	29	21	20	.001	.46	6.4
4,4'-DDE	29	25	23	0.13	.61	15	29	24	23	.008	1.1	28
2,4'-DDT	0	--	--	--	--	--	16	7	6	.22	1.7	5.7
4,4'-DDT	0	--	--	--	--	--	29	20	19	.05	3.4	38
DDT+DDE+DDD[1]	29	26	24	.5	1.7	38	29	25	24	.008	2.9	80
Dieldrin	29	20	13	.77	1.6	35	29	21	18	.003	.07	3.6
Endosulfan I	26	15	13	.73	1.8	11	29	2	2	.04	--	.69
Endosulfan II	29	5	4	1.0	1.8	5.2	16	1	0	--	--	--
Endosulfan sulfate	0	--	--	--	--	--	16	0	0	--	--	--
Endrin	29	4	3	1.5	2.8	9.1	29	1	1	--	.31	--
Endrin aldehyde	0	--	--	--	--	--	16	0	0	--	--	--
Endrin ketone	0	--	--	--	--	--	16	0	0	--	--	--
HCB	29	16	0	--	--	--	16	5	3	.08	.16	.33
α-HCH	29	9	0	--	--	--	16	2	1	--	0.02	--
β-HCH	29	0	0	--	--	--	0	--	--	--	--	--
δ-HCH	29	0	0	--	--	--	16	0	0	--	--	--
Heptachlor	29	0	0	--	--	--	29	0	0	--	--	--
Heptachlor epoxide	29	3	0	--	--	--	29	2	1	--	.15	--
Lindane (γ-HCH)	29	1	0	--	--	--	29	2	1	--	.06	--
2,4'-Methoxychlor	29	2	0	--	--	--	16	1	1	--	≤3.4	--
4,4'-Methoxychlor	11	0	0	--	--	--	29	1	1	--	≤.32	--
Mirex	29	3	0	--	--	--	29	2	2	.004	--	.008
trans-Nonachlor	0	--	--	--	--	--	16	3	1	--	.21	--
cis- and trans-Permethrin	0	--	--	--	--	--	16	0	0	--	--	--
Perthane	29	2	0	--	--	--	13	0	0	--	--	--
PCB Aroclor 1260	0	--	--	--	--	--	15	2	2	E 1.2	--	8.8
Gross PCN	0	--	--	--	--	--	13	0	0	--	--	--
Toxaphene	0	--	--	--	--	--	14	1	1	--	160	--

[1]In filtered-water samples, DDT+DDE+DDD represents the sum of 4,4'-DDT, 4,4'-DDE, and 4,4'-DDD. In suspended-sediment samples, however, DDT+DDE+DDD represents the sum of 2,4'-DDT, 4,4'-DDT, 4,4'-DDE, 2,4'-DDD, and 4,4'-DDD.

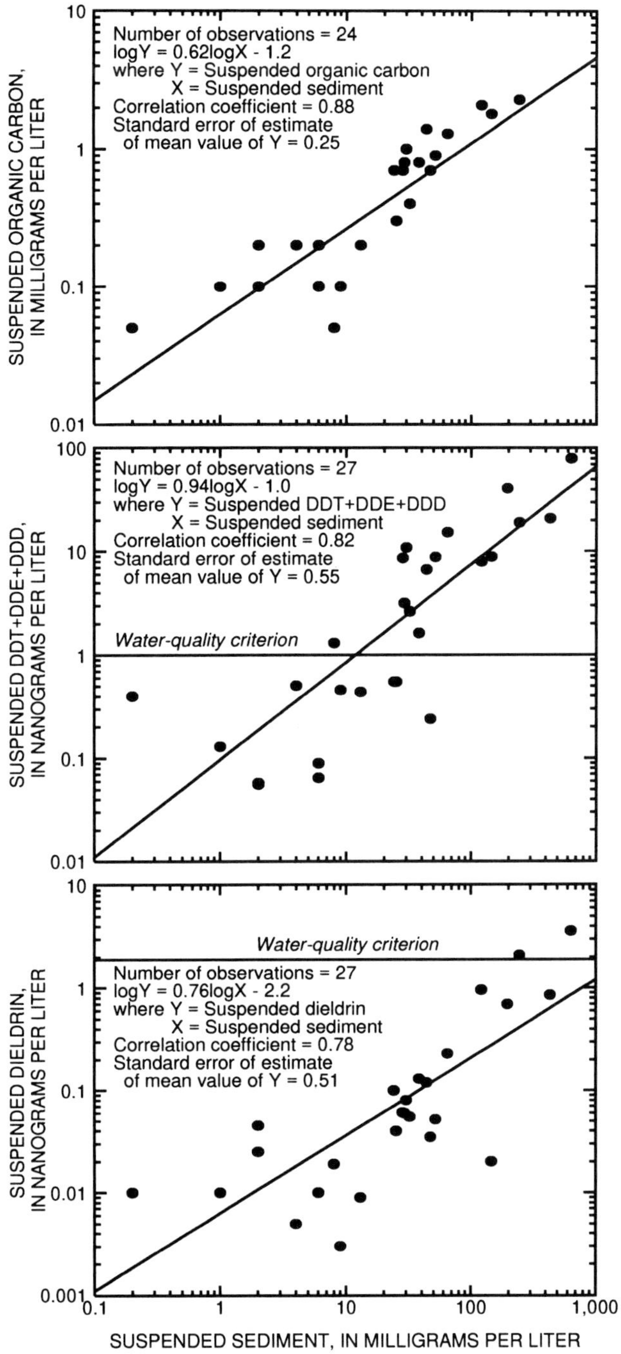

Figure 19. Relation of concentrations of suspended sediment and suspended organic carbon, 4,4'-dichlorodiphenyltrichloroethane (DDT) plus 4,4'-dichlorodiphenyldichloroethylene (DDE) plus 4,4'-dichlorodiphenyldichloroethane (DDD), and dieldrin in unfiltered-water samples from selected stations, Yakima River Basin, Washington, June 25–30, 1989. (Values less than the minimum reporting level are graphically represented and statistically summarized as one-half the minimum reporting level; greater-than values were excluded. Water-quality criterion, the chronic-toxicity water-quality criterion for the protection of freshwater aquatic life established by the U.S. Environmental Protection Agency [1986].)

DDD, dieldrin, and (or) endosulfan I were the East-side tributaries and South Drain (table 60, at back of report). These stations also have the highest suspended-sediment concentrations. High concentrations of organochlorine compounds in East-side tributaries can be attributed to (1) historical applications of DDT and dieldrin prior to bans and restrictions in the 1970s, (2) continued application of endosulfan I to orchards (table 59, at back of report), (3) highly erosive soils of the Warden-Esquatzel association in the East-side area, (4) steeper subbasin slopes that promote erosion, (5) rill irrigation, and (6) the strong tendency for these compounds to sorb to soils and stream sediment. The increased transport of DDT+DDE+DDD in South Drain can be attributed to increased erosion caused by frequent tillage and rill irrigation of hop fields. Controlling erosion of contaminated soils from agricultural fields will help to reduce concentrations of DDT+DDE+DDD and dieldrin in streams in the Yakima River Basin.

Because of relatively high sediment-water partition coefficients (for example, K_{oc} = 243,000 mL/g for DDT, 4,400,000 for DDE, 770,000 for DDD, and 1,700 for dieldrin), organochlorine compounds are highly sorbed to soils in the agricultural fields (table 61, at back of report). As concentrations of sediment decrease from fields (2,000,000 mg/L of soil and soil-pore water) to ditches and then to streams (typically less than 500 mg/L), some of the sorbed compound will undergo dissolution from the suspended soil particles into the water column. As K_{oc} decreases and (or) as suspended organic carbon decreases, equilibrium-partitioning calculations (equation 1) indicate that a smaller portion of the mass of these compounds will remain sorbed to the suspended sediment. Results of compound dissolution from soil particles into stream water are apparent in data collected in 1991 from major agricultural return flows and the Yakima River at Kiona (data summary shown in table 27). Assuming equilibrium partitioning in agricultural soils, more than 97 percent of the dieldrin should be sorbed to soils and the remaining should be dissolved in the soil-pore water. Assuming equilibrium partitioning in streams, less than 2 percent of the dieldrin should be sorbed to stream sediment (table 28). The measured concentrations in streams show that a large portion (median of 80 percent) of the dieldrin has undergone dissolution from the eroded soils. According to equilibrium calculations, however, more than 98 percent of the dieldrin should be dissolved, indicating that either

Table 27. Summary of concentrations of organochlorine compounds in water samples, Yakima River Basin, Washington, May-September 1991

[Concentrations reported in nanograms per liter; results from duplicate analyses were averaged for this summary; 23 samples from 6 stations were analyzed in May-September 1991; sample volumes as large as 5 liters were collected for this analysis; DDD, 4,4'-dichlorodiphenyldichloroethane; DDE, 4,4'-dichlorodiphenyldichloroethylene; DDT, 4,4'-dichlorodiphenyltrichlorethane; HCH, hexachlorocyclohexane; PCB, polychlorinated biphenyls; PCN, polychlorinated naphthalenes; --, insufficient data to calculate statistic]

Compound	Summary for filtered-water samples: Number of quantifiable concentrations	Quantifiable concentrations: Minimum	Median	Maximum	Summary for suspended-sediment samples: Number of quantifiable concentrations	Quantifiable concentrations: Minimum	Median	Maximum
Aldrin	0	--	--	--	0	--	--	--
Chlordane (technical)	0	--	--	--	1	--	1	--
DDD	9	1	2	5	15	0.1	1	4
DDE	23	1	5	34	22	1	2.5	29
DDT	12	1	4	25	19	.2	4	37
Dieldrin	21	1	4	16	17	.1	1	5
Endosulfan I	5	2	2	3	9	.1	1	2
Endrin	0	--	--	--	0	--	--	--
Heptachlor	0	--	--	--	0	--	--	--
Heptachlor epoxide[1]	1	--	1	--	1	--	.1	--
Lindane (γ-HCH)	4	.3	2	7	4	.1	.2	.5
4,4'-Methoxychlor	6	--	--	--	6	--	--	--
Mirex	0	--	--	--	0	--	--	--
PCB	4	100	100	150	10	10	23	70
PCN	0	--	--	--	0	--	--	--
Perthane	0	--	--	--	0	--	--	--
Toxaphene	0	--	--	--	0	--	--	--

[1]For heptachlor epoxide, only 21 samples from 6 stations were analyzed.

dieldrin has not equilibrated with the water and sediment phases or possibly that the K_{oc} value of 1,700 mL/g (table 61, at back of report) is too low for sediment in the Yakima River Basin. K_{oc} values have been shown to vary for different types of sediment (Smith and others, 1988).

Concentrations of DDT+DDE+DDD in filtered-water and suspended-sediment samples are listed in table 29 and shown in figure 20 for 1991 data.The data show that DDT+DDE+DDD is associated with both the dissolved and suspended phases. Measured concentrations of suspended DDT+DDE+DDD generally are lower than the calculated concentrations (equation 1), with a median difference of 20 percent. In the 1991 sampling, a 1-μm (micrometer) pore-size filter was used to separate the dissolved and suspended phases. Some particles smaller than 1 μm in diameter would have been included in the analysis of the dissolved phase and may account for this bias, however, a similar bias was not observed for dieldrin results. A closer examination of the equilibrium-partitioning calculations shows that most of the bias may be attributed to DDE partitioning, suggesting that the assumed K_{oc} value of 4,400,000 mL/g (table 61, at back or report) may be inordinately high for Yakima River Basin sediment. Regardless of bias, the measured data show that the dissolved phase, which includes the colloidal phase, is an important transport phase for organochlorine compounds (Pankow and McKenzie, 1991).

Two noteworthy differences between measured and calculated concentrations of DDT+DDE+DDD occurred in samples from Granger Drain and Sulphur Creek Wasteway on June 20, 1991, during a period of storm runoff. For both samples, the measured suspended DDT+DDE+DDD concentrations are much smaller than the calculated concentration (table 29). Both of these samples had concentrations of dissolved organic carbon (22 mg/L at Granger Drain and 15 mg/L at Sulphur Creek Wasteway; Rinella, McKenzie, Crawford, and others, 1992) that were 5 to 7 times higher than most of the other samples collected in 1991. This increase in dissolved organic carbon may have enhanced partitioning of DDT+DDE+DDD into the dissolved phase (Witkowski and others, 1987).

As for the carbamate and organophosphorus insecticides and the triazine and chlorophenoxy-acid herbicides, the number of detections of organochlorine compounds in streams was expected to coincide with pesticide use. Accordingly, endosulfan was detected in

Table 28. Partitioning of dieldrin in filtered water and suspended sediment for selected stations, Yakima River Basin, Washington, May-September 1991

[For theoretical equilibrium-partitioning calculations, the sediment-water partition coefficient equals 1,700 milliliters per gram; concentrations reported in nanograms per liter; *, concentration of dieldrin in filtered water or suspended sediment was reported as a less-than (<) value]

Station number	Station name	Date	Dieldrin concentration: Filtered water	Dieldrin concentration: Suspended sediment	Percent of dieldrin mass associated with suspended sediment: Observed	Percent of dieldrin mass associated with suspended sediment: Theoretical equilibrium partitioning (calculated)
12496550	Buckskin Slough	05–22–91	<1	<1	*	0.1
		06–12–91	4	<1	*	.1
12500420	Moxee Drain	05–23–91	<1	<1	*	.3
		06–17–91	2	1	33	.6
		06–20–91	1	.2	17	1.7
		07–10–91	2	.1	5	.2
		09–04–91	4	1	20	.4
12505450	Granger Drain	05–23–91	3	<1	*	.5
		06–17–91	8	2	20	.9
		06–20–91	9	2	18	1.8
		07–10–91	6	5	45	1.5
		09–04–91	6	1	14	.4
12508850	Sulphur Creek Wasteway	05–24–91	2	1	33	.1
		06–19–91	6	1	14	.3
		06–20–91	16	2	11	1.8
		07–11–91	7	2	22	.6
		09–05–91	5	2	28	.3
12509499	Chandler Canal	05–22–91	1	<1	*	.1
		06–13–91	1	<1	*	.4
12510500	Yakima River at Kiona	05–21–91	3	1	25	.1
		07–09–91	3	<1	*	.1
		08–06–91	<1	.4	*	.2
		08–06–91	1	<1	*	.2
		09–25–91	5	1	17	.1

areas of the basin where it was applied heavily. In 1989, endosulfan was applied primarily to apple, pear, and peach orchards in the Tieton, East-side, and West-side areas (table 59, at back of report). Endosulfan was minimally used in the Kittitas area where it was not detected in the runoff (table 60, at back of report). Historical use of DDT and dieldrin in the Yakima River Basin was not estimated in this study; however, their detections in June 1989 at most of the stations throughout the agricultural areas (fig. 21; table 60, at back of report) indicate that their past use was extensive.

The acarcide dicofol may have been a relatively recent source of DDT+DDE+DDD in the basin (Johnson and others, 1986). DDT+DDE+DDD and DDT-related compounds are impurities that have been detected in dicofol. In the United States, these impurities averaged 5.75 percent in 1984. In 1985, they were reduced to 2.5 percent with plans to further reduce them to 0.1 percent by January 1, 1987 (Johnson and others, 1986). As of 1992, the concentration of DDT+DDE+DDD in dicofol was about 0.5 percent (U.S. Soil Conservation Service and others, 1992).

Water-quality criteria for the protection of aquatic life (U.S. Environmental Protection Agency, 1986; table 62, at back of report) for DDT+DDE+DDD, dieldrin, PCB, endosulfan, endrin, chlordane, and toxaphene have been exceeded in agricultural areas throughout the basin. Exceedances of chronic-toxicity criteria for the protection of freshwater aquatic life are listed in table 30. Criteria for acute toxicity were not exceeded in this study. Of the 31 stations sampled during 1988–91, 21 stations had exceedances for DDT+DDE+DDD, 12 for dieldrin, 5 for PCB, 3 for endosulfan, 3 for endrin, 1 for chlordane, and 1 for toxaphene. As expected, sampling stations without exceedances include streams that have been minimally affected by human activity. Additionally, organochlorine-compound concentrations in several shallow wells in agricultural areas did not exceed the criteria. Considering that ground-water discharge may constitute a considerable portion of streamflow in agricultural return

Table 29. Partitioning of dichlorodiphenyltrichloroethane (DDT) plus dichlorodiphenyldichloroethylene (DDE) plus dichlorodiphenyldichloroethane (DDD) in filtered water and suspended sediment for selected stations, Yakima River Basin, Washington, May-September 1991

[Concentrations reported in nanograms per liter; in filtered-water samples, DDT+DDE+DDD represents the sum of 4,4'-DDT, 4,4'-DDE, and 4,4'-DDD; in suspended-sediment samples, however, DDT+DDE+DDD represents the sum of 2,4'-DDT, 4,4'-DDT, 4,4'-DDE, 2,4'-DDD, and 4,4'-DDD]

Station number	Station name	Date	DDT+DDE+DDD concentration		Percent of DDT+DDE+DDD mass associated with suspended sediment	
			Filtered water	Suspended sediment	Observed	Theoretical equilibrium partitioning (calculated)
12496550	Buckskin Slough	05–22–91	5	4	44	53
		06–12–91	13	3	19	36
12500420	Moxee Drain	05–23–91	6	6	50	72
		06–17–91	6	12	67	73
		06–20–91	2	2.1	51	90
		07–10–91	9	4	31	58
		09–04–91	5	3	38	74
12505450	Granger Drain	05–23–91	9	10	53	80
		06–17–91	19	32	63	80
		06–20–91	61	39	39	86
		07–10–91	12	67	85	82
		09–04–91	12	38	76	69
12508850	Sulphur Creek Wasteway	05–24–91	6	19	76	52
		06–19–91	25	11	30	43
		06–20–91	38	2.3	6	86
		07–11–91	9	21	70	71
		09–05–91	6	29	83	63
12509499	Chandler Canal	05–22–91	4	1	20	76
		06–13–91	.3	2.7	90	90
12510500	Yakima River at Kiona	05–21–91	4	2	33	65
		07–09–91	1	1.2	54	68
		08–06–91	4	1	20	82
		08–06–91	4	2	33	87
		09–25–91	2	1	33	70

flows during fall and winter low flows, the lack of exceedances of organochlorine compounds in ground water provides important information concerning seasonal transport of these compounds in surface water.

Prior to its ban in 1979 (table 24), PCB compounds were used chiefly in insulation for electrical wires in electric condensers, so their occurrence is often associated with urban, industrial, and electrical-power-generation activities. Analyses of PCB compounds in 1-L unfiltered-water samples collected in 1988–89 show that all of the 77 samples had concentrations lower than the minimum reporting level of 100 ηg/L (table 25). In June 1989, PCB concentrations in suspended-sediment samples from Granger Drain and Sulphur Creek Wasteway were 8.8 and 1.2 ηg/L, respectively (table 26). Analyses of filtered water and suspended sediment collected from May-September 1991 (table 27) show that PCB (filtered plus suspended) exceeded the chronic-toxicity criterion of 14 ηg/L for the protection of freshwater aquatic life (U.S. Environmental Protection Agency, 1986) at Moxee Drain, Granger Drain, Sulphur Creek Wasteway, Buckskin Slough, and Yakima River at Kiona. Large amounts of suspended sediment were collected and analyzed to reduce conventional minimum reporting levels. In these analyses, the minimum reporting level for PCB associated with suspended sediment typically was reduced to 10 ηg/L, whereas the level for filtered water remained at 100 ηg/L. PCB was not detected in unfiltered-water samples from the Yakima River at Kiona in 1988–89, but it was detected in several filtered-water samples from the Yakima River at Kiona in 1991 at concentrations (100–150 ηg/L) that exceeded the chronic-toxicity criterion. In 1990, the USGS began using gas chromatography (GC) with capillary columns in place of GC with packed columns. Capillary columns provide increased analytical resolu-

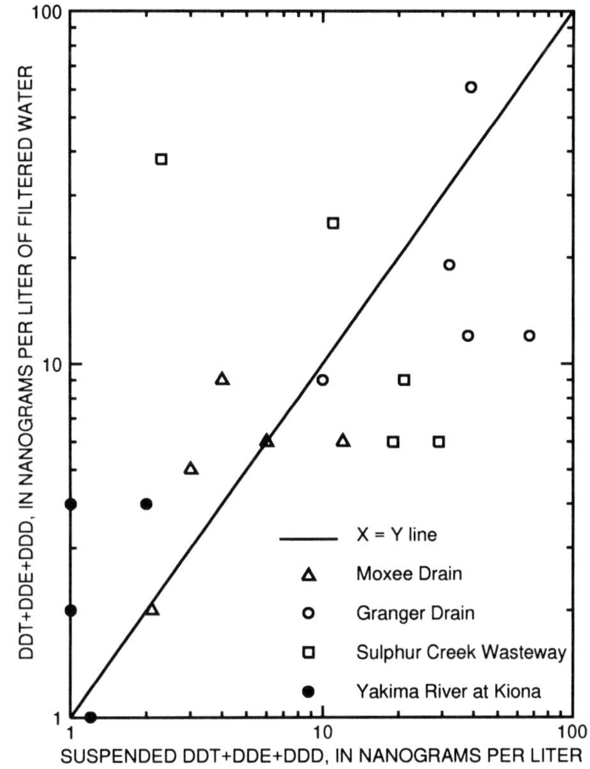

Figure 20. Relation of concentrations of 4,4'-dichlorodiphenyltrichloroethane (DDT) plus 4,4'-dichlorodiphenyldichloroethylene (DDE) plus 4,4'-dichlorodiphenyldichloroethane (DDD) in filtered-water and suspended-sediment samples from major agricultural-return flows and the Yakima River at Kiona, Yakima River Basin, Washington, May-September 1991. (Values less than the minimum reporting level are excluded.)

tion and may account for the detections of PCB in the Yakima River at Kiona in 1991. Results from the 1991 sampling indicate that PCB contamination is a concern in the lower Yakima River Basin.

Human-health criteria for several organochlorine compounds have been established for human ingestion of water and fish (table 62, at back of report). On the basis of current water-quality criteria (U.S. Environmental Protection Agency, 1991a and 1992d), 18 of 29 stations (62 percent) sampled in June 1989 in the Yakima River Basin had concentrations of DDT, DDE, or DDD that exceeded the human health criteria of 0.59, 0.59, or 0.83 ηg/L, respectively, for an increase in cancer risk of 1 in 1,000,000 for people exposed over a 70-year lifetime. None of the stations sampled in June 1989 had concentrations of DDT, DDE, or DDD that exceeded human health criteria of 59, 59, and 83 ηg/L for an increase in cancer risk of 1 in 10,000 people.

In the Yakima River Basin, most of the health risk associated with DDT+DDE+DDD results from ingesting contaminated resident fish and not from drinking the water. In June 1989, the concentration of DDE in the city of Yakima's municipal drinking water was 0.36 ηg/L; DDT and DDD concentrations were lower than minimum reporting levels (Rinella, McKenzie, Crawford, and others, 1992). Assuming that 2 L of water are ingested each day and that resident fish from the Yakima River Basin are not consumed, the increased lifetime (70 year) cancer risk from DDE is 3 in 1,000,000,000 people (U.S. Environmental Protection Agency, 1992d).

At several stations, concentrations of dieldrin and PCBs also exceeded current human-health criteria for ingestion of water and fish for an increased lifetime risk of 1 in 1,000,000. In June 1989, dieldrin concentrations exceeded the criterion of 0.14 ηg/L (table 62, at back of report) at 15 of 29 stations (52 percent) that were sampled. In 1991, low-level analyses of PCBs in streams show that the human-health criterion was exceeded at Buckskin Slough, Moxee Drain, Granger Drain, Sulphur Creek Wasteway, and the Yakima River at Kiona. Eleven of 21 samples from these stations had concentrations that exceeded the criterion of 0.044 ηg/L for an increase in the lifetime cancer risk of 1 in 1,000,000. Six of these samples had concentrations that exceeded levels corresponding to an increased risk of 1 in 1,000. The number of stations sampled in 1991 was inadequate for accurately defining the spatial distribution of PCB in the Yakima River Basin. The data, however, do suggest concern for ingesting water and fish from the lower Yakima River. Currently, the lower Yakima River is not used for drinking water, but the main route of contaminant exposure is through fish consumption.

Long-term data are available to examine changes in concentrations of DDT+DDE+DDD and dieldrin in unfiltered-water samples collected from the Yakima River at Kiona (fig. 22). Concentrations of DDT+DDE+DDD and dieldrin have decreased by as much as one order of magnitude since the ban of DDT in 1973 and the suspended use of dieldrin on food products in 1974. After 1974, concentrations in samples from the Yakima River at Kiona were below minimum reporting levels for about 10 years, until Johnson and others (1986) detected both compounds at concentrations just below 10 ηg/L (fig. 22). Exceedances of chronic-toxicity criteria were confirmed in the 1988–89 NAWQA sampling. The decrease in concentrations

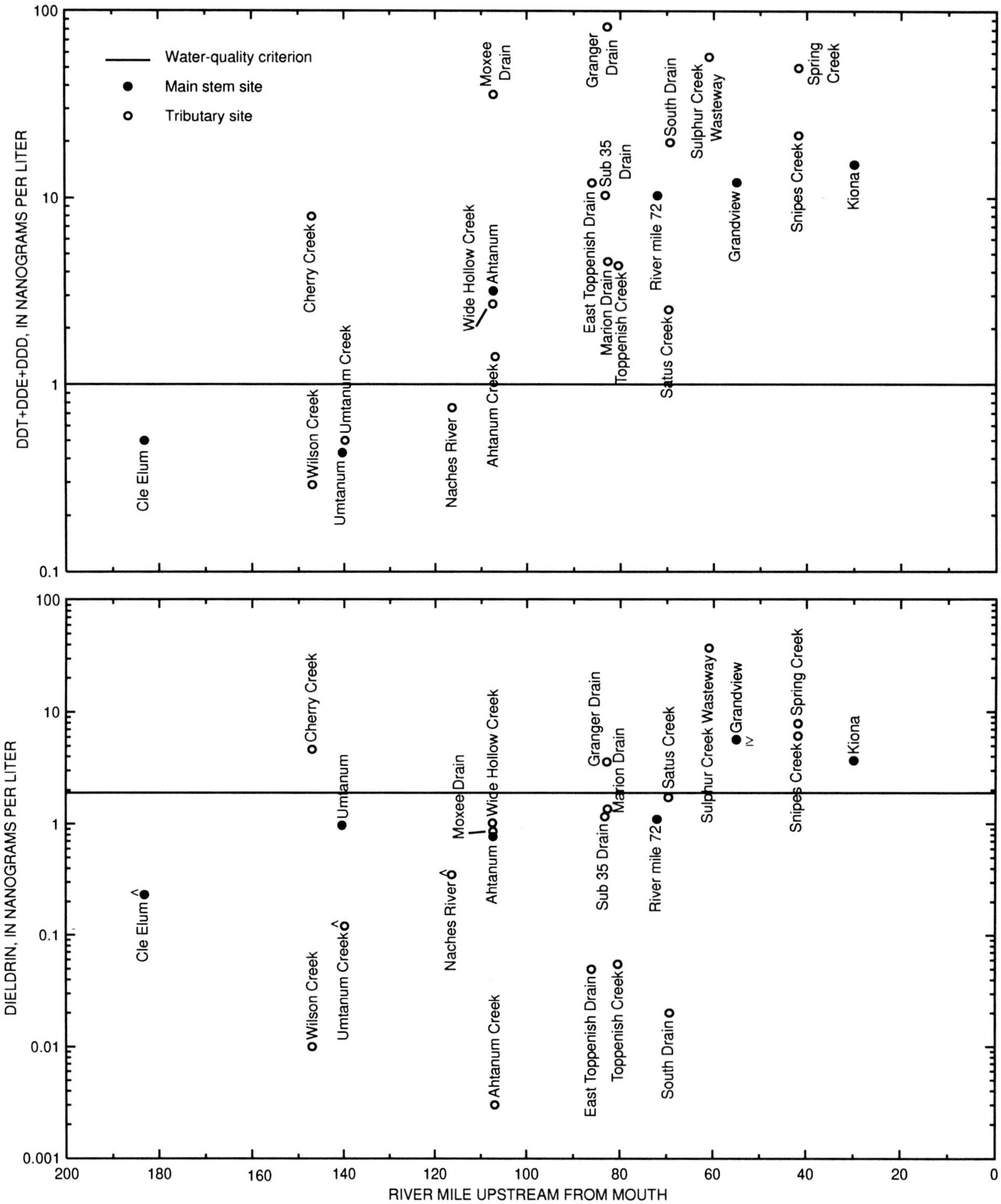

Figure 21. Concentrations of 4,4'-dichlorodiphenyltrichloroethane (DDT) plus 4,4'-dichlorodiphenyldichloroethylene (DDE) plus 4,4'-dichlorodiphenyldichloroethane (DDD) and dieldrin in unfiltered-water samples from selected stations, Yakima River Basin, Washington, June 25–30, 1989. (Concentrations represent the sum of filtered-water and suspended-sediment analyses. Water-quality criterion, the chronic-toxicity water-quality criterion for the protection of freshwater aquatic life established by the U.S. Environmental Protection Agency [1986]; <. data point represents a value less than the minimum reporting level; ≤, data point represents a value quantified as greater than or equal to a concentration.)

Table 30. Exceedances of water-quality criteria for the protection of freshwater aquatic life for organochlorine compounds in water samples, Yakima River Basin, Washington, 1988–91

[Data are shown only when criteria are equaled or exceeded and are based on analyses of either concentrations in unfiltered water or concentrations in both filtered water and suspended sediment; only one determination is included in listing for replicate samples collected within a 1-hour period; see table 62 for criteria and guidelines; in water samples, DDT+DDE+DDD represents the sum of 4,4'-dichlorodiphenyltrichloroethane (DDT) plus 4,4'-dichlorodiphenyldichloroethylene (DDE) plus 4,4'-dichlorodiphenyldichloroethane (DDD); in suspended-sediment sample, however, DDT+DDE+DDD represents the sum of 2,4'-DDT, 4,4'-DDT, 2,4'-DDE, 4,4'-DDE, 2,4'-DDD, and 4,4'-DDD; PCB, polychlorinated biphenyls; --, no exceedance detected]

Station number	Station name	Number of samples equaling or exceeding criteria and number of samples collected						
		Chlordane	DDT+DDE+DDD	Dieldrin	Endosulfan	Endrin	PCB[1]	Toxaphene
12484480	Cherry Creek at Thrall	1 of 7	7 of 7	6 of 7	--	--	--	--
12496510	Pacific Power and Light Company Wasteway	--	1 of 1	--	--	--	--	--
12496550	Buckskin Slough below Gleed Ditch	--	2 of 2	1 of 2	--	--	1 of 2	--
12500430	Moxee Drain at Thorp Road[2]	--	16 of 16	9 of 16	2 of 16	--	3 of 5	--
12500445	Wide Hollow Creek near mouth	--	7 of 7	4 of 7	3 of 7	--	--	--
12500450	Yakima River above Ahtanum Creek	--	2 of 2	--	--	1 of 2	--	--
12502500	Ahtanum Creek at Union Gap	--	1 of 1	--	--	--	--	--
12505350	East Toppenish Drain at Wilson Road	--	1 of 1	--	--	--	--	--
12505410	Sub 35 Drain at Parton Road	--	1 of 1	--	--	--	--	--
12505460	Granger Drain at mouth[3]	--	15 of 15	13 of 15	3 of 15	--	2 of 5	1 of 15
12505510	Marion Drain at Indian Church Road	--	2 of 2	--	--	--	--	--
12507508	Toppenish Creek at Indian Church Road	--	6 of 7	2 of 7	--	--	--	--
12507585	Yakima River at river mile 72	--	1 of 1	--	--	--	--	--
12508620	Satus Creek at gage at Satus	--	2 of 2	2 of 2	--	--	--	--
12508630	South Drain near Satus	--	1 of 1	--	--	--	--	--
12508850	Sulphur Creek Wasteway	--	14 of 14	13 of 14	--	--	2 of 5	--
12509050	Yakima River at Euclid Bridge at river mile 55	--	2 of 2	2 of 2	--	1 of 2	--	--
12509499	Chandler Canal at Bunn Road	--	2 of 2	--	--	--	--	--
12509710	Spring Creek at mouth	--	2 of 2	2 of 2	--	1 of 2	--	--
12509829	Snipes Creek at mouth	--	2 of 2	2 of 2	--	--	--	--
12510500	Yakima River at Kiona	--	14 of 15	8 of 15	--	--	3 of 4	--

[1]Samples collected during summer 1991.

[2]Includes data collected from Moxee Drain at Birchfield Road.

[3]Includes data collected upstream at Granger Drain at Granger.

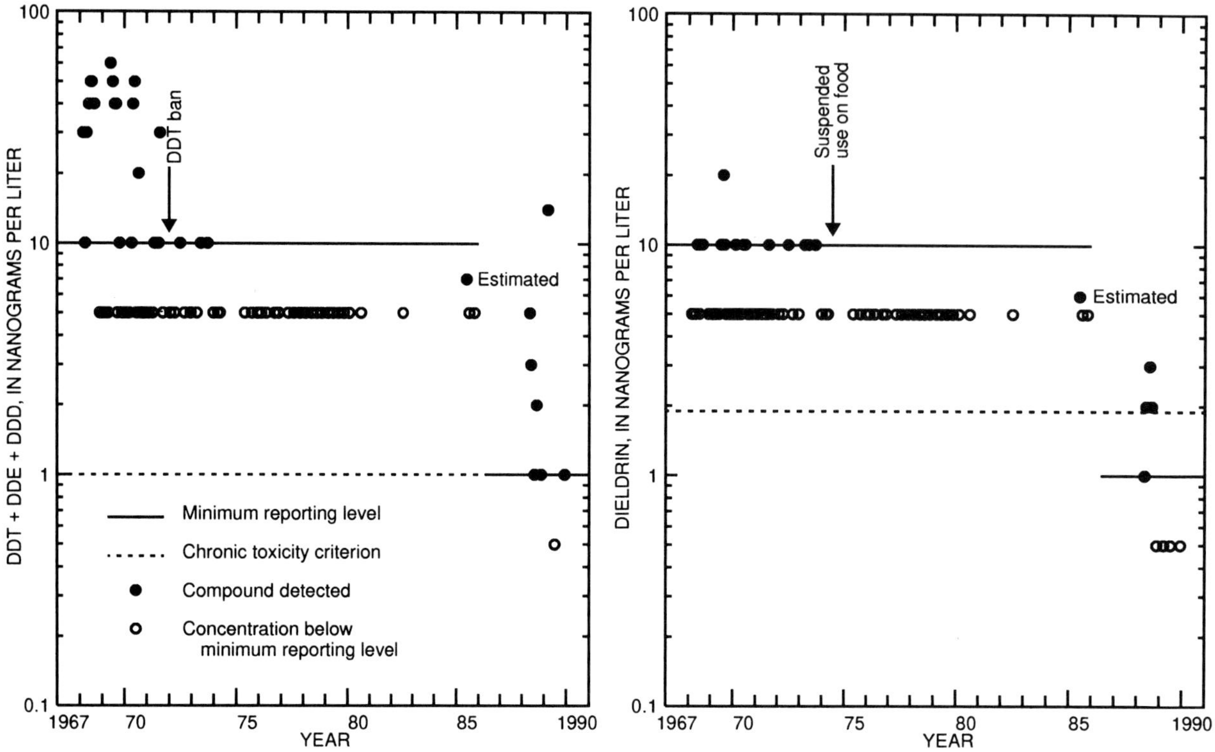

Figure 22. Concentrations of 4,4'-dichlorodiphenyltrichloroethane (DDT) plus 4,4'-dichlorodiphenyldichloroethylene (DDE) plus 4,4'-dichlorodiphenyldichloroethane (DDD) and dieldrin in unfiltered-water samples from the Yakima River at Kiona, Yakima River Basin, Washington, 1968–89. (Values less than the minimum reporting level are graphically represented as one-half the minimum reporting level. Chronic-toxicity criterion, the chronic-toxicity water-quality criterion for the protection of freshwater aquatic life established by the U.S. Environmental Protection Agency [1986].)

of DDT+DDE+DDD over time coincides with, and probably resulted from, the ban in 1973. Data are not available, however, in the Yakima River Basin to show how concentrations of DDT+DDE+DDD have changed in agricultural soils, suspended sediment, and atmospheric deposition over time. The relation (fig. 23) between the concentrations of DDT+DDE+DDD and suspended sediment in the basin in June 1989 suggests that DDT+DDE+DDD concentrations in unfiltered water may have decreased, in part, as a result of decreased suspended-sediment concentrations. Historical concentrations of suspended sediment, however, are not available to evaluate this hypothesis.

Volatile Organic Compounds

Volatile organic compounds that were analyzed in this study include halogenated aliphatic and monocyclic aromatic hydrocarbons. Many of these compounds are highly toxic and are used as solvents, degreasers, refrigerants, fumigants, dry-cleaning solvents, and in the manufacturing of pharmaceuticals and other synthetic organic compounds. These compounds are characterized by high vapor pressures, high solubilities, and relatively low sediment-water partition coefficients. The primary fate for many of these compounds in streams is volatilization followed by photolytic degradation in the atmosphere (Smith and others, 1988).

Eleven unfiltered-water samples were collected from 7 sampling stations in the basin and were analyzed for 47 volatile compounds (Rinella, McKenzie, Crawford, and others, 1992). Samples were collected from the city of Yakima's municipal drinking-water supply (finished chlorinated water), several agricultural return flows (Moxee Drain, Wide Hollow Creek, Granger Drain, and Sulphur Creek Wasteway) that also convey urban and industrial discharges, Pacific Power & Light Company Wasteway, and the Yakima River at Kiona. Three compounds were detected: chloroform (22,000 ηg/L) in the water supply, tetrachloroethene (200–300 ηg/L) in four samples from Wide Hollow Creek near the mouth, and trichlorofluoromethane (900 ηg/L) in Sulphur Creek Wasteway near Sunnyside.

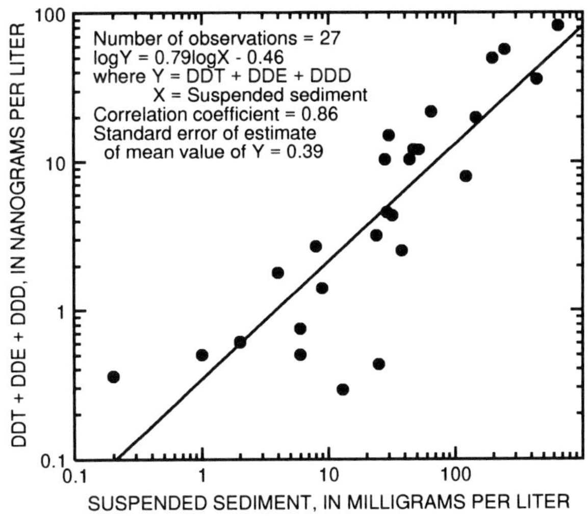

Figure 23. Relation of concentrations of suspended sediment and 4,4'-dichlorodiphenyltrichloroethane (DDT) plus 4,4'-dichlorodiphenyldichloroethylene (DDE) plus 4,4'-dichlorodiphenyldichloroethane (DDD) in unfiltered-water samples from selected stations, Yakima River Basin, Washington, June 25–30, 1989. (Values less than the minimum reporting level are excluded.)

Chloroform is widely distributed in the environment and occurs in municipal drinking-water supplies primarily as a result of the chlorination process. A survey of 80 cities in the United States by the EPA detected chloroform in every water system in concentrations ranging from less than 300 to 311,000 ηg/L (Sittig, 1981). The concentration of 22,000 ηg/L in Yakima's drinking water is about one-fifth of EPA's maximum contaminant level of 100,000 ηg/L for drinking water (table 62, at back or report), although it exceeds EPA's health advisory of 6,000 ηg/L for a lifetime cancer risk of 1 in 1,000,000 people (U.S. Environmental Protection Agency, 1992a). Only one sample of the city of Yakima's drinking water was collected for analysis in this study, so the temporal distribution of chloroform concentrations could not be determined.

Tetrachloroethene was detected at Wide Hollow Creek during irrigation season in August 1988, low flows in November 1988, and storm runoff in March 1989. Concentrations ranged from 200 to 300 ηg/L, well below EPA's chronic-toxicity criterion of 840,000 ηg/L for the protection of freshwater aquatic life (U.S. Environmental Protection Agency, 1986). Possible sources upstream from the sampling station in Wide Hollow Creek Subbasin include light industry, urban runoff, the Yakima airport, and agricultural activities. Tetrachloroethene is a widely used solvent in dry-cleaning agents, degreasers, and fumigants.

Trichlorofluoromethane, a refrigerant and cleaning agent probably resulting from light-industrial and (or) urban activity in the vicinity of the city of Sunnyside was detected at 900 ηg/L in Sulphur Creek Wasteway during irrigation season in August 1988. Even though Sulphur Creek Wasteway is not used currently as a drinking-water supply, this concentration is well below EPA's lifetime health advisory of 2,000,000 ηg/L for protection from adverse, noncarcinogenic, health effects (U.S. Environmental Protection Agency, 1992a).

Sources and Loads

The application of pesticides on agricultural crops, in urban areas, and along road sides and channel banks of canals and agricultural return flows is a major source of organic compounds to streams in the Yakima River Basin. Estimates for 1989 indicate that about 3 million kilograms of active pesticide ingredients were applied to increase the production and quality of agricultural crops (table 59, at back of report). This use is equivalent to an annual average pesticide application of about 1,600 kilograms per square kilometer of irrigated crop (14 pounds per acre). Oil is used as a dormant spray (insecticide) on fruit trees and accounts for about 30 percent of the annual pesticide application in the basin.

Estimates for pesticide use have been compiled for several major agricultural areas (table 59, at back of report). These areas include the Kittitas, Tieton, Moxee, East-side, and West-side areas and consist of about 96,000, 31,000, 25,000, 160,000, and 120,000 acres of irrigated land, respectively. For examining relations between application and runoff loads of pesticides, agricultural return flows in these areas were sampled synoptically during June 25–30,1989, near peak irrigation. This time frame was selected because the 1988 data showed that concentrations of hydrophilic and hydrophobic compounds increased following pesticide application near or during peak irrigation.

In the June 1989 synoptic sampling, about 100 compounds were analyzed at 29 sampling stations (Rinella, McKenzie, Crawford, and others, 1992). The following 20 compounds were quantified most frequently (concentrations at each station are shown in table 60, at back of report): cis- and trans-chlordane,

DDT+DDE+DDD, dieldrin, and endosulfan I (organochlorine compounds); chlorpyrifos, diazinon, dimethoate, malathion, parathion, phorate, and phosphamidon (organophosphorus compounds); atrazine, prometon, and simazine (triazine compounds); alachlor and metolachlor (acetamide compounds); and 2,4-D, dicamba, EPTC, and propargite (miscellaneous compounds). With the exception of DDT+DDE+DDD, dieldrin, chlordane, and prometon, these compounds were among the most abundantly used compounds in the basin in 1989, which probably accounts for their frequent detections (table 3 and table 59, at back of report). Even though the production and sale of DDT, dieldrin, and chlordane were banned, suspended, or cancelled in 1973, 1974, and 1987, respectively, these three persistent compounds were detected frequently (28 of 29 stations for DDT+DDE+DDD, 20 stations for dieldrin, and 11 stations for chlordane), suggesting historical broad usage throughout the basin.

In this study, instantaneous pesticide loads in streams during the June 1989 synoptic sampling were compared to annual pesticide applications in 1989 (table 31). The term "instantaneous" implies that a sample generally was collected within a 30- to 60-minute time period (up to 3 hours for large sample volumes). As mentioned earlier, the synoptic study occurred in June because pesticide concentrations in streams in the Yakima River Basin tend to increase at or near peak irrigation following springtime pesticide applications.

A major objective for calculating compound loads is to estimate the mass transport of organic compounds from major tributaries into the Yakima River. Instantaneous loads (shown in table 60, at back of report) also may be used to identify potential sources and sinks of suspended sediment and organic compounds. For example, the instantaneous load of suspended sediment in Sulphur Creek Wasteway, which flows into the Yakima River at RM 61, in June 1989 was 160,000 kg/d. The load in the Yakima River at Kiona (RM 29.9) was 110,000 kg/d. The difference in loads suggests that sediment deposition is occurring between RM 61 and RM 29.9, probably in the slower moving reach upstream from Prosser Dam (RM 47.1). Another explanation for the difference may be that tributary loads were not constant during the 5-day sampling period in June 1989; therefore, the data may not be appropriate for mass balancing using steady-state assumptions.

The predominant source area for many of the frequently occurring organic-compounds in Yakima River Basin streams during the June 1989 sampling is the East-side area (area east of the Yakima River from Moxee Valley downstream to Benton City near Kiona and includes the Moxee, Granger, Sunnyside, and Whitstran areas; table 60, at back of report). This area has the largest acreage of irrigated land in this study, and generally receives the largest application of pesticides in the basin (table 59, at back of report). Sulphur Creek Wasteway was the predominant source of suspended sediment and many of the pesticides to the Yakima Rive. When compared to streamflows in the other agricultural return flows in June 1989, Sulphur Creek Wasteway also had the largest discharge, which contributed to the larger pesticide loads. Sulphur Creek Wasteway had among the highest concentrations of suspended sediment, alachlor, DDT+DDE+DDD, dieldrin, dimethoate, endosulfan I, prometon, propargite, and simazine.

Owing to the highly erosive soil characteristics of the Warden-Esquatzel association, steep subbasin slopes, and rill irrigation of tilled crops, the suspended-sediment load from the East-side area is five or more times larger than the loads from the other areas. Similarly, several of the more hydrophobic compounds (DDT+DDE+DDD, dieldrin, endosulfan I, phorate, and propargite) have loads from the East-side area that are four or more times larger than their corresponding loads from other areas. Results from the sampling of 29 stations in June 1989 show that concentrations of DDT+DDE+DDD, dieldrin, diazinon, and malathion in unfiltered water (suspended plus dissolved phases) and concentrations of phosphamidon, propargite, atrazine, and simazine in filtered water increase as concentrations of suspended sediment increase; rank correlations between suspended sediment and each of the pesticides were significant ($\rho \leq 0.02$). The correlations suggest that these hydrophobic and hydrophilic compounds are being transported, at least initially, in the suspended phase from the agricultural fields. The data summary for the June 1989 synoptic sampling shows that tributaries to the Yakima River with the highest concentrations of suspended sediment have among the highest concentrations of many of the hydrophilic and hydrophobic compounds (table 32). As expected, background stations upstream from agricultural activities have among the lowest concentrations of suspended sediment and pesticides

Table 31. Mean daily applications of selected pesticides in 1989 and instantaneous pesticide and suspended-sediment loads in stream runoff for selected areas from June 25–30, 1989, Yakima River Basin, Washington

[Applications and runoff load reported in grams per day; Ratio, the ratio of runoff load to application, in percent; B, value is below minimum reporting level; East side, the area east of Yakima River from Moxee Valley downstream to Benton City near Kiona including the Moxee, Granger, and Sunnyside areas; West side, the area west of Yakima River downstream from Ahtanum Ridge to the city of Mabton including the Wapato, Toppenish, and Mabton areas; Yakima River Basin, the Yakima River Basin downstream to Benton City near Kiona excluding the Kennewick Irrigation District; EPTC, *S*-ethyl dipropylthiocarbamate; 2,4-D, (2,4-dichlorophenoxy) acetic acid]

Constituent	Kittitas area	Tieton area	Moxee Subbasin	East side	West side	Yakima River Basin
Organochlorine compound						
Endosulfan I						
Application	2,000	30,000	7,000	45,000	15,000	100,000
Runoff load[1]	B	.34	1.9	6.2	.87	4.4
Ratio	B	.0011	.027	.014	.0058	.0044
Organophosphorus compounds						
Chlorpyrifos						
Application	3,000	45,000	10,000	80,000	40,000	200,000
Runoff load[1]	.09	B	.08	.54	.55	35
Ratio	.0030	B	.00080	.00068	.0014	.018
Diazinon						
Application	5,000	4,000	40,000	100,000	50,000	200,000
Runoff load[1]	B	.66	65	120.5	56.3	140
Ratio	B	.016	.16	.12	.11	.070
Dimethoate						
Application	1,500	2,000	900	38,000	9,500	54,000
Runoff load[2]	B	.46	B	210	3.1	B
Ratio	B	.023	B	.55	.033	B
Malathion						
Application	90,000	45,000	25,000	200,000	80,000	450,000
Runoff load[1]	B	1.35	B	8.80	14.9	260
Ratio	B	.0030	B	.0044	.019	.058
Parathion						
Application	3,000	20,000	15,000	80,000	25,000	150,000
Runoff load[1]	B	B	.02	1.50	63.7	B
Ratio	B	B	.00013	.0019	.25	B
Phorate						
Application	1,000	0	4,500	13,000	10,000	25,000
Runoff load[1]	1.7	B	21	21	4.6	B
Ratio	.17	B	.47	.16	.046	B
Phosphamidon						
Application	1,500	35,000	8,000	35,000	15,000	100,000
Runoff load[2]	B	.95	6.8	16.1	10.6	19
Ratio	B	.0027	.085	.046	.071	.019
Acetamide, sulfite, thiocarbamate, and triazine compounds						
Alachlor						
Application	4,000	0	150	9,000	25,000	45,000
Runoff load[2]	B	B	B	24	25.9	44
Ratio	B	B	B	.27	.10	.098
Atrazine						
Application	1,500	90	150	10,000	10,000	25,000
Runoff load[2]	31.5	.55	1.4	39.7	81.1	120
Ratio	2.1	.61	.93	.40	.81	.48

Table 31. Mean daily applications of selected pesticides in 1989 and instantaneous pesticide and suspended-sediment loads in stream runoff for selected areas from June 25–30, 1989, Yakima River Basin, Washington—Continued

Constituent	Kittitas area	Tieton area	Moxee Subbasin	East side	West side	Yakima River Basin
Acetamide, sulfite, thiocarbamate, and triazine compounds—Continued						
EPTC						
Application	4,500	900	1,000	24,000	15,000	45,000
Runoff load[2]	11.6	B	B	.84	3.48	12
Ratio	.26	B	B	.0035	.023	.027
Metolachlor						
Application	400	0	45	4,500	3,500	9,000
Runoff load[2]	B	B	B	B	3.23	7
Ratio	B	B	B	B	.092	.078
Prometon						
Application	90	100	70	200	100	600
Runoff load[2]	3.23	1.97	.60	19.5	.26	11
Ratio	3.6	2.0	.86	9.8	.26	1.8
Propargite						
Application	400	5,000	3,500	70,000	40,000	130,000
Runoff load[2]	B	B	33	207	48	26
Ratio	B	B	.94	.30	.12	.020
Simazine						
Application	1,000	15,000	3,500	40,000	15,000	80,000
Runoff load[2]	4.9	.95	B	65.0	20.0	66
Ratio	.49	.0063	B	.16	.13	.082
Benzoic compounds and chlorophenoxy-acid herbicides						
Dicamba						
Application	40,000	3,000	5,000	25,000	20,000	100,000
Runoff load[2]	49	B	B	13.2	B	B
Ratio	.12	B	B	.053	B	B
2,4-D						
Application	250,000	40,000	35,000	150,000	100,000	700,000
Runoff load[2]	89	B	19	89.9	35.1	330
Ratio	.036	B	.054	.060	.035	.047
Suspended-sediment load in stream runoff, in kilograms per day	41,600	1,150	317,000	59,500	180,000	110,000

[1]Based on concentrations in unfiltered water (suspended plus dissolved phases).
[2]Based on concentrations in filtered water.

in the Yakima River Basin. Main-stem stations generally have low to intermediate concentrations, depending on whether the station is upstream or downstream from tributaries containing agricultural return flow, respectively.

The relation between mass transport of 17 pesticides in streams (instantaneous loads in June 1989 listed in table 31) and pesticide use in 1989 is shown in figure 24. The annual amount of pesticide applied was divided by 365 days (to give the mean daily application) in order to compare runoff and application in the same units (grams per day). As pesticide applications increase, pesticide loads in streams generally increase; the rank correlation was highly significant ($\rho < 0.0001$). Typically, however, the measured pesticide loads in streams vary in range by more than three orders of magnitude for a particular pesticide-application level. An attempt was made to model the loads using a multiple linear regression, with independent variables of pesticide application, area of irrigated crops, half-life of the compound in soil, the K_{oc} values, and suspended-sediment concentration. These variables could

Table 32. Summary of pesticide concentrations in water samples for selected areas, Yakima River Basin, Washington, June 25–30, 1989

[Concentrations reported in nanograms per liter, except suspended sediment which is in milligrams per liter (mg/L); Background and drinking water stations include: Cooper Creek, Yakima River at Cle Elum, Umtanum Creek, Pacific Power and Light Company Wasteway, Satus Creek above Wilson-Charley, city of Yakima finished water, one well at Sunnyside, and one well at Harrah; Tributaries with medium suspended-sediment concentrations (less than 40 mg/L) include: Wilson Creek, Wide Hollow Creek, Ahtanum Creek, Sub 35 Drain, Marion Drain, Toppenish Creek, and Satus Creek; Tributaries with large suspended-sediment concentrations include: Cherry Creek, Moxee Drain, East Toppenish Drain, Granger Drain, South Drain, Sulphur Creek, Spring Creek, and Snipes Creek; main-stem stations include: Yakima River at Umtanum, Union Gap, river mile 72 (upstream from Satus Creek), Grandview, and Kiona; <, less than; nd, not detected and ranked as zero values; DDT+DDE+DDD, the sum of 2,4'-dichlorodiphenyltrichloroethane (DDT), 4,4'-DDT, 2,4'-dichlorodiphenyldichloroethylene (DDE), 4,4'-DDE, 2,4'-dichlorodiphenyldichloroethane (DDD), and 4,4'-DDD; EPTC, *S*-ethyl dipropylthiocarbamate; 2,4-D, (2,4-dichlorophenoxy) acetic acid]

Constituent	Background and drinking water stations				Tributaries with medium suspended-sediment concentrations				Tributaries with large suspended-sediment concentrations				Naches River and main-stem stations downstream from Umtanum			
	Number of samples	Minimum	Median	Maximum	Number of samples	Minimum	Median	Maximum	Number of samples	Minimum	Median	Maximum	Number of samples	Minimum	Median	Maximum
Suspended sediment	6	<1	2	6	7	8	28	38	8	52	170	643	6	6	28	47
Organochlorine compounds																
Chlordane[1], cis- and trans-	8	nd	nd	.15	7	nd	.41	3.0	8	nd	.26	4	6	nd	nd	nd
DDT+DDE+DDD[1]	8	nd	.50	1.8	7	.29	2.7	10	8	8.0	29	83	6	.43	6.8	15
Dieldrin[1]	8	nd	nd	nd	7	.003	1.0	1.7	8	.02	4.1	37	8	nd	1.0	5.7
Endosulfan I[1]	8	nd	nd	8.0	7	nd	1.2	2.6	8	nd	nd	12	8	nd	.78	1.2
Organophosphorus compounds																
Chlorpyrifos[1]	8	nd	nd	1.0	7	nd	nd	.20	8	nd	.25	3.7	8	nd	nd	9.6
Diazinon[1]	8	nd	nd	.67	7	nd	22	55	8	nd	49	410	6	nd	22	120
Dimethoate[2]	8	nd	nd	nd	7	nd	nd	7.7	8	nd	nd	330	6	nd	nd	7.3
Malathion[1]	8	nd	nd	5.6	7	nd	13	31	8	nd	nd	48	6	nd	11	70
Parathion[1]	8	nd	nd	nd	7	nd	nd	180	8	nd	.17	23	6	nd	nd	30
Phorate[1]	8	nd	nd	nd	7	nd	nd	19	8	nd	nd	130	6	nd	nd	3.6
Phosphamidon[2]	8	nd	nd	nd	7	nd	5.3	16	8	nd	17	56	6	nd	nd	22
Acetamide, sulfite, thiocarbamate, and sulfite compounds																
Alachlor[2]	8	nd	nd	nd	7	nd	13	98	8	nd	2.2	33	6	nd	5.5	19
Atrazine[2]	8	nd	nd	57	7	nd	34	88	8	8.5	48	130	6	.26	18	61
EPTC[2]	8	nd	nd	.01	7	nd	2.8	6.3	8	nd	2.4	37	6	nd	2.4	7.3
Metolachlor[2]	8	nd	nd	nd	7	nd	nd	4.1	8	nd	nd	6.0	6	nd	.46	1.9
Prometon[2]	8	nd	nd	nd	7	nd	2.7	22	8	nd	.7	32	6	nd	1.9	5.2
Propargite[2]	8	nd	nd	nd	7	nd	11	30	8	nd	42	260	6	nd	5.1	44
Simazine[2]	8	nd	nd	6.4	7	4.2	17	43	8	nd	10	130	6	.24	10	33
Chlorophenoxy-acid and benzoic compounds																
2,4-D[2]	8	nd	nd	nd	6	nd	5	140	8	nd	70	290	6	nd	35	90
Dicamba[2]	8	nd	nd	nd	6	nd	nd	50	8	nd	10	100	6	nd	nd	10

[1]Concentrations in unfiltered water (suspended plus dissolved phases).

[2]Concentrations in filtered water.

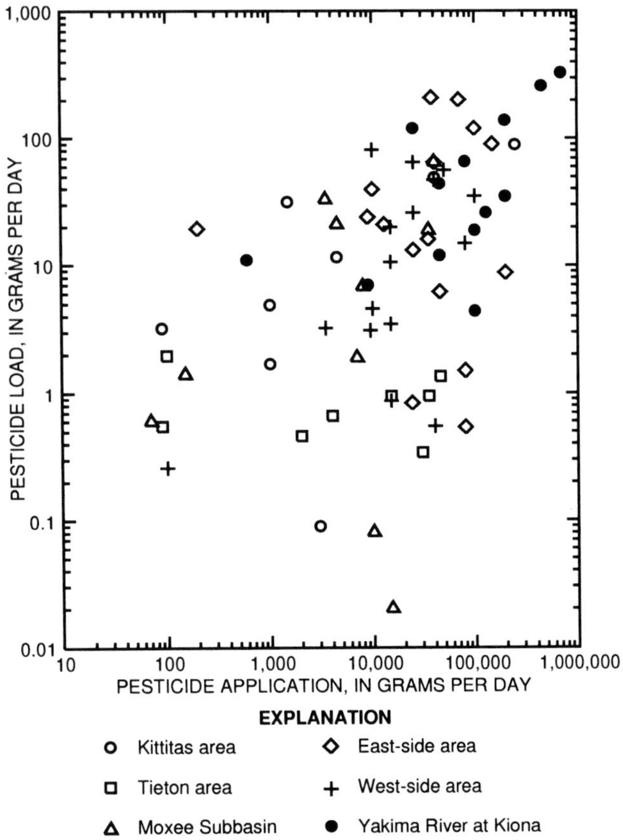

Figure 24. Relation of mean daily application of 17 selected pesticides in 1989 and instantaneous pesticide load in runoff from selected agricultural areas, Yakima River Basin, Washington, June 25–30, 1989. (Data listed in table 31. Values less than the minimum reporting level are excluded.)

explain only 37 percent of the variability observed in the instantaneous pesticide loads. Results from this type of regression may be improved by modeling annual pesticide loads rather than instantaneous loads. Deterministic modeling also could be used to help understand transport processes in the basin. Deterministic modeling, however, requires information on many factors that govern transport, including frequency and amount of pesticide application; method, frequency, and volume of irrigation-water application; rates of erosion; soil-flushing processes; sorption and dissolution of pesticides to and from soils, suspended sediment, and bed sediment; pesticide half-lives in soils and streams; biological uptake and metabolism of the pesticides; and hydrological and physical characteristics (time of travel, stream velocity and turbulence relative to sediment suspension and contaminant volatility, light penetration relative to photolysis, channel characteristics, and water use). The amount and complexity of information needed for deterministic modeling in a basinwide study was well beyond the scope and funding of this pilot study.

For several compounds (2,4-D, diazinon, atrazine, and simazine), relations between pesticide application and mass transport in runoff exist that could be used for estimating pesticide loads in runoff in larger basins. For example, a linear regression between application and instantaneous load of 2,4-D using data from the Kittitas, Tieton, Moxee, East-side, and West-side areas (table 31) can be used to estimate that the instantaneous load of 2,4-D associated with a mean daily application of 700,000 g/d (grams per day) for the Yakima River Basin is about 330 g/d, which corresponds well to the observed instantaneous load in June 1989 of 330 g/d (fig. 25). Considering the large temporal variability expected for instantaneous pesticide concentrations, the agreement between the extrapolated and measured loads is remarkably good. These relations for 2,4-D, diazinon, atrazine, and simazine (fig. 25) suggest that data may be collected in source areas and extrapolated to make estimates of instantaneous pesticide loads in receiving streams. The other compounds listed in table 31, however, did not show significant relations between pesticide application and loading in stream runoff. Similar analyses of additional compounds may reveal other such relations.

Ratios of instantaneous pesticide loads in stream runoff to pesticide application are listed in table 31 and figure 26. The instantaneous loads generally represent less than 0.6 percent of the mean daily pesticide application. These ratios are low even though the data used for calculating instantaneous pesticide loads were collected during the peak-irrigation period when pesticide concentrations and streamflows were expected to be high. In Midwestern streams in the United States, the median annual-mass-transport rates (April 1991 through March 1992) for alachlor, metolachlor, atrazine, and simazine from the Mississippi River Basin, as a percent of mass applied, were 0.2, 0.8, 1.4, and 2.5 percent, respectively (D.A. Goolsby and W.A. Battaglin, 1993). These mass-transport rates measured in Midwestern streams are similar to the corresponding rates measured in the Yakima River Basin (table 31).

Two compounds, prometon and atrazine, have notably larger ratios than other compounds in the Yakima River Basin (fig. 26). In June 1989, prometon was detected at 13 stations, with concentrations as high as 32 ηg/L in Granger Drain; atrazine was detected at 23 stations, with concentrations as high as

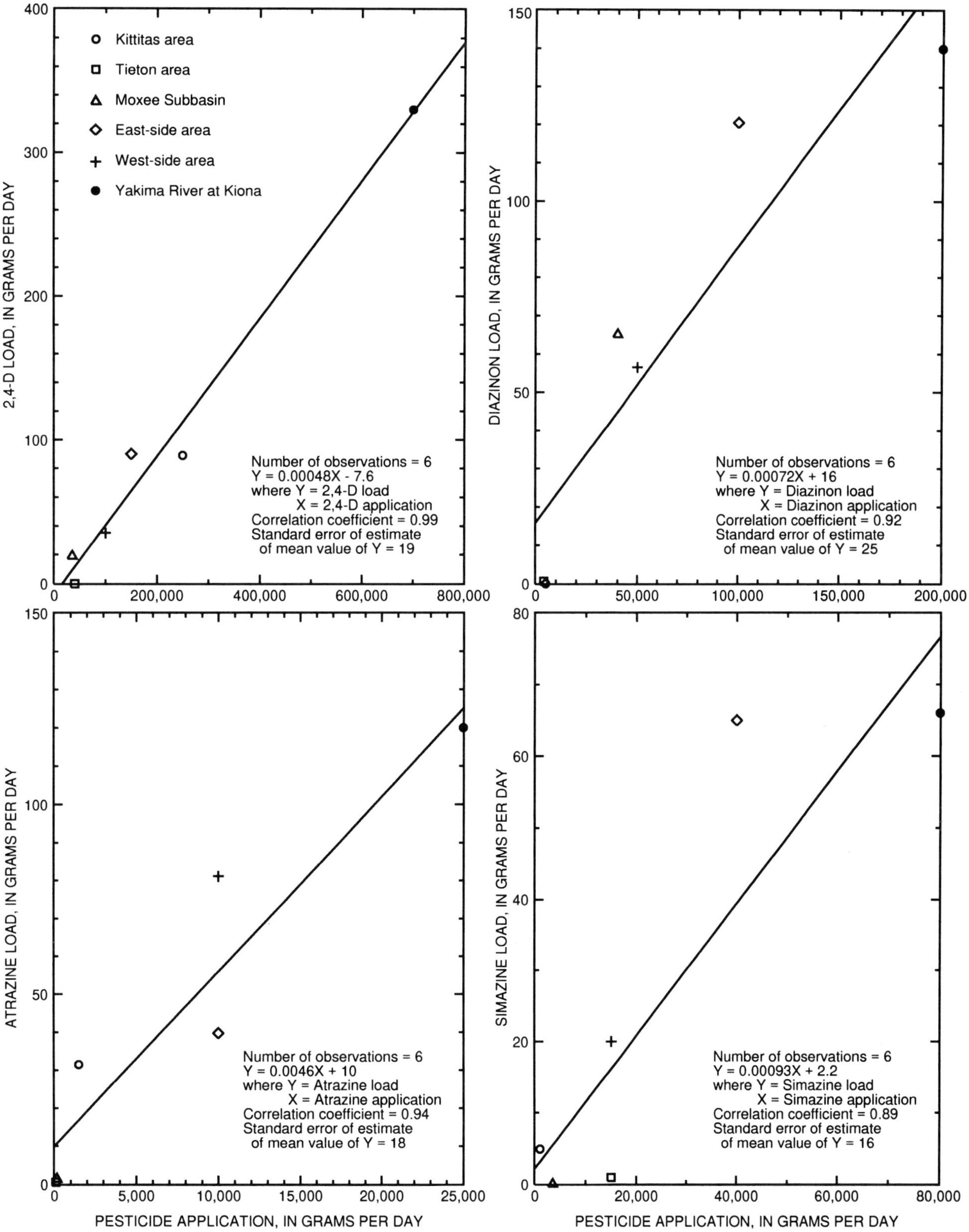

Figure 25. Relation of mean daily application of selected individual pesticides in 1989 and instantaneous pesticide load runoff from selected agricultural areas, Yakima River Basin, Washington, June 25–30, 1989. (Data reported in table 31 2,4-D, (2,4-dichlorophenoxy) acetic acid.)

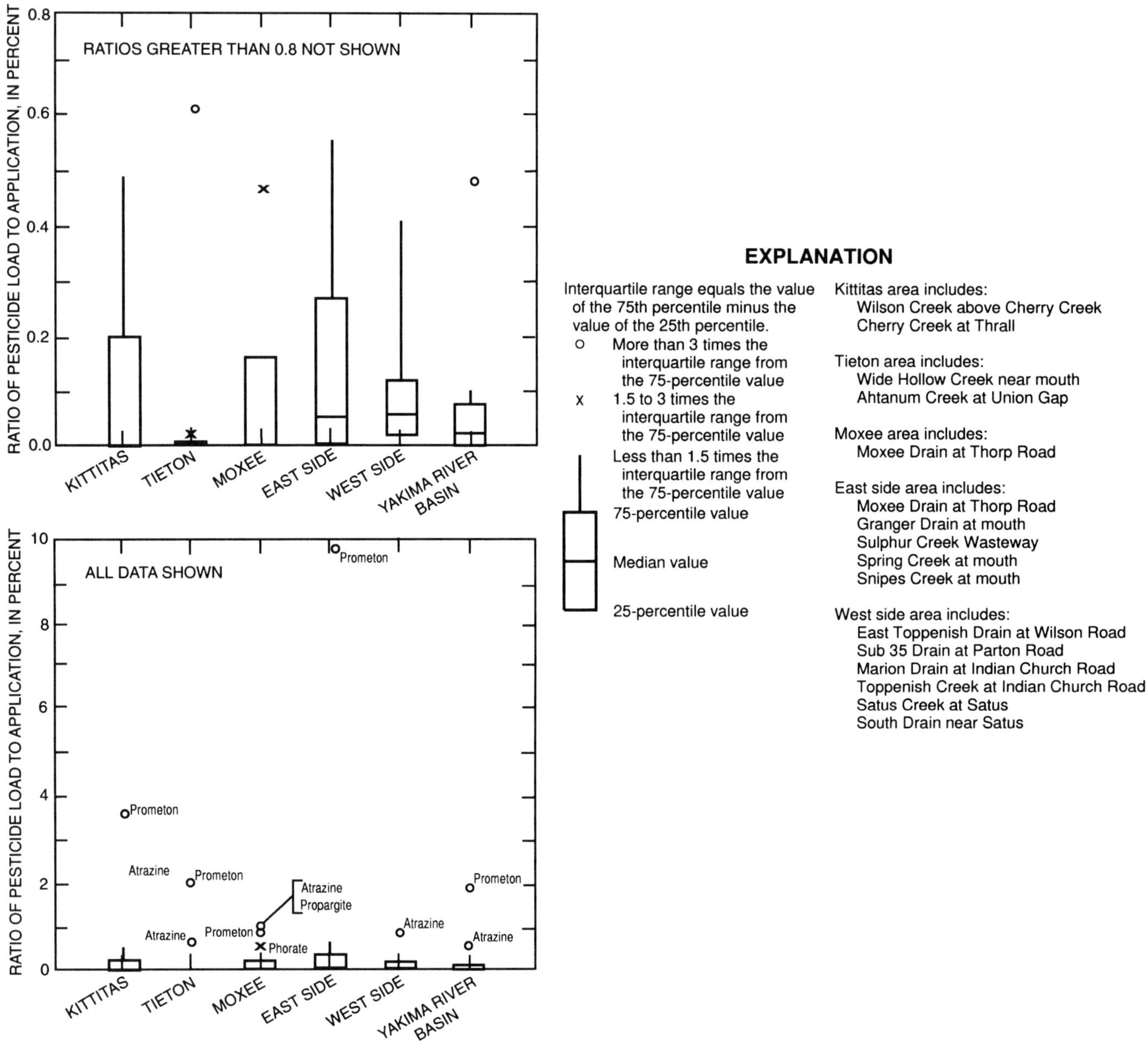

Figure 26. Ratio of instantaneous loads of 17 selected pesticides in June 1989 runoff to mean-daily pesticide application in 1989, Yakima River Basin, Washington. (Data reported in table 31.)

130 ηg/L in East Toppenish Drain and South Drain. Both pesticides are organonitrogen herbicides used on broadleaf weeds and grasses. Atrazine is used mostly on corn, and prometon is widely available for controlling weeds, grasses, and brush (Bob Parker, Extension Weed Scientist, Irrigated Agriculture Research and Extension Center, Washington State University, Prosser, Washington, written commun., August 5, 1992). A high ratio for atrazine also was determined for Midwestern streams (D.A. Goolsby and W.A. Battaglin, U.S. Geological Survey, Denver, Colorado, written commun., November 1992). The reason for the high ratio in Yakima River Basin streams is unknown. Prometon ratios may appear large if the amount of application was underestimated in this study. According to the agricultural marketing specialist for prometon in the Yakima River Basin, about 500 kg were sold in 1991 (Tom Edwards, Senior Marketing Specialist, Ciba Geigy Corporation, Yakima, Washington, oral commun., November 1992), which is in close agreement to the 600-kg estimate for the basin; it is possible, however, that additional prometon used in the basin was

purchased outside of the basin. In addition, prometon has a relatively long half-life in soil (as long as 500 days). Consequently, it may build up in soils and contribute to the increased ratios during peak irrigation.

Dissolved Organic Carbon in the Yakima River Basin

By George R. Aiken

Dissolved organic carbon (DOC) is commonly a major pool of organic matter in ecosystems. DOC is operationally defined as organic carbon in the filtrate (dissolved and colloidal phases) that has passed through a 0.45-μm pore-size filter. Many organic compounds, including many pesticides discussed in this report, can interact with dissolved or colloidal phases of organic matter and can be transported in streams far beyond distances that would be predicted by analyses of their structural and activity relations. To the extent possible, the chemical characteristics of various classes of DOC are presented in this report relative to the behavior, transport, and distribution of pesticides in streams in the Yakima River Basin. The assessment of DOC in our Nation's streams, however, is in its early stages, so that results presented in this section primarily reflect research characterizations of DOC in the Yakima River Basin.

In addition to reacting with pesticides and other contaminants, DOC in streams and other aqueous systems often controls geochemical processes by acting as a proton donor/acceptor and pH buffer. DOC also affects the degradation of various chemical constituents and can affect mineral dissolution/precipitation reactions. Numerous studies have recognized the importance of natural DOC in the mobilization of hydrophobic organic species, metals (lead, cadmium, copper, zinc, mercury, and chromium), and radionuclides (plutonium, uranium, and cobalt) into the dissolved phase. In addition, DOC is a substrate for many microbial metabolic processes. The foregoing factors have led researchers to modify conceptual models of chemical behavior from simple two-phase systems (solid/water) to complex multiphase systems (solids/water/colloids/cosolutes), which necessitate consideration of constituent-DOC interactions (Huling, 1989; McCarthy and Zachara, 1989).

In most aquatic ecosystems, DOC is derived from the soils, terrestrial plants, and algae and other microorganisms growing in the aquatic environment (Steinberg and Muenster, 1985). Major fractions of DOC in natural water are biologically stable organic compounds that are operationally defined as aquatic fulvic acids (FAs) and hydrophilic acids (HPIAs) (Aiken and others, 1992). Aquatic FAs are yellow, heterogeneous, organic acids that usually make up 20 to 80 percent of the DOC. HPIAs have a greater abundance of acid-functional groups than do FAs and generally constitute about 10 to 20 percent of the DOC. Operational definitions used in this report to describe these DOC fractions are given in table 33.

Results of a preliminary study of DOC in the Yakima River are presented in this section. Objectives of the study were to define chemical characteristics of dissolved FA and HPIA in the Yakima River, and to provide information of chemical reactivity essential to understanding pesticide transport in streams. Water samples were collected from two Yakima River stations in June 1987 during irrigation season. The stations were located upstream (Yakima River near Cle Elum—RM 183.1) and downstream (Yakima River at Kiona—RM 29.9) from agricultural return flows in order to assess DOC changes resulting from agricultural activities.

The Cle Elum station is downstream from three large, relatively unpolluted reservoirs: Keechelus, Kachess, and Cle Elum Lakes. At the Cle Elum station, the river is cool and swift with a low concentration of suspended sediment (table 34); some periphyton were observed near the shore. In contrast, the Kiona station is located near the terminus of the basin, where agricultural return flows account for most of the flow in the Yakima River during irrigation season. At the Kiona station, the river is warmer, more turbid, and slower moving than at Cle Elum, and higher phytoplankton concentrations were noticeable.

Distribution and General Properties of Dissolved Organic Carbon

Large-volume water samples (up to 240 gallons) were collected for analyses of the FA and HPIA fractions in filtered water. The fractions were isolated from

Table 33. Operational definitions of different fractions of organic acids comprising dissolved organic carbon obtained using XAD-8 and XAD-4 resins

[[k'_{cutoff}, the value of the capacity factor (k') for a solute at which 50 percent is retained at the hydrophobic-hydrophilic break (Aiken and others, 1992)]

Acid fraction	Definition
Hydrophobic (includes fulvic-acid fraction)	That portion of the dissolved organic carbon that sorbs on a column of XAD-8 resin at pH 2 under conditions where $k'_{cutoff} = 50$ for the column and is eluted at pH 13. This fraction can contain aliphatic carboxylic acids of five to nine carbons, one- and two-ring phenols, and aquatic humic substances.
Hydrophilic	That portion of the dissolved organic carbon contained in the XAD-8 resin effluent at pH 2 that sorbs on a column of XAD-4 resin under conditions where $k'_{cutoff} = 50$ for the column, and that are eluted at pH 13. This fraction can contain polyfunctional organic acids and aliphatic acids with five or fewer carbon atoms.

Table 34. Water-quality data for the Yakima River at Cle Elum and Kiona, Yakima River Basin, Washington, 1987

[Measurements were made 4 to 5 weeks prior to sampling for dissolved organic carbon]

Parameter	Yakima River at Cle Elum June 11, 1987	Yakima River at Kiona June 18, 1987
Streamflow, in cubic feet per second	2,460	1,240
Specific conductance, in microsiemens per centimeter	48	281
pH	7.1	8.4
Temperature, in degrees Celsius	10.5	21.5
Turbidity, in nephelometric turbidity units	.5	8.0
Suspended sediment, in milligrams per liter	2	25

samples using the XAD-8 method of Thurman and Malcolm (1981), which was modified for this study by Aiken and others (1992). In samples collected on July 22 and 23, 1987, DOC concentrations were higher at Kiona (2.4 mg/L) than at Cle Elum (1.6 mg/L) (table 35). Fixed-station data collected during 1987–91 also indicate that DOC concentrations at Kiona were consistently higher than those at Cle Elum (fig. 27). DOC concentrations at Cle Elum were relatively constant, probably because of the leveling effect of the upstream reservoirs, which attenuates seasonal differences. In contrast, DOC concentrations at Kiona were more variable. For example, on June 18, 1987, a high DOC concentration (7.5 mg/L) was observed in a sample from the Kiona station. Similarly, other high DOC concentrations were detected sporadically from the Kiona station in 1988.

Large increases in DOC concentrations at Kiona may have resulted from agricultural stormwater runoff. DOC data from May to October 1991 for two major agricultural return flows, Moxee Drain and Granger Drain, are presented in figure 28. These drains flow into the Yakima River upstream from Kiona at RMs 107.3 and 82.8, respectively. In both drains, DOC concentrations were generally higher than those at Kiona, and they were substantially higher (>20 mg/L) during stormwater-runoff periods. These high DOC levels may significantly affect the transport phase of pesticides in streams in the Yakima River Basin by increasing dissolution of the hydrophobic pesticides from the soil and sediment phases.

At both Cle Elum and Kiona, FAs were found to be a significant fraction of the total DOC—26 percent and 24 percent, respectively (table 35). These FA concentrations are, however, at the low end of the concentration ranges expected for streams located in temperate climates. In most natural waters, aquatic FAs are a major fraction (20–80 percent) of DOC. The HPIA fraction of the total DOC from both stations represents a small percentage of the DOC—8 percent and 12 percent from Cle Elum and Kiona, respectively. The fractions of HPIA measured in the Yakima River are similar o those found in other aquatic samples; however, few data are currently available on the distribution of HPIA fractions in aquatic environments (Aiken and others, 1992).

Table 35. Fractionation of dissolved organic carbon in filtered-water samples, Yakima River at Cle Elum and Kiona, Yakima River Basin, Washington, 1987

Station number	Station name	Date	Dissolved organic carbon, in milligrams of carbon per liter	Fulvic acid, in percent	Hydrophilic acid, in percent
12479500	Yakima River at Cle Elum	07–22–87	1.6	26	8
12510500	Yakima River at Kiona	07–23–87	2.4	24	12

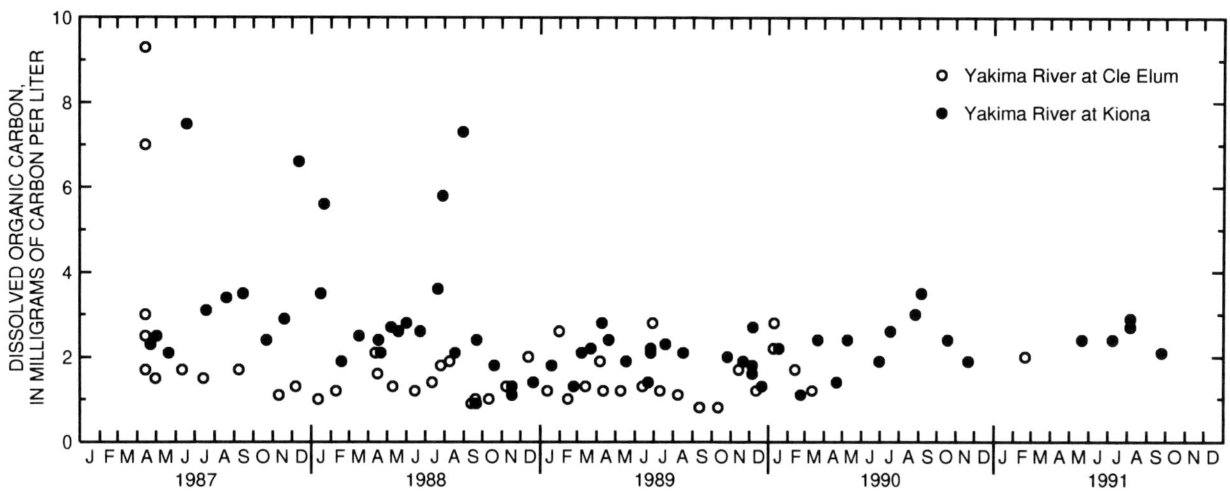

Figure 27. Concentrations of dissolved organic carbon in the Yakima River at Cle Elum and Kiona, Yakima River Basin, Washington, 1987–91.

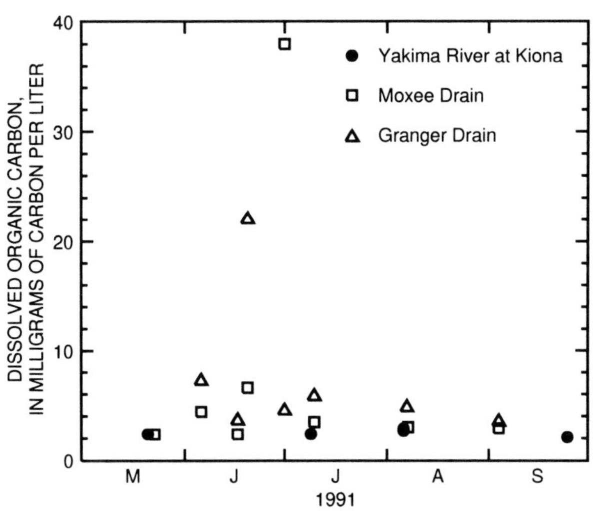

Figure 28. Concentrations of dissolved organic carbon in the Yakima River at Kiona, Moxee Drain at Birchfield Road near Union Gap, and Granger Drain at Granger, Yakima River Basin, Washington, May-September 1991.

Influence of Inorganic Constituents

Concentrations of dissolved inorganic constituents (table 36) may be affecting DOC concentrations in the Yakima River. Aiken and Malcolm (1987) concluded that the presence of divalent cations in soils and river water decreases the solubility of higher molecular weight components of humic substances by suppressing charge on the organic polyanions. As expected for headwater streams, the Cle Elum station had a low specific conductance (48 microsiemens per centimeter at 25 degrees Celsius) and low concentrations of major inorganic constituents. In general, the weathering of soils in the reaches upstream from the Cle Elum station does not contribute large amounts of dissolved inorganic constituents to the river. In contrast, the water at Kiona was found to have larger concentrations of sodium (Na^{+}), and the divalent cations, calcium (Ca^{2+}) and magnesium (Mg^{2+}), were greater at the Kiona station than at Cle Elum. Increased cation concentrations at Kiona reflect changes in soil chemistry between the two stations resulting from land-use activities, water-use practices (nonpoint and point sources), and evapotranspiration, particularly in the irrigated areas. Even though concentrations of divalent cations were greater at Kiona, DOC concentrations at Kiona were considerably higher than those at Cle Elum. Therefore, higher DOC input from urban and

Table 36. Concentrations of inorganic constituents in filtered-water samples, Yakima River at Cle Elum and Kiona, Yakima River Basin, Washington, 1987
[Concentrations reported in milligrams per liter; <, less than]

Element	Yakima River at Cle Elum July 22, 1987	Yakima River at Kiona July 23, 1987
Boron	<0.01	0.0103
Barium	.0024	.024
Calcium	5.1	28
Cadmium	<.01	.0012
Copper	.0038	.0065
Iron	.056	.052
Lithium	.0081	.011
Magnesium	2.5	10
Manganese	.0006	.0072
Molybdenum	<.01	.210
Sodium	1.5	16
Nickel	.013	.026
Lead	.035	.015
Silicon	4.7	22
Strontium	.024	.13
Zinc	.0006	.0019

agricultural areas appears to be overriding charge suppression effects on the organic polyanions.

Elemental Composition

FA and HPIA fractions from Yakima River samples were analyzed for elemental composition (table 37) to provide information on the nature and source of the organic isolates (Steelink, 1985). In general, the FA and HPIA samples are composed primarily of carbon, hydrogen, oxygen, nitrogen, and sulfur. With exception of the nitrogen and sulfur concentrations, the elemental composition of the FA and HPIA fractions were similar at the Cle Elum and Kiona stations. The Kiona sample had higher concentrations of nitrogen and sulfur in the FA and HPIA fractions. Increased nitrogen concentrations in the acid fractions in the Kiona samples may have been caused by increased biological activity resulting from nutrient contributions from agricultural return flows. McKnight and others (1991) previously have shown that FA and HPIA fractions derived from algal and bacterial processes were enriched in nitrogen relative to acid fractions derived from higher plants.

Molecular Weight

Vapor pressure osometry (VPO) is a complicated colligative-property procedure (Aiken and Gillam, 1989) used to determine the molecular weight of humic substances that consist of a complex mixture of polyelectrolytic acids (Wershaw and Aiken, 1985). The molecular weight of the FA in the Kiona sample (650 daltons) was less than that in the Cle Elum sample (707 daltons) (table 37). These molecular weights are midrange values compared to FA molecular weights for a wide range of aquatic systems reported by Aiken and Malcolm (1987). The difference in FA average molecular weights between the Cle Elum and Kiona samples is consistent with the difference in the concentrations of inorganic Ca^{2+} and Mg^{2+} in the water samples from the two stations. According to Aiken and Malcolm (1987), molecular weights of FA fractions are inversely related to cation concentrations in the original mixing water. Charges on the humic substances in the soil environment are neutralized by Ca^{2+} and Mg^{2+}. As a consequence, humic substances have a decreased solubility (Hayes and Swift, 1978). Aiken and Malcolm (1987) further concluded that the divalent cations in soils and river water decrease the solubility of higher molecular-weight components of the humic substances by suppressing charges on the organic polyanions. Therefore, the higher concentrations of Ca^{2+} and Mg^{2+} cations in the water at Kiona probably limit the solubility of the higher molecular-weight molecules of humic substances.

Acid-Base Titrations and Carbon-13 Nuclear-Magnetic-Resonance Analyses

Assessment of the acidic character of organic fractions in the Yakima River is complicated because DOC substances are complex heterogeneous mixtures of molecules. According to Bowles and others (1989), FAs contain carboxyl and phenolic functional groups with pK_a (acid-dissociation constants) values that extend over a wide range. As a consequence, titration curves of FA do not resemble those of simple monoprotic or polyprotic acids. Nonetheless, estimates for the concentrations of free carboxyl and phenolic functional groups can be made. Acid-base titration curves for the FA and HPIA fractions in samples from the Yakima River at Cle Elum are presented in figure 29, and table 38 contains results of those titrations.

The titration of base from pH 2.8 to 8.0 is attributable to the neutralization of carboxyl groups, whereas

Table 37. Elemental-composition, ash-content, and molecular-weight data for fulvic-acid and hydrophilic-acid fractions in filtered-water samples, Yakima River at Cle Elum and Kiona, Yakima River Basin, Washington, 1987

[Elemental-composition and ash-content data reported in percent; na, not analyzed]

Station number	Station name	Date	Fraction	Elemental composition (ash free) Carbon	Hydrogen	Oxygen	Nitrogen	Sulphur	Ash content	Molecular weight, in daltons
12479500	Yakima River at Cle Elum	07–22–87	Fulvic acid	57.2	4.9	35.9	1.0	0.6	8.5	707
			Hydrophilic acid	52.2	4.6	40.8	1.5	.6	3.1	na
12510500	Yakima River at Kiona	07–23–87	Fulvic acid	56.1	5.0	35.5	2.2	1.0	1.1	650
			Hydrophilic acid	50.5	4.4	40.6	3.0	1.2	3.9	411

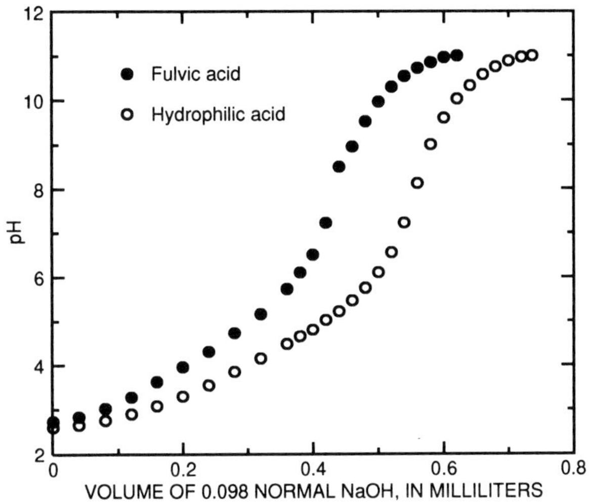

Figure 29. Acid-base titration curves for fulvic-acid and hydrophilic-acid fractions in filtered-water samples from the Yakima River at Cle Elum, Yakima River Basin, Washington, July 22, 1987. (NaOH, sodium hydroxide.)

base consumption from pH 8.0 to 11.0 is considered to represent neutralization of phenolic groups. The HPIA fraction has a larger relative abundance of functional carboxyl groups than does the FA fraction.

Concentrations of carboxylic acid in the HPIA samples from Kiona and Cle Elum were identical, whereas the concentration of carboxylic acid in the FA sample from Kiona was slightly higher than the concentration in the Cle Elum sample. Comparison of the phenolic concentrations in the FA and HPIA samples shows that the Cle Elum samples contain higher concentrations of phenolic groups than does the Kiona sample. Several mechanisms can be proposed to account for the difference in phenolic concentrations between water samples from the two stations: (1) Yakima River water at Kiona could have a lower phenolic concentration in the FA and HPIA fractions because of the interactions with divalent cations (mentioned earlier), and (2) increased amounts of organic matter derived from biota in the agricultural return flows could be diluting phenolic concentrations. McKnight and others (1991) suggested that microbially derived DOC fractions of FA and HPIA are depleted in phenolic concentrations. In the Yakima River at Kiona, DOC probably is composed mainly of algae and bacteria, which may result in depletion of phenolic concentrations. If the DOC was dominated by lignin, phenolic concentrations would be larger.

Carbon-13 nuclear-magnetic-resonance (^{13}C-NMR) techniques have been used to study humic substances. Analytical techniques and their application to environmental studies have been described by Wershaw and Mikita (1987), Preston (1987), and Steelink and others (1990). Quantitative-solution-state ^{13}C-NMR spectra for FA and HPIA from the Yakima River at Kiona are presented in figure 30. Comparison of these spectra (fig. 31) indicates that differences were detected between the FA and HPIA fractions. For each station, the HPIA fraction is less aromatic than the FA fraction, and the HPIA fraction has a greater amount of carboxyl and aliphatic II carbon than the FA fraction.

In comparing samples from each location, the Kiona FA sample is less aromatic than the Cle Elum FA sample. These ^{13}C-NMR results for the Kiona FA sample are supported by the lower phenolic concentration determined by the acid-base titration analysis (table 38). Another mechanism causing the selective depletion of the aromatic carbon is photooxidation. The photooxidative degradation of humic materials in various environments has been studied by Hayes and Swift (1978); McKnight and others (1988); Chen and Bada (1989); and Mopper and others (1991). DOC may be photooxidized in the river as the water moves downgradient, which may, in part, be responsible for the loss of aromatic and phenolic components in the FA

Table 38. Results of acid-base titrations of fulvic-acid and hydrophilic-acid fractions in filtered-water samples, Yakima River at Cle Elum and Kiona, Yakima River Basin, Washington, 1987
[DOC, dissolved organic carbon]

Station number	Station name	Date	Fraction	Carboxyl, in milliequivalents per gram of DOC	Phenolic, in milliequivalents per gram of DOC
12479500	Yakima River at Cle Elum	07–22–87	Fulvic acid	5.2	1.4
			Hydrophilic acid	5.9	1.1
12510500	Yakima River at Kiona	07–23–87	Fulvic acid	5.4	.8
			Hydrophilic acid	5.9	.9

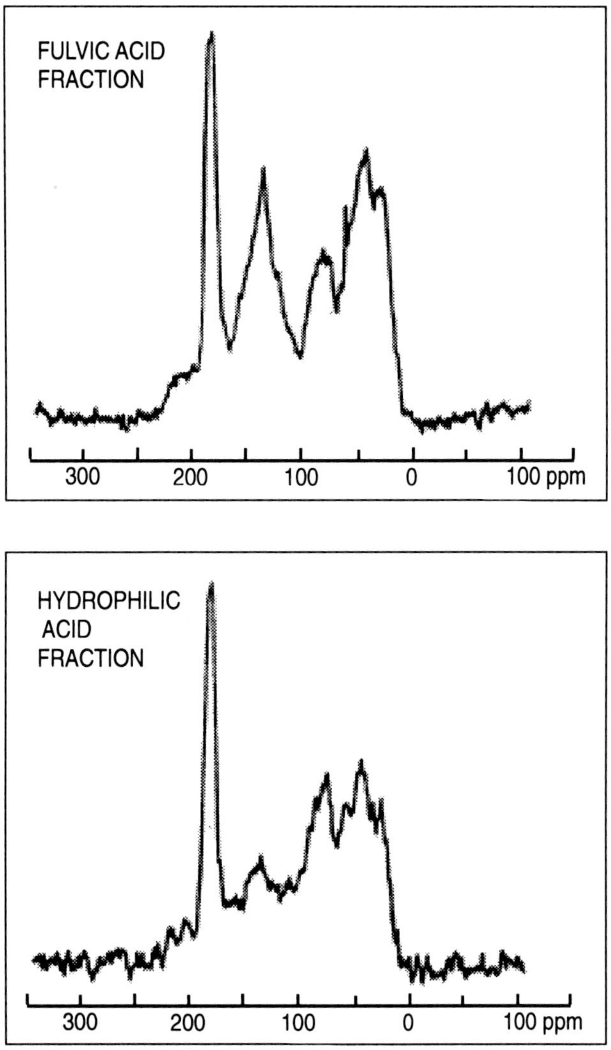

Figure 30. Carbon-13 nuclear-magnetic-resonance spectra for fulvic-acid and hydrophilic-acid fractions in filtered-water samples from the Yakima River at Kiona, Yakima River Basin, Washington, July 23, 1987. (ppm, parts per million of magnetic-field strength.)

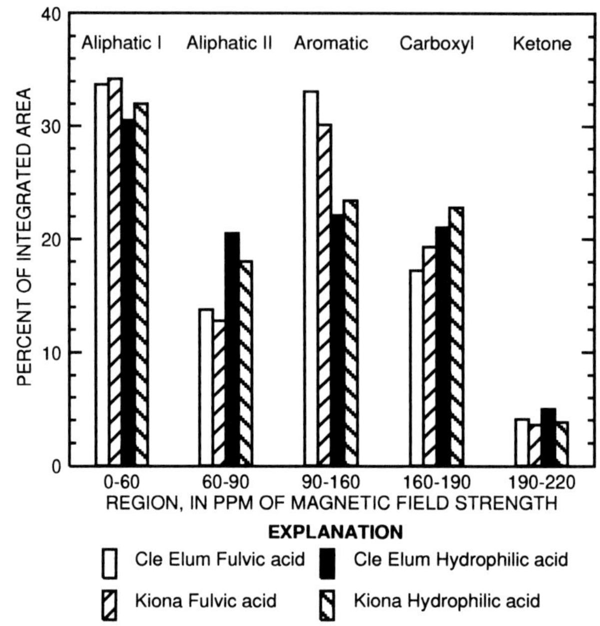

Figure 31. Quantitative carbon-13 nuclear-magnetic-resonance-integration data for fulvic-acid and hydrophilic-acid fractions in filtered-water samples from the Yakima River at Cle Elum and Kiona, Yakima River Basin, Washington, July 22–23, 1987. (ppm, parts per million.)

fraction. The photooxidative process likely would be enhanced by (1) increased water surface area resulting from the network of relatively shallow canals, ditches, and agricultural return flows that transport irrigation water throughout the area and (2) the low gradient and slow velocities in the main stem downstream from the city of Yakima.

Influences of Dissolved Organic Carbon on Contaminant Transport

The solubility of hydrophobic pesticides and other organic compounds in water has been shown to increase 2 to 3 times at concentrations of humic substances that typically occur in natural waters (Thurman, 1985; Chiou and others, 1986). Carter and Suffet (1982) found that the binding of DDT to humic substances varied depending on the source of the DOC. In addition, they found the binding of DDT to vary with pH, ionic strength, and the inorganic ions that were present. Humic substances also have surfactant qualities with hydrophilic and hydrophobic ends of the humic molecule. These molecules are believed to form water-soluble micelles that enhance solubilization and stream transport of hydrophobic organic compounds in the dissolved phase (Thurman, 1985).

Source materials for humic substances change with different land and water uses, so that humic substances derived from these different source areas have distinctive characteristics that can affect water-solubility-enhancement effects of humic materials (McKnight and others, 1991 and 1992). In upstream forested subbasins that influence water quality in the Yakima River at Cle Elum, lignaceous plants are significant precursor materials of humic substances. Consequently, these substances have high concentrations of aromatic-carbon and phenolic constituents and low concentrations of nitrogen. In the nutrient-rich River reach at Kiona, the DOC pool consists of microbially derived humic substances that contain higher nitrogen concentrations and lower aromatic-carbon and phenolic concentrations than those in the Yakima River at Cle Elum. Higher concentrations of divalent inorganic ions at the Kiona station result in low-molecular-weight (LMW) humic substances. These LMW substances are less effective than higher-molecular-weight humic substances in the Yakima River at Cle Elum in enhancing the solubility of hydrophobic pesticides in streams. The foregoing statement, however, neglects solubility-enhancement effects of surfactants that are commonly added to pesticide mixtures. These surfactants resulting from extensive pesticide applications in the lower Yakima River Basin may, in part, be controlling and enhancing the solubility of hydrophobic pesticides in the Yakima River at Kiona and in other streams that receive agricultural runoff.

The role of storm-driven transport of DOC in agricultural return flows and streams requires further study to better understand the transport of hydrophobic pesticides in the Yakima River Basin. Storm-runoff samples from Granger Drain and Sulphur Creek Wasteway on June 20, 1991, had DOC concentrations that were five to seven times higher than those in non-storm-runoff samples. This increase in DOC concentrations appears to have enhanced the partitioning of DDT+DDE+DDD into the dissolved phase, because measured concentrations of DDT+DDE+DDD in filtered water were much higher than calculated values based on equilibrium-partitioning theory (table 29). During storm runoff, DOC is washed into streams from a variety of natural (for example, humic substances) and man-affected (for example, surfactants) sources. Surfactant materials are particularly effective in interacting with pesticides and may play a major role in pesticide transport and bioavailability.

Bed Sediment and Agricultural Soil

When stream energy decreases and is unable to suspend and transport sediment in the water column, the sediment settles and is deposited on the streambed. A tendency for sediment to sorb and desorb nonionic organic compounds is primarily attributable to partitioning between the water phase and organic-matter phase of the sediment. In this study, the following persistent compounds were analyzed in bed sediment: organochlorine insecticides and other semivolatile organic compounds, including aliphatic compounds, cresols, ethers, monocyclic- and polycyclic-aromatic hydrocarbons, phenols, and phthalate esters. They are characterized by low water solubilities and large K_{oc} values (U.S. Environmental Protection Agency, 1982). Many of these compounds are toxic and have been designated by the U.S. Environmental Protection Agency as "Priority Pollutants" (Chapman and others, 1982).

Bed-sediment analyses are useful for evaluating the occurrence, transport, and fate of organic contaminants in the Yakima River Basin. Bed sediment functions as a reservoir for organic contaminants and constantly is sorbing and releasing compounds in the process of equilibrating with variations in contaminant concentrations in the overlying water column (Witkowski and others, 1987). In this study, bed-sediment samples were collected from 59 stations, with some samples analyzed individually and others composited for analyses by crop type (see table 8 and fig. 7 for station locations). Bed-sediment samples generally were collected from depositional areas in order to collect a higher proportion of fine-grained sediment that were expected to be organically enriched. As a result, contaminant concentrations detected in the bed sediment are representative of those concentrations in the depositional area and not necessarily the entire stream reach.

Previous studies have shown that several organic compounds sorb preferentially to finer sized sediment (silt and clay) because of the typically high organic-carbon concentration in these sediments (Witkowski and others, 1987). In contrast, Goerlitz and Law (Witkowski and others, 1987, p. 23) concluded that all particle-size fractions were important because of the presence of organic sorbents that are sand sized. To further investigate the importance of particle size, two size fractions were analyzed in this study: sediment with sieve diameters finer than 2,000 μm and sediment finer than 62 μm. In the finer than 2,000-μm samples, the percent of sediment finer than 62 μm in diameter ranged from 6 to 87 percent, with a median of 56 percent.

A comparison of the concentrations of three organochlorine compounds that were detected frequently in the bed sediment is shown in figure 32. These concentrations were normalized for variations in organic-carbon concentrations because these compounds tend to partition to organic matter (U.S. Environmental Protection Agency, 1988b). The diagonal lines (X = Y relation) in figure 32 indicate where concentrations in the two size fractions are equal. Points above the line indicate preferential sorption of compounds to finer sized sediment, and points below the line indicate preferential sorption to coarser sized sediment. Concentrations of DDT+DDE+DDD (4,4'-DDT+ 4,4'DDE+4,4'DDD) and dieldrin (both normalized and not normalized for organic carbon) shown in figure 32 are similar in each size fraction, usually within a factor of 10. In contrast, concentrations of endosulfan I are dissimilar in each size fraction, plotting well above or well below the X = Y line. These comparisons show that particle-size effects may not be important in controlling the sorption of these organochlorine compounds in bed sediment in the Yakima River Basin.

Results from the sampling show that several organochlorine compounds and semivolatile compounds occur in bed sediment throughout the basin (Rinella, McKenzie, Crawford, and others, 1992). The following organochlorine insecticides were detected in more than 50 percent of the samples: dieldrin, DDE, DDT, and DDD (table 39). The following semivolatile compounds were detected in more than 30 percent of the samples: phenol, bis(2-ethylhexyl) phthalate, naphthalene, and phenanthrene (table 40). The occurrence of several phthalate compounds can be expected because of the broad use of phthalate compounds as plasticizers in domestic and industrial products. A study by Johnson and others (1986) in 1985 showed that DDT, DDE, DDD, and dieldrin were the predominant organochlorine compounds detected in bed sediment. In a national study of pesticides in rivers, the most frequently detected organochlorine compounds in bed sediment were DDE (17 percent of the samples), DDD (12 percent), dieldrin (12 percent), chlordane (9.9 percent), and DDT (8.5 percent) (Gilliom and others, 1985). With the exception of chlordane, the national results are consistent with the frequent occurrence of these compounds bed-sediment samples from the Yakima River Basin.

Concentrations of DDT+DDE+DDD in bed-sediment samples collected upstream from nonagricultural areas had concentrations ranging from less than 0.3 to 1.4 μg/kg (micrograms per kilogram), dry weight, with a median concentration of 0.4. These concentrations are lower than those measured in 1966 in surface-soil samples collected in the Oregon Cascade Range in areas that did not receive direct DDT application (mean concentration was 16 μg/kg, dry weight) (Moore and Loper, 1980). The primary source of DDT in Oregon's Cascade Range at untreated locations was believed to be atmospheric input. Considering that DDT, DDE, and DDD compounds have degraded since the ban of DDT in 1973, the lower concentrations in the nonagricultural areas in the Yakima River Basin in 1987–90 are consistent with the higher concentrations measured in the untreated areas of the Oregon Cascade Range in 1966. This comparison suggests that DDT+DDE+DDD concentrations in streams located upstream from agricultural activities are not enriched over levels that might be expected by atmospheric deposition.

To explore relations among concentrations of organochlorine insecticides and other semivolatile compounds with agricultural activities, samples were collected from 3 agricultural fields and from 31 ditches (table 41). The bed-sediment samples from the ditches were composited into seven samples by crop type. DDT+DDE+DDD was detected in ditches draining orchards or fields of apples, grapes, pears, and potatoes, but not of asparagus, corn, or hops. In this study, soil samples from two hop fields and one apple orchard were collected and analyzed. DDT+DDE+DDD was detected in both the A and B soil horizons in each of the soil samples (table 41); however, the data are limited in number and spatial coverage to be able to conclude an association between the crop type and amount of DDT+DDE+DDD contamination. In addition, crop history was not assessed in this study to determine how DDT applications and crop types have changed in these areas over time. The discrepancy

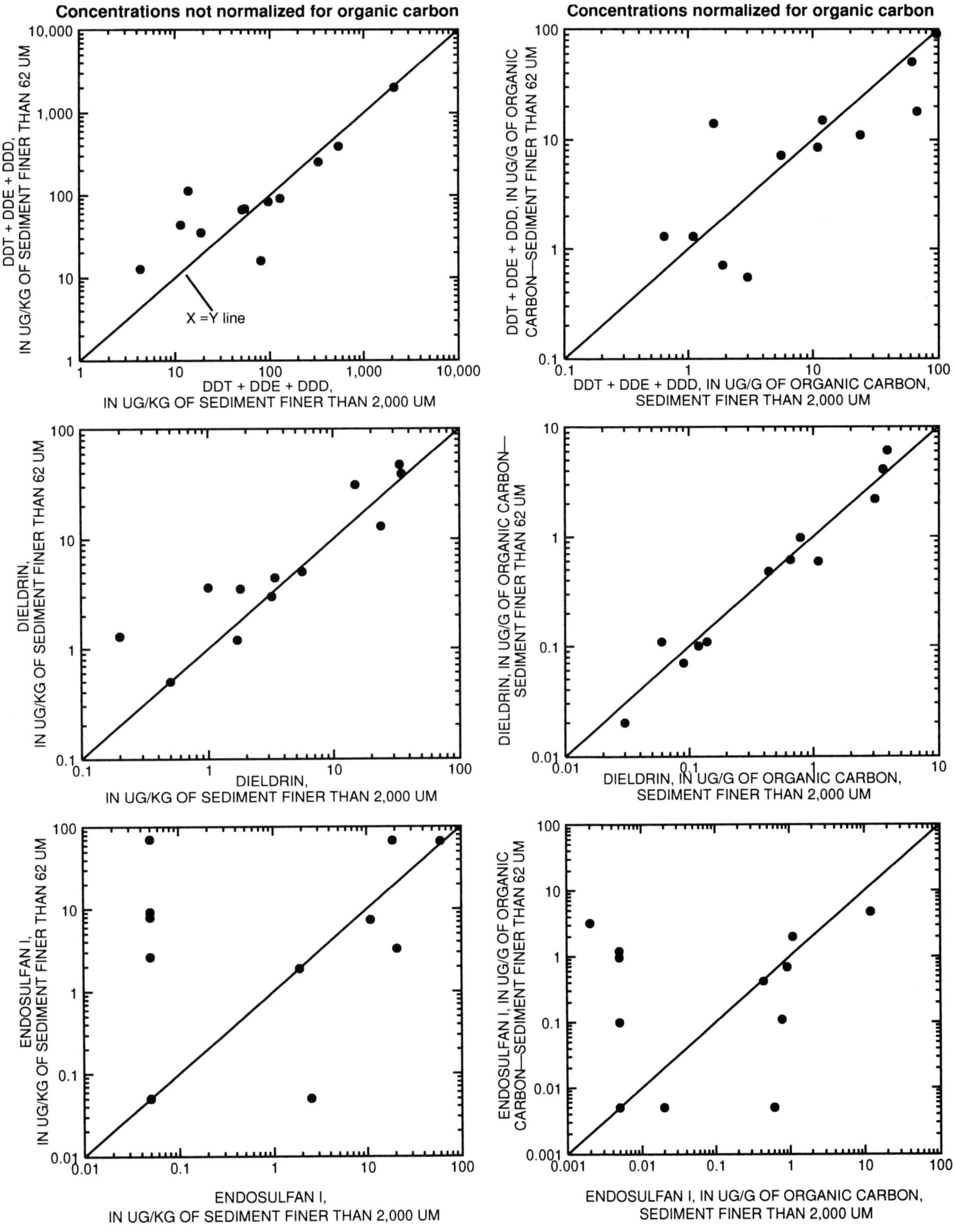

Figure 32. Relation of concentrations of selected organic compounds in two particle-size fractions of bed-sediment samples from selected stations, Yakima River Basin, Washington, 1988. (DDT+DDE+DDD, 4,4'-dichlorodiphenyltrichloroethane [DDT] plus 4,4'-dichlorodiphenyldichloroethylene [DDE] plus 4,4'-dichlorodiphenyldichloroethane [DDD]; UG/KG, micrograms per kilogram, UM, micrometers; UG/G, micrograms per gram.)

Table 39. Summary of concentrations of organochlorine compounds in bed-sediment samples, Yakima River Basin, Washington, 1987–90

[Minimum reporting levels and concentrations reported in micrograms per kilogram, dry weight; --, insufficient data to calculate statistic; DDD, 4,4'-dichlorodiphenyldichloroethane; DDE, 4,4'-dichlorodiphenyldichloroethylene; DDT, 4,4'-dichlorodiphenyltrichloroethane; HCH, hexachlorocyclohexane; PCB, polychlorinated biphenyls; PCN, polychlorinated naphthalenes]

Compound	Particle size finer than, in micrometers	Number of samples	Minimum reporting level(s)	Number of quantifiable concentrations	Quantifiable concentrations		
					Minimum	Median	Maximum
Aldrin	2,000	20	0.1	1	--	0.4	--
	62	19	.1, .5	2	0.2	--	0.4
Gross chlordane	2,000	20	1, 10	3	1	4	9
	62	19	1, 5	5	3	5	15
Dieldrin	2,000	20	.1	14	.2	2.5	35
	62	19	.5	12	.5	4.0	47
DDD	2,000	20	.1	15	.1	13	48
	62	19	.1	18	.1	11	440
DDE	2,000	20	.1	16	.1	32	1,700
	62	19	.1	19	.2	16	1,400
DDT	2,000	20	.1	13	.3	11	370
	62	19	.1, .5	14	.1	9.2	180
Endosulfan I	2,000	20	.1	6	1.9	15	61
	62	19	.1, .5	9	1.9	7.9	71
Endrin	2,000	20	.1	6	.1	.4	5.3
	62	19	.1, .5	4	.1	1.8	17
Heptachlor	2,000	20	.1	0	--	--	--
	62	19	.1, .5	6	.1	.1	.1
Heptachlor epoxide	2,000	20	.1	4	.1	.2	1.3
	62	19	.1, .5	4	.1	.2	2.5
Lindane (γ-HCH)	2,000	20	.1	2	.1	--	.3
	62	19	.1, .5	1	--	.8	--
4,4'-Methoxychlor	2,000	20	.1, 1	0	--	--	--
	62	19	1, .5	4	.6	1.4	3.2
Mirex	2,000	20	.1	0	--	--	--
	62	19	.1, .5	0	--	--	--
PCB	2,000	20	1	0	--	--	--
	62	19	1, 5	0	--	--	--
PCN	2,000	20	1	0	--	--	--
	62	19	1, 5	0	--	--	--
Perthane	2,000	20	1	1	--	2	--
	62	19	1, 5	0	--	--	--
Toxaphene	2,000	20	10	0	--	--	--
	62	19	10, 50	0	--	--	--

between the soil and bed-sediment data for hops indicates that additional data are needed to characterize relations among compound concentrations in bed sediment and soils associated with different agricultural crops in the Yakima River Basin. Except for a few miscellaneous detections (phenol, anthracene, fluoranthene, phenanthrene, and pyrene), semivolatile compounds were not detected in the bed-sediment samples.

Concentrations of selected organochlorine insecticides and other semivolatile organic compounds (PAHs and phthalate) were normalized for organic-carbon concentrations (dividing compound concentrations by the concentration of organic carbon) so that these data could be compared to bed-sediment criteria for the protection of benthic fauna (U.S. Environmental Protection Agency, 1988b; Nowell and Resek, 1994; David J. Hansen, U.S. Environmental Protection Agency, written commun., 1990) (tables 42 and 43). Highest concentrations of semivolatile organic compounds were detected in agricultural return flows that also received point-source discharges and urban runoff (table 43). Concentrations of semivolatile organic compounds, however, did not exceed sediment-quality criteria at any of the 24 stations that were sampled throughout the basin. As expected, agricultural soils and stations receiving agricultural return flow had the highest concentrations of DDT+DDE+DDD, chlordane,

Table 40. Summary of concentrations of semivolatile compounds in bed-sediment samples, Yakima River Basin, Washington, 1987–90

[Minimum reporting levels and concentrations reported in micrograms per kilogram, dry weight; --, insufficient data to calculate statistic; PCB, polychlorinated biphenyls]

Compound	Particle size finer than, in micrometers	Number of samples	Minimum reporting level(s)	Number of quantifiable concentrations	Quantifiable concentrations		
					Minimum	Median	Maximum
Pesticide							
Isophorone	2,000	21	200, 400	1	--	110	--
	62	15	200	0	--	--	--
PCB-related compound							
2-Chloronaphthalene	2,000	21	200, 400	0	--	--	--
	62	15	200	0	--	--	--
Halogenated aliphatic compounds							
Hexachlorobutadiene	2,000	21	200, 400	0	--	--	--
	62	15	200	0	--	--	--
Hexachlorocyclopentadiene	2,000	21	200, 400	0	--	--	--
	62	15	200	0	--	--	--
Hexachloroethane	2,000	21	200, 400	0	--	--	--
	62	15	200	0	--	--	--
Ethers							
Bis(2-chloroethoxy)	2,000	21	200, 400	0	--	--	--
methane	62	15	200	0	--	--	--
Bis(2-chloroethyl) ether	2,000	21	200, 400	1	--	160	--
	62	15	200	0	--	--	--
Bis(2-chloroisopropyl) ether	2,000	21	200, 400	1	--	120	--
	62	15	200	0	--	--	--
4-Bromophenyl phenyl ether	2,000	21	200, 400	0	--	--	--
	62	15	200	0	--	--	--
4-Chlorophenyl phenyl ether	2,000	21	200, 400	0	--	--	--
	62	15	200	0	--	--	--
Monocyclic aromatic compounds							
1,2-Dichlorobenzene	2,000	21	200, 400	0	--	--	--
	62	15	200	0	--	--	--
1,3-Dichlorobenzene	2,000	21	200, 400	0	--	--	--
	62	15	200	0	--	--	--
1,4-Dichlorobenzene	2,000	21	200, 400	0	--	--	--
	62	15	200	0	--	--	--
2,4-Dinitrotoluene	2,000	21	200, 400	0	--	--	--
	62	15	200	0	--	--	--
2,6-Dinitrotoluene	2,000	21	200, 400	1	--	160	--
	62	15	200	0	--	--	--
Hexachlorobenzene	2,000	21	200, 400	0	--	--	--
	62	15	200	0	--	--	--
Nitrobenzene	2,000	21	200, 400	1	--	120	--
	62	15	200	0	--	--	--
1,2,4-Trichlorobenzene	2,000	21	200, 400	0	--	--	--
	62	15	200	0	--	--	--
Phenols and cresols							
2-Chlorophenol	2,000	21	200, 400	0	--	--	--
	62	15	200	0	--	--	--

Table 40. Summary of concentrations of semivolatile compounds in bed-sediment samples, Yakima River Basin, Washington, 1987–90—Continued

Compound	Particle size finer than, in micrometers	Number of samples	Minimum reporting level(s)	Number of quantifiable concentrations	Quantifiable concentrations		
					Minimum	Median	Maximum
Phenols and cresols—Continued							
2,4-Dichlorophenol	2,000	21	200, 400	0	--	--	--
	62	15	200	0	--	--	--
2,4-Dimethylphenol	2,000	21	200, 400	0	--	--	--
	62	15	200	0	--	--	--
4,6-Dinitro-*o*-cresol	2,000	21	600, 1200	0	--	--	--
	62	15	600	0	--	--	--
2,4-Dinitrophenol	2,000	21	600, 1200	0	--	--	--
	62	15	600	0	--	--	--
2-Nitrophenol	2,000	21	200, 400	0	--	--	--
	62	15	200	0	--	--	--
4-Nitrophenol	2,000	21	600, 1200	1	--	250	--
	62	15	600	0	--	--	--
p-Chloro-*m*-cresol	2,000	21	600, 1200	0	--	--	--
	62	15	600	0	--	--	--
Pentachlorophenol	2,000	21	600, 1,200	0	--	--	--
	62	15	600	0	--	--	--
Phenol	2,000	21	200, 400	8	3	110	870
	62	15	200	6	35	110	310
2,4,6-Trichlorophenol	2,000	21	600, 1200	0	--	--	--
	62	15	600	0	--	--	--
Phthalate esters							
Bis(2-ethylhexyl) phthalate	2,000	21	200	13	4	340	5,200
	62	15	200	10	180	480	790
Diethyl phthalate	2,000	21	200, 400	5	79	81	90
	62	15	200	3	80	81	83
Dimethyl phthalate	2,000	21	200, 400	0	--	--	--
	62	15	200	1	--	99	--
Di-*n*-butyl phthalate	2,000	21	200, 400	6	3	120	160
	62	15	200	4	100	120	120
Di-*n*-octyl phthalate	2,000	21	400, 800	4	150	170	190
	62	15	400	1	--	170	--
n-Butyl benzyl phthalate	2,000	21	200, 400	5	120	150	360
	62	15	200	1	--	160	--
Polycyclic aromatic compounds							
Acenaphthene	2,000	21	200, 400	0	--	--	--
	62	15	200	0	--	--	--
Acenaphthylene	2,000	21	200, 400	2	81	--	87
	62	15	200	0	--	--	--
Anthracene	2,000	21	200, 400	2	86	--	99
	62	15	200	0	--	--	--
Benzo [a] anthracene	2,000	21	400, 800	1	--	130	--
	62	15	400	0	--	--	--
Benzo [a] pyrene	2,000	21	400, 800	1	--	160	--
	62	15	400	0	--	--	--

Table 40. Summary of concentrations of semivolatile compounds in bed-sediment samples, Yakima River Basin, Washington, 1987–90—Continued

Compound	Particle size finer than, in micrometers	Number of samples	Minimum reporting level(s)	Number of quantifiable concentrations	Quantifiable concentrations		
					Minimum	Median	Maximum
Polycyclic aromatic compounds—Continued							
Benzo [b] fluoranthene	2,000	21	400, 800	1	--	200	--
	62	15	400	0	--	--	--
Benzo [k] fluoranthene	2,000	21	400, 800	1	--	140	--
	62	15	400	0	--	--	--
Benzo [ghi] perylene	2,000	21	400, 800	1	--	160	--
	62	15	400	0	--	--	--
Chrysene	2,000	21	400, 800	2	81	--	160
	62	15	400	0	--	--	--
Dibenzo [ah] anthracene	2,000	21	400, 800	0	--	--	--
	62	15	400	0	--	--	--
Fluoranthene	2,000	21	200, 400	3	120	120	200
	62	15	200	3	4	5	120
Fluorene	2,000	21	200, 400	0	--	--	--
	62	15	200	0	--	--	--
Indeno [1,2,3-cd] pyrene	2,000	21	400, 800	0	--	--	--
	62	15	400	0	--	--	--
Naphthalene	2,000	21	200, 400	3	30	96	100
	62	15	200	6	2	14	95
Phenanthrene	2,000	21	200, 400	4	90	93	140
	62	15	200	8	4	33	83
Pyrene	2,000	21	200, 400	3	120	120	190
	62	15	200	4	3	37	110
Nitrosamines							
N-nitrosodimethylamine	2,000	21	200, 400	1	--	120	--
	62	15	200	0	--	--	--
N-nitrosodi-*n*-propylamine	2,000	21	200, 400	0	--	--	--
	62	15	200	0	--	--	--
N-nitrosodiphenylamine	2,000	21	200, 400	2	43	--	220
	62	15	200	1	--	160	--

dieldrin, endosulfan I, and endrin (tables 42 and 44). Even though DDT was banned in 1972, chlordane was cancelled for all uses in 1987, dieldrin was suspended on food products in 1974 and banned in 1985, and endrin was cancelled in 1984 (U.S. Environmental Protection Agency, 1991b; table 24), these persistent organochlorine insecticides had concentrations that exceeded interim sediment-quality criteria for the protection of benthic fauna (table 42). Exceedances of bed-sediment criteria occurred only in agricultural return flows and in the Yakima River at RM 68 and at Kiona (RM 29.9). Most of these stations had exceedances for DDT+DDE+DDD and dieldrin. According to the revised sediment-quality criteria (Nowell and Resek, 1994), dieldrin and endrin concentrations did not exceed the criteria at any stations sampled in the basin. Selected data from this study and data collected by Johnson and others (1986) in 1985 are listed in table 45. Except for Moxee Drain, the concentrations are similar. The differences observed at Moxee Drain may, in part, have been the result of sampling two different locations.

Aquatic Biota

Biological samples were collected from 33 sampling stations. Each sample was a composite of either

Table 41. Concentrations of selected organochlorine compounds in agricultural soils and bed-sediment samples from ditches that drain specific crop types, Yakima River Basin, Washington, 1987

[Concentrations reported in micrograms per kilogram, dry weight; A horizon, upper 6 to 10 inches of soil; B horizon, subsurface soil layer from about 6 to 12 inches; DDT+DDE+DDD, 4,4'-dichlorodiphenyltrichloroethane (DDT) plus 4,4'-dichlorodiphenyldichloroethylene (DDE) plus 4,4'-dichlorodiphenyldichloroethane (DDD); <, less than; see table 8 for location of stations; see Rinella, McKenzie, Crawford, and others, 1992 for additional determinations]

Crop type	Dieldrin	DDT+DDE+DDD	Endosulfan I	Endrin
Agricultural soils				
Apple orchard				
A horizon	<0.1	1,390	<0.1	<0.1
B horizon	<.1	108	<.1	<.1
Hops				
A horizon	.8	317	<.1	.3
B horizon	.1	52	<.1	<.1
Hops				
A horizon	.1	475	<.1	.2
Composited samples from ditches				
Apple orchard	<10	682	17	<10
Asparagus	<10	<10	<10	<10
Corn	<10	<10	<10	<10
Grape	<10	81	<10	<10
Hops	<10	<10	<10	<10
Pear orchard	<10	269	<10	<10
Potato	38	82	<10	<10

4 to 10 whole fish, 5 to 59 mollusks (soft parts only), approximately 20 whole crayfish, or the apical 5 centimeters of selected submersed aquatic plants. This report primarily contains an analysis of whole-fish samples that were collected in late October and early November 1989 and 1990. Fish-fillet samples also were collected in 1991 and used in this report to assess human-health cancer risks associated with consumption of these fish. Whole-fish samples were sent to a contract laboratory, certified by the USFWS, where 25 organochlorine compounds and 14 PAHs were measured. Fish-fillet samples were analyzed at the USGS National Water-Quality Laboratory for selected organochlorine compounds. Organophosphorus compounds, which were analyzed in water samples, were not analyzed in biological samples because these compounds either do not readily bioaccumulate or they are rapidly metabolized (McCarthy and Shugart, 1990). Crayfish and aquatic plants proved to be poor accumulators of organic contaminants. Data from these organisms are used in this report only for assessing PAHs. In addition, four of the samples were analyzed for dioxin (2,3,7,8-tetrachlorodibenzo-*p*-dioxin). Station locations, sample collection, sample processing, and analytical methods for whole fish are detailed in Rinella, McKenzie, Crawford, and others (1992).

Contaminant data from these samples are interpreted in four different ways in this study. First, individual compounds are characterized by their presence or absence and their frequency of occurrence. Second, individual contaminants are compared among sampling stations within the basin to identify stations or groups of stations where a particular compound is elevated in the biota. Third, comparisons are made against three previous studies—one focuses on the Yakima River Basin and the other two are of national scope. Finally, the data from this study are compared with existing environmental-quality guidelines for contaminant concentrations in fish to determine whether the detected concentrations pose a threat to humans or wildlife.

The three previous studies in the Yakima River Basin were done by the Washington Department of Ecology (WDOE), USFWS, and EPA. Sampling for the WDOE study of fish and invertebrates was done in 1985 and focused exclusively on the Yakima River Basin (Johnson and others, 1986). Water, sediment, and fish, including both resident and migratory fish, were collected and analyzed in this study. Most important for comparison with the present work are the samples of resident fish from four main-stem Yakima River stations.

Table 42. Summary of sediment-quality criteria and concentrations of selected organochlorine compounds in bed-sediment and soil samples, Yakima River Basin, Washington, 1987–90

[Concentrations reported in micrograms per gram of organic carbon; concentrations within the sediment-criteria confidence intervals can be considered either safe or hazardous with respect to that compound with certainties between 2.5 and 97.5 percent (U.S. Environmental Protection Agency, 1988b); polychlorinated biphenyls and toxaphene were not detected; DDT+DDE+DDD, 4,4-dichlorodiphenyltrichloroethane (DDT) plus 4,4-dichlorodiphenyldichloroethylene (DDE) plus 4,4-dichlorodiphenyldichloroethane (DDD); HCH, hexachlorocyclohexane; --, not applicable; *, concentration exceeds lower confidence level; <, less than; #, concentration is calculated using organic-carbon data collected during a trace-element study on a different sampling date; **, concentration exceeds upper confidence level; ns, not sieved]

Station number	Station name	Date	Particle size finer than, in micrometers	DDT+ DDE+ DDD[1]	Chlordane (technical)	Dieldrin	Endosulfan I	Endrin	Heptachlor[1]	Lindane (γ-HCH)
SEDIMENT-QUALITY CRITERIA										
Interim sediment-quality criteria (U.S. Environmental Protection Agency, 1988b; Nowell and Resek, 1994)										
Mean		--	--	0.83	0.31	0.13	0.33	0.05	0.11	0.43
95-percent confidence interval[2]		--	--	.18–3.8	.036–2.8	.01–1.8	.034–3.1	.006–.44	.015–.84	.070–2.6
Revised sediment-quality criteria (Nowell and Resek, 1994)										
Mean		--	--	--	--	11.1	--	4.2	--	--
95-percent confidence interval[2]		--	--	--	--	5.17–23.8	--	2.0–9.1	--	--
CONCENTRATIONS IN BED SEDIMENT										
Upper main-stem stations										
12484500	Yakima River at Umtanum	05–04–89	2,000	.31*	<.06	.06*	<.01	<.01	<.01	<.01
# 12500450	Yakima River above Ahtanum Creek	09–16–87	180	<14	<4.9	<4.9	<4.9	<4.9	<4.9	<4.9
Lower main-stem stations										
# 12508625	Yakima River at river mile 68	09–15–87	180	9.7**	<6.4	<6.4	<6.4	<6.4	<6.4	<6.4
12510500	Yakima River at Kiona	08–31–88	2,000	1.9*	<.44	.09*	<.04	<.04	<.04	<.04
		08–31–88	62	.71*	.28*	.07*	<.01	<.01	<.01	<.01
Nonagricultural stations										
12478300	Cle Elum River above Cle Elum Lake	08–31–88	2,000	<.17	<.56	<.06	<.06	<.06	<.06	<.06
12479720	Jungle Creek near mouth	05–02–89	2,000	<.08	<.26	<.03	<.03	<.03	<.03	<.03
# 12480000	Teanaway River below forks	11–06–89	2,000	<.12	<.39	<.04	<.04	<.04	<.04	<.04
# 12481900	Taneum Creek at Taneum Meadow	11–03–90	62	.09	<1.1	<.11	<.11	<.11	.02*	<.11
# 12483190	South Fork Manastash Creek	11–02–90	62	.20*	<1.7	<.17	<.17	<.17	.03*	<.17
# 12483750	Naneum Creek below High Creek	11–02–90	62	.09	<.55	<.06	<.06	<.06	.01	<.06
12484550	Umtanum Creek near mouth	05–03–89	2,000	.04	<.10	<.01	<.01	<.01	<.01	<.01
# 12487200	Little Naches River at mouth	10–30–90	62	.12	<1.2	<.12	<.12	<.12	.02*	<.12
# 12488250	American River at Hells Crossing	11–07–89	2,000	.05	<.27	<.03	<.03	<.03	<.03	<.03
# 12489100	Rattlesnake Creek above North Fork Rattlesnake Creek	11–07–89	2,000	.03	<.34	<.03	<.03	<.03	<.03	<.03
# 12500900	South Fork Ahtanum Creek near Tampico	11–06–90	62	.27*	<.97	<.10	<.10	<.10	.02*	<.10
# 12508500	Satus Creek below Dry Creek	11–07–90	62	.78*	<4.4	<.44	<.44	<.44	<.44	<.44

Table 42. Summary of sediment-quality criteria and concentrations of selected organochlorine compounds (normalized for organic carbon in bed-sediment and soil samples, Yakima River Basin, Washington, 1987–90—Continued

Station number	Station name	Date	Particle size finer than, in micrometers	DDT+ DDE+ DDD[1]	Chlordane (technical)	Dieldrin	Endosulfan I	Endrin	Heptachlor[1]	Lindane (γ-HCH)
Agricultural stations										
12484480	Cherry Creek at Thrall	08–31–88	2,000	5.6**	.93*	3.6**	<0.01	0.02*	<0.01	<0.01
		08–31–88	62	7.2**	1.0 *	4.1**	<.01	.02*	<.01	<.01
12500430	Moxee Drain at Thorp Road	08–30–88	2,000	11**	<.08	.14*	.92*	.03*	<.01	<.01
		08–30–88	62	8.5**	.28*	.11*	.69*	.01*	<.01	<.01
12500437	Wide Hollow Creek at West Valley Middle School	05–01–89	2,000	8.7**	<.42	.04*	<.004	.02*	<.004	<.004
12500445	Wide Hollow Creek near mouth	08–30–88	2,000	3.0*	<.04	.12*	.78*	.02*	<.004	<.004
		08–30–88	62	.55*	.52*	.10*	.11*	<.004	<.004	<.004
12503640	Unnamed drain at Lateral and Riggs	09–01–88	2,000	62**	<.12	3.9**	<.01	<.01	<.01	<.01
	Roads	09–01–88	62	51**	<.13	6.1**	1.2*	<.01	<.01	<.01
12505460	Granger Drain at mouth	09–02–88	2,000	24**	<.24	.44*	.61*	<.02	<.02	.02
12505460	Granger Drain at mouth	09–02–88	62	11**	<.14	.48*	<.01	<.01	<.01	<.01
12507508	Toppenish Creek at Indian Church	09–01–88	2,000	1.1*	.06*	.03*	<.01	<.01	<.01	<.01
	Road	09–01–88	62	1.3*	<.04	.02*	.10*	<.004	<.004	<.004
12508850	Sulphur Creek Wasteway	08–31–88	2,000	1.6*	<.12	.66*	<.01	<.01	<.01	<.01
		08–31–88	62	14**	<.12	.61*	.96*	<.01	<.01	<.01
		08–31–88	2,000	12**	.93*	.79*	.44*	<.02	<.02	<.02
		08–31–88	62	15**	.89*	.98*	.42*	<.02	<.02	<.02
461652119522000	Unnamed Drain at County Line	09–01–88	2,000	69**	<.20	3.1**	12**	.02*	<.02	<.02
	Road near Grandview	09–01–88	62	18**	<.07	2.2**	4.8**	<.01	<.01	<.01
461744119522400	Unnamed Drain to East Turbine	09–01–88	2,000	.64*	<.06	.06*	1.1*	.30*	<.01	.02
	Lateral near Grandview	09–01–88	62	1.3*	<.03	.11*	2.0*	.50**	<.003	.02
463501120353300	Wide Hollow Creek at Ahtanum	09–01–88	2,000	97**	<.05	1.1*	<.005	<.005	<.005	<.005
	Boulevard	09–01–88	62	92**	<.05	.59*	3.2**	.16*	<.005	<.005
CONCENTRATIONS IN AGRICULTURAL SOILS										
463202120223600	Apple orchard near Moxee City									
	A Horizon, 0–6 inches	07–01–89	ns	99**	<.07	<.01	<.01	<.01	<.01	<.01
	B Horizon, 6–9 inches	07–01–89	ns	21**	<.19	<.02	<.02	<.02	<.02	<.02
463232120234900	Hop field near Moxee City									
	A Horizon, 0–9 inches	07–01–89	ns	20**	.25*	.05*	<.01	.02*	<.01	<.01
	B Horizon, 9–12 inches	07–01–89	ns	8.5**	.16*	.02*	<.02	<.02	<.02	<.02
463247120222300	Hop field near Moxee City									
	A Horizon, 1–10 inches	07–01–89	ns	34**	.14*	.01*	<.01	.01*	<.01	<.01

[1]Criteria for DDT+DDE+DDD and heptachlor are derived from final residue values; other criteria are derived from final chronic values.The final chronic value is defined as the contaminant concentration in water that will not result in unacceptable effects on aquatic organisms in the water column during chronic (24-hour) exposure. The final residue value is defined as the contaminant concentration in water that will protect wildlife that consume aquatic organisms and that will not result in tissue concentrations that exceed U.S. Food and Drug Administration action levels.

[2]The upper value of the confidence interval represents the concentration which, with 97.5-percent certainty, will result in hazardous long-term effects on the benthic fauna.

Table 43. Summary of sediment-quality criteria and concentrations of semivolatile organic compounds in bed-sediment samples, Yakima River Basin, Washington, 1987–90

[Concentrations reported in micrograms per gram of organic carbon; concentrations within the sediment-criteria confidence intervals can be considered either safe or hazardous with respect to that compound with certainties between 2.5 and 97.5 percent (U.S. Environmental Protection Agency, 1988b); --, not applicable; <, less than; #, concentration is calculated using organic-carbon data collected during a trace-element study on a different sampling date]

Station number	Station name	Date	Particle size finer than, in micrometers	Acenaph-thene	Benzo[a]-pyrene	Benzo[a]-anthracene	Bis(2-ethylhexyl) phthalate	Fluoran-thene	Phenan-threne	Pyrene
SEDIMENT-QUALITY CRITERIA										
Interim sediment-quality criteria (U.S. Environmental Protection Agency, 1988b; Nowell and Resek, 1994)										
Mean		--	--	732	1,063	1,317	172,000	1,883	174	1,311
95-percent confidence interval[1]		--	--	180–3,030	225–5,018	217–7,999	--	423–8,375	21.8–1,347	265–6,465
Revised sediment-quality criteria (Nowell and Resek, 1994)										
Mean		--	--	134	--	--	--	513	184	--
95-percent confidence interval[1]		--	--	62.4–288	--	--	--	239–1,100	85.5–394	--
CONCENTRATIONS IN BED SEDIMENT										
Upper main-stem stations										
12484500	Yakima River at Umtanum	05–04–89	2,000	<12	<25	<25	19	<12	<12	<12
		# 11–05–90	62	<49	<98	<98	<49	<49	<49	<49
# 12500450	Yakima River above Ahtanum Creek at Union Gap	09–16–87	180	<95	<190	<190	<95	<95	<95	<95
Lower main-stem stations										
# 12508625	Yakima River below Satus Creek at river mile 68 near Satus	09–15–87	180	<120	<250	<250	<120	<120	<120	<120
12510500	Yakima River at Kiona	08–31–88	2,000	<87	<170	<170	<87	<87	<87	<87
		08–31–88	62	<11	<22	<22	34	<11	<11	<11
		# 11–10–90	62	<11	<22	<22	<11	.22	.22	.17
Nonagricultural stations										
# 12478100	Waptus River at mouth near Roslyn	11–06–89	2,000	<61	<120	<120	120	<61	<61	<61
12478300	Cle Elum River above Cle Elum Lake near Roslyn	08–31–88	2,000	<110	<220	<220	<110	<110	<110	<110
12479720	Jungle Creek near mouth near Cle Elum	05–02–89	2,000	<51	<100	<100	62	<51	<51	<51
# 12480000	Teanaway River below forks near Cle Elum	11–06–89	2,000	<77	<150	<150	<77	<77	<77	<77
12484550	Umtanum Creek near mouth at Umtanum	05–03–89	2,000	<20	<40	<40	520	<20	<20	<20
# 12488250	American River at Hells Crossing near Nile	11–07–89	2,000	<53	<100	<100	<53	<53	<53	<53

Table 43. Summary of sediment-quality criteria and concentrations of semivolatile organic compounds (normalized for organic carbon) in bed-sediment samples, Yakima River Basin, Washington, 1987–90—Continued

Station number	Station name	Date	Particle size finer than, in micrometers	Acenaph-thene	Benzo[a]-pyrene	Benzo[a]-anthracene	Bis(2-ethylhexyl) phthalate	Fluoran-thene	Phenan-threne	Pyrene
Nonagricultural stations—Continued										
# 12489100	Rattlesnake Creek above North Fork Rattlesnake Creek near Nile	11–07–89	2,000	<61	<120	<120	<61	<61	<61	<61
Agricultural stations										
12484480	Cherry Creek at Thrall	08–31–88	2,000	<21	<41	<41	260	12	9.9	12
		08–31–88	62	<21	<42	<42	57	<21	8.6	<21
12500430	Moxee Drain at Thorp Road near Union Gap	08–30–88	2,000	<17	<34	<34	28	10	7.6	10
		08–30–88	62	<18	<37	<37	17	<18	<18	<18
12500437	Wide Hollow Creek at West Valley Middle School near Ahtanum	05–01–89	2,000	<8.3	<17	<17	10	<8.3	<8.3	<8.3
12500445	Wide Hollow Creek near mouth at Union Gap	08–30–88	2,000	<7.4	<15	<15	16	<7.4	3.3	<7.4
		08–30–88	62	<6.9	<14	<14	20	4.2	2.4	2.4
12503640	Unnamed drain at Lateral & Riggs Roads near Wapato	09–01–88	2,000	<23	<46	<46	.46	<23	<23	<23
		09–01–88	62	<26	<52	<52	<26	<26	<26	<26
12505460	Granger Drain at mouth near Granger	09–02–88	2,000	<49	<98	<98	<49	<49	<49	<49
		09–02–88	62	<27	<55	<55	<27	<27	<27	<27
12507508	Toppenish Creek at Indian Church Road near Granger	09–01–88	2,000	<12	<24	<24	20	<12	<12	<12
		09–01–88	62	<7.4	<15	<15	29	<7.4	.74	<7.4
12508850	Sulphur Creek Wasteway near Sunnyside	08–31–88	2,000	<24	<47	<47	<24	<24	<24	<24
		08–31–88	62	<24	<49	<49	43	<24	<24	<24
		08–31–88	2,000	<46	<93	<93	120	<46	<46	<46
		08–31–88	62	<44	<89	<89	96	<44	18	24
# 12509710	Spring Creek at mouth at Whitstran	11–08–90	62	<330	<670	<670	<330	8.3	15	6.7
461652119522000	Unnamed drain at County Line Road near Grandview	09–01–88	2,000	<41	<82	<82	<41	<41	<41	<41
		09–01–88	62	<14	<29	<29	24	<14	2.7	<14
461744119522400	Unnamed drain to East Turbine Lateral near Grandview	09–01–88	2,000	<11	<22	<22	14	<11	<11	<11
			62	<5.9	<12	<12	5.9	<5.9	<5.9	<5.9
# 463359120281400	Unnamed urban runoff at Union Gap	08–28–87	180	<64	<130	<130	<64	18	11	19
463501120353300	Wide Hollow Creek at Ahtanum Boulevard near Ahtanum	09–01–88	2,000	<9.1	7.3	5.9	18	9.1	6.4	8.7
			62	<9.1	<18	<18	24	<9.1	1.3	<9.1

[1]The upper value of the confidence interval represents the concentration which, with 97.5-percent certainty, will result in hazardous long-term effects on the benthic fauna.

Table 44. Summary of concentrations of organochlorine compounds in bed-sediment samples collected in agricultural and nonagricultural areas, Yakima River Basin, Washington, 1987–90

[Concentrations reported in micrograms per gram of organic carbon, except organic-carbon concentrations which are in grams per kilogram; nd, not detected; --, insufficient data to calculate statistic; lower main-stem sites are located downstream from Union Gap; DDD, 4,4'-dichlorodiphenyldichloroethane; DDE, 4,4'-dichlorodiphenyldichloroethylene; DDT, 4,4'-dichlorodiphenyltrichloroethane]

Compound	Station location/ sampling media	Number of samples	Minimum concentration	Concentration at indicated percentile			Maximum concentration
				25	50	75	
Chlordane (technical)	Upper main stem/bed sediment	2	nd	--	--	--	nd
	Lower main stem/bed sediment	3	nd	--	nd	--	.3
	Nonagriculture/bed sediment	12	nd	nd	nd	nd	nd
	Agriculture/bed sediment	23	nd	nd	nd	.3	1.0
	Agriculture/soil	5	nd	nd	.1	.2	.2
DDD	Upper main stem/bed sediment	2	nd	--	--	--	.1
	Lower main stem/bed sediment	3	nd	--	.2	--	.4
	Nonagriculture/bed sediment	12	nd	nd	.02	.1	.1
	Agriculture/bed sediment	23	nd	1.0	1.9	4.9	20
	Agriculture/soil	5	.2	.4	1.1	2.9	4.0
DDE	Upper main stem/bed sediment	2	nd	--	--	--	.2
	Lower main stem/bed sediment	3	.6	--	1.3	--	9.4
	Nonagriculture/bed sediment	12	nd	.007	.04	.1	.7
	Agriculture/bed sediment	23	nd	.9	7.6	12	78
	Agriculture/soil	5	4.7	6.7	10	29	49
DDT	Upper main stem/bed sediment	2	nd	--	--	--	.1
	Lower main stem/bed sediment	3	nd	--	nd	--	.1
	Nonagriculture/bed sediment	12	nd	nd	nd	.008	.04
	Agriculture/bed sediment	23	nd	.3	1.3	3.3	21
	Agriculture/soil	5	3.5	6.1	11	34	46
DDT+DDE+DDD	Upper main stem/bed sediment	2	nd	--	--	--	.3
	Lower main stem/bed sediment	3	.7	--	1.9	--	9.4
	Nonagriculture/bed sediment	12	nd	.008	.1	.2	.8
	Agriculture/bed sediment	23	.6	1.6	11	24	97
	Agriculture/soil	5	8.4	14	20	66	99
Dieldrin	Upper main stem/bed sediment	2	nd	--	--	--	.1
	Lower main stem/bed sediment	3	nd	--	.1	--	.1
	Nonagriculture/bed sediment	12	nd	nd	nd	nd	nd
	Agriculture/bed sediment	23	.02	.1	.6	2.2	6.1
	Agriculture/soil	5	nd	nd	.007	.03	.05
Endosulfan I	Upper main stem/bed sediment	2	nd	--	--	--	nd
	Lower main stem/bed sediment	3	nd	--	nd	--	nd
	Nonagriculture/bed sediment	12	nd	nd	nd	nd	nd
	Agriculture/bed sediment	23	nd	nd	.4	1.1	12
	Agriculture/soil	5	nd	nd	nd	nd	nd
Endrin	Upper main stem/bed sediment	2	nd	--	--	--	nd
	Lower main stem/bed sediment	3	nd	--	nd	--	nd
	Nonagriculture/bed sediment	12	nd	nd	nd	nd	nd
	Agriculture/bed sediment	23	nd	nd	nd	.02	.5
	Agriculture/soil	5	nd	nd	nd	.02	.02
Organic carbon	Upper main stem/bed sediment	2	2.1	--	--	--	16
	Lower main stem/bed sediment	3	1.6	--	2.3	--	18
	Nonagriculture/bed sediment	12	1.2	2.7	3.8	5.1	9.9
	Agriculture/bed sediment	23	4.1	7.7	11	22	34
	Agriculture/soil	5	5.2	5.7	14	15	16

Table 45. Concentrations of 4,4'-dichlorodiphenyltrichloroethane (DDT) plus 4,4'-dichlorodiphenyldichloroethylene (DDE) plus 4,4'-dichlorodiphenyldichloroethane (DDD) and dieldrin in bed-sediment samples collected in 1985 and 1987–90, Yakima River Basin, Washington

[Concentrations reported in micrograms per gram of organic carbon; M, sample collected at mouth; RM, river mile; --, no data]

Sampling station	DDT+DDE+DDD		Dieldrin	
	1985[1]	1987–90	1985[1]	1987–90
Moxee Drain at Thorp Road	0.8 M	8.5–11	0.02 M	0.11–.14
Granger Drain at mouth	12	11–24	.23	.44–.48
Sulphur Creek Wasteway	16–23	1.6–15	.77–.99	.61–.98
Yakima River at Kiona (RM 29.9)	--	.71–1.9	--	.07–.09
Yakima River at Horn Rapids Dam (RM 18.0)	4	--	.17	--

[1]Johnson and others (1986); 1985 data includes concentrations of 2,4'-DDT which generally accounts for less than 10 percent of the sum of DDT+DDE+DDD.

From 1974–1990, the USFWS managed the National Contaminant Biomonitoring Program (NCBP), in which pesticides and metals were measured in fish and birds. In this program, samples were collected about every 3 years, with the 1984 sampling period being the most recent for which a report has been issued (Schmitt and others, 1990). Sampling locations for the NCBP were almost exclusively on major rivers. Only one of the 112 NCBP sampling locations, the Yakima River at Granger, was in the Yakima River Basin. The NCBP provides a nationwide data base against which contaminant concentrations in Yakima River Basin fish can be compared.

The EPA study was a one-time sampling effort, initially called the National Bioaccumulation Study and now known as the National Study of Chemical Residues in Fish (NSCRF). NSCRF sample collections were made from 1986 to 1989 (U.S. Environmental Protection Agency, 1992c) at 314 sampling locations nationwide, many of which were selected because of their proximity to known sources of pollution.

Lipid Normalization

The accumulation of an organic compound in biota is a function of several factors including the (1) physical and chemical properties of the compound, (2) concentration of the compound in water, and (3) uptake and elimination of the compound by the organism. One additional factor is the lipid (fat) concentration of the organism. Nonionic hydrophobic compounds tend to sorb to lipid deposits in organisms. As a consequence, the lipid concentration of an organism can greatly influence the concentration of organic contaminants in that organism. Previous researchers differ in their views as to the utility of normalizing organic-contaminant concentrations to lipid content (compound concentration divided by the lipid concentration) in biological samples. For example, Phillips (1980, p. 39–90) concluded that both wet-weight concentrations and lipid-weight concentrations should be used in evaluating organic contaminants in biological organisms. The author indicated that the variability in organochlorine concentrations among individuals of aquatic biota decreased when the concentrations were normalized to lipid content. Phillips (1980) also suggested two disadvantages of lipid normalization: (1) failure to account for the different types of lipid that make up the total lipid pool in organisms, and (2) the assumption that the amount of contaminant in the organism is dependent on the level of contaminant in the environment rather than the amount of lipid in the organism. More recently, Schmitt and others (1990) chose not to use lipid normalization for their analysis of NCBP data because normalization failed to help in the interpretation of previous NCBP data, and because Huckins and others (1988) reported that lipid normalization added variability to their data rather than reducing it.

In this study, wet-weight concentrations of most organochlorine compounds for samples of whole fish are significantly correlated with lipid concentrations in the fish (table 46). As a result, lipid-normalized concentrations of organochlorine compounds in whole fish are used for comparisons of compound concentrations among stations and for comparisons among fish species at a station. Wet-weight concentrations in whole fish were used for comparisons with existing action levels and water-quality criteria and for comparisons with data from other studies.

In mollusk samples, lipid concentrations were poorly correlated with contaminant concentrations (table 46), probably because lipid concentrations varied little among the samples, and because mollusk samples

Table 46. Spearman's rank-order correlation coefficients between concentrations of lipids and organochlorine compounds in biological samples, Yakima River Basin, Washington, 1989–90
[DDT, dichlorodiphenyltrichloroethane; DDE, dichlorodiphenyldichloroethylene; DDD, dichlorodiphenyldichloroethane; DDT+DDE+DDD, the sum of 2,4'-DDT, 4,4'-DDT, 2,4'-DDE, 4,4'-DDE, 2,4'-DDD, and 4,4'-DDD; chlordane-related compounds, the sum of cis-chlordane, trans-chlordane, cis-nonachlor, trans-nonachlor, oxychlordane, and heptachlor epoxide; **, statistically significant at the 0.01 level of confidence; ND, compound not detected in any sample]

Organochlorine compound	Spearman's correlation coefficients between lipid and contaminant concentrations	
	Fish	Mollusks
2,4'-DDT	0.20	-0.06
4,4'-DDT	.69**	.51
2,4'-DDD	.67**	-.08
4,4'-DDD	.74**	.00
2,4'-DDE	.49**	.02
4,4'-DDE	.79**	-.02
DDT+DDE+DDD	.79**	.04
Dicofol	.31	-.08
Total polychlorinated biphenyls (PCBs)	.55**	ND
Hexachlorobenzene (HCB)	-.10	ND
α-hexachlorocyclohexane (HCH)	ND	ND
β-HCH	ND	ND
δ-HCH	ND	ND
Lindane (γ-HCH)	ND	ND
cis-Chlordane	.73**	.00
trans-Chlordane	ND	ND
cis-Nonachlor	ND	ND
trans-Nonachlor	.71**	.00
Oxychlordane	.44**	ND
Heptachlor epoxide	.44**	ND
Chlordane-related compounds	.77**	.00
Toxaphene	.62**	ND
Dieldrin	.84**	.05
Endrin	ND	ND
Mirex	ND	ND
Kepone	ND	ND
Octachlorostyrene	ND	ND

were collected from only five stations in the basin. The importance, however, of normalizing contaminant concentrations for lipid content in both mollusk and fish samples is shown in figure 33. After normalizing concentrations of DDT+DDE+DDD for lipid content in Asiatic clams (*Corbicula fluminea*), smallmouth bass (*Micropterus dolomieu*), mountain whitefish (*Prosopium williamsoni*), and largescale sucker (*Catostomus macrocheilus*) collected from the Yakima River at Kiona, the DDT+DDE+DDD concentrations were remarkably similar among the species, ranging from 19.6 to 25.9 μg/g of lipid. This similarity provides justification for normalizing lipophilic-contaminant concentrations to lipid content in both mollusk and fish samples. Crayfish and aquatic-plant data are not used in this report for among-station comparisons, so contaminant concentrations referenced for these biota are not normalized for lipid concentrations.

Contaminant Occurrence

In this study, small numbers of fish, mollusk, crayfish, and aquatic-plant samples were collected from a small number of sampling stations in the Yakima River Basin over a short time period. Given these data limitations, extrapolation of study findings to other areas in the basin should be made with caution. Study results, however, are adequate to determine

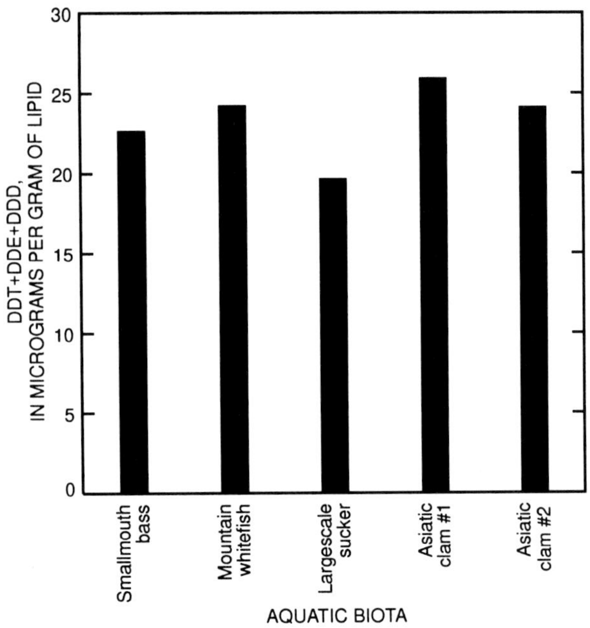

Figure 33. Lipid-normalized concentrations of dichlorodiphenyltrichloroethane (DDT) plus dichlorodiphenyldichloroethylene (DDE) plus dichlorodiphenyldichloroethane (DDD) in fish and mollusk samples from the Yakima River at Kiona, Washington, October 1989. (DDT+DDE+DDD, the sum of 2,4'-DDT, 4,4'-DDT, 2,4'-DDE, 4,4'-DDE, 2,4'-DDD, and 4,4'-DDD.)

approximate contaminant concentrations in several species of aquatic biota and to establish baseline data for assessing long-term changes at a station.

Organochlorine Compounds

Of the 25 organochlorine compounds analyzed in biological samples from the Yakima River Basin, one DDT metabolite, DDE, was found in fish samples at all stations in 1989 and at 68 percent (17 of 25) of the sampling stations in 1990 (table 47). DDE is the most widely occurring organic compound detected in the study. DDT, DDD, cis-chlordane, trans-nonachlor, and dieldrin also were detected frequently each year, with each compound being found in fish at 32 percent or more of the stations. Dicofol, PCBs, and toxaphene were found less frequently in fish samples, from 16 to 42 percent of the stations each year. HCB, HCH (α-, β-, γ-, and δ-isomers), cis-nonachlor, endrin, mirex, kepone, and octachlorostyrene were not detected at any of the sampling stations in either fish, mollusk, or plant samples in either year (table 47).

Of the compounds not detected in this study, endrin and the four HCHs were analyzed in the WDOE study of fish and mussels in the Yakima River Basin (Johnson and others, 1986). Endrin was detected in only two samples, both of mixed species of suckers from the Yakima River at Buena. Buena is located midway between two sampling stations used in this study, the Yakima River at Parker and the Yakima River below Toppenish Creek; the Buena station is upstream from Toppenish Creek and Granger Drain. HCHs were not detected in the WDOE study. In the most recent report from the NCBP, the USFWS did not detect these compounds in their 1984 sample from the Yakima River at Granger, the only station in the Yakima River Basin sampled in the NCBP (Schmitt and others, 1990). In the NSCRF, one whole-fish sample was collected from the Yakima River Basin. This was a sample of carp (*Cyprinus carpio*) from the Yakima River at Richland, Washington, which is about 25 miles downstream from Kiona, the farthest downstream station in the current study. Concentrations of lindane (γ-HCH), endrin, cis-nonachlor, mirex, or octachlorostyrene were not detected in the sample. α-HCH was detected, but at a concentration below the minimum reporting level of 0.0025 μg/g (U.S. Environmental Protection Agency, 1992c). β-HCH, δ-HCH, and kepone were not analyzed in the EPA study.

Dioxin was not detected in any of the four aquatic-biota samples collected in the Yakima River Basin, possibly because the minimum reporting level of the analysis was too high (0.01 nanograms per gram [ηg/g]). Dioxin concentrations as small as 0.001 to 0.003 ηg/g may be of environmental significance. Accordingly, the few data collected in this study do not provide assurance that dioxin is not a concern in the Yakima River Basin. In EPA's NSCRF study, however, 2,3,7,8-tetrachlorodibenzo-*p*-dioxin was detected at a concentration of 0.00079 ηg/g in a whole-carp sample from the Yakima River at Richland.

Polycyclic Aromatic Hydrocarbons

Fourteen PAHs were analyzed in 33 biological samples collected from 14 stations in the Yakima River Basin. Fifteen of the samples were rooted aquatic plants, 11 were Asiatic clams, 3 were western pearlshell clams (*Margaritifera falcata*), 3 were crayfish (*Pacifastacus* spp.), and 1 was rainbow trout (*Oncorhynchus clarki*). Nine tributary stations and 5 main-stem stations were represented among the 33 samples.

Naphthalene and fluoranthene were detected in samples from 6 of the 14 stations (table 48). Phenan-

Table 47. Occurrence of organochlorine compounds in biological samples, Yakima River Basin, Washington, 1989–90

[DDD, dichlorodiphenyldichloroethane; DDE, dichlorodiphenyldichloroethylene; DDT, dichlorodiphenyltrichloroethane]

	Number of sampling stations having contaminant occurrences					
	Fish		Mollusks		Plants	
Organochlorine compound	1989	1990	1989	1990	1989	1990
2,4'-DDD	9	11	4	2	0	0
4,4'-DDD	12	14	4	2	0	0
2,4'-DDE	1	10	3	2	0	0
4,4'-DDE	19	17	4	2	7	0
2,4'-DDT	2	3	1	2	0	0
4,4'-DDT	12	13	4	2	0	0
Dicofol	8	4	0	1	0	0
Total polychlorinated biphenyls (PCBs)	3	6	0	0	0	0
Hexachlorobenzene (HCB)	0	0	0	0	0	0
Hexachlorocyclohexanes (HCH) (α, β, δ, and γ compounds)	0	0	0	0	0	0
cis-Chlordane	6	12	0	2	0	0
trans-Chlordane	0	4	0	0	0	0
cis-Nonachlor	0	0	0	0	0	0
trans-Nonachlor	6	12	0	2	0	0
Oxychlordane	0	6	0	0	0	0
Heptachlor epoxide	0	8	0	0	0	0
Toxaphene	4	5	0	0	0	0
Dieldrin	16	11	4	3	0	0
Endrin	0	0	0	0	0	0
Mirex	0	0	0	0	0	0
Kepone	0	0	0	0	0	0
Octachlorostyrene	0	0	0	0	0	0
	Number of stations sampled					
	19	25	5	3	14	0

threne and benzo(e)pyrene were detected at five stations. Fluorene, pyrene, 1,2-benzanthracene, and chrysene were detected at three, two, one, and one stations, respectively (tables 48 and 49). Anthracene, benzo(g,h,i)perylene, benzo(b)fluoranthene, benzo(k)fluoranthene, benzo(a)pyrene, and 1,2,5,6-dibenzanthracene were not detected in any sample.

Contaminant Concentrations and Spatial Distribution

Organochlorine Compounds

DDT and Its Metabolites

In addition to being the most ubiquitous compounds detected in aquatic biota (detected in 48 of 58 samples—83 percent), concentrations of DDT and its metabolites were among the highest detected in the group of organochlorine compounds (table 50). The highest concentration of DDT+DDE+DDD was detected in a sample of a largescale sucker from Sulphur Creek Wasteway (fig. 7). This sample had a concentration of 4.8 µg/g, wet weight. The median DDT+DDE+DDD concentration in the Yakima River Basin fish samples was 0.67 µg/g, excluding the sculpin (*Cottus* spp.) data. The highest concentrations of DDT+DDE+DDD in biota were detected at main-stem and tributary stations downstream from the city of Yakima, where agriculture is the primary land use. These compounds also were detected in samples from upstream stations, but at lower concentrations.

Concentrations of DDT+DDE+DDD in resident fish from the Yakima River Basin collected during this

Table 48. Occurrence of noncarcinogenic polycyclic aromatic hydrocarbons (PAHs) in biological samples, Yakima River Basin, Washington, 1989–90

[Minimum reporting level = 0.01 micrograms per gram, wet weight; --, not detected; yes, detected]

Station name	Were PAHs detected in aquatic plants, mollusks, fish, or crayfish?							
	Anthracene	Benzo(e)pyrene	Benzo(g,h,i)perylene	Fluoranthene	Fluorene	Naphthalene	Phenanthrene	Pyrene
Tributary stations								
Cherry Creek at Thrall	--	--	--	--	--	--	--	--
Umtanum Creek near mouth at Umtanum	--	--	--	--	--	--	--	--
Little Naches River at mouth near Cliffdell	--	--	--	--	--	yes	--	--
Moxee Drain at Thorp Road near Union Gap	--	yes	--	--	--	--	--	--
Wide Hollow Creek at West Valley Middle School near Ahtanum	--	--	--	--	--	--	--	--
Wide Hollow Creek at old sewage treatment plant at Union Gap	--	yes	--	yes	--	--	--	--
Toppenish Creek at Indian Church Road near Granger	--	yes	--	--	--	--	--	--
Satus Creek at gage at Satus	--	--	--	--	--	--	--	--
Spring Creek at mouth at Whitstran	--	yes	--	yes	yes	yes	yes	yes
Main-stem stations								
Yakima River at Parker	--	--	--	--	--	--	--	--
Yakima River at Umtanum	--	--	--	yes	--	yes	yes	--
Yakima River at river mile 72 above Satus Creek near Sunnyside	--	--	--	yes	yes	yes	yes	--
Yakima River at Euclid Bridge at river mile 55 near Grandview	--	yes	--	yes	--	yes	yes	yes
Yakima River at Kiona	--	--	--	yes	yes	yes	yes	--

study are elevated compared with nationwide data sets (table 50). When compared with data collected by USFWS in 1984 in their NCBP (Schmitt and others, 1990), the arithmetic mean DDT+DDE+DDD concentration in resident fish (1.12 µg/g, wet weight) from the Yakima River Basin is more than four times larger than the geometric mean concentration in the NCBP. When compared with data collected by EPA in their NSCRF, the arithmetic mean DDE concentration in the resident fish (0.85 µg/g, wet weight) from the Yakima River Basin is almost three times larger than EPA's arithmetic mean.

In the WDOE study (Johnson and others, 1986), concentrations of DDT+DDE+DDD increased in a downstream direction, with the largest concentrations detected in a fish sample from the Yakima River at Kiona. In the WDOE study, the composite largescale-sucker sample from Kiona had a DDT+DDE+DDD concentration of 1.1 µg/g, wet weight, compared with a mean concentration of 2.5 µg/g, wet weight, for sucker samples from Kiona in the present study.

Largescale suckers were sampled at 13 stations in the present study. Lipid-normalized concentrations of DDT+DDE+DDD in these samples ranged from 0.9 µg/g-lipid to 65.5 µg/g-lipid (fig. 34). Suckers collected at three main-stem stations had relatively high lipid-normalized concentrations of DDT+DDE+DDD: Yakima River near Grandview (RM 55), Yakima River at RM 72, and Yakima River at Kiona (RM 29.9). Suckers from two Yakima River tributaries, Sulphur Creek Wasteway and Granger Drain, which flow into the Yakima River at RMs 61.0 and 82.8, respectively, also had high concentrations of DDT+DDE+DDD. Whole largescale-sucker samples from Moxee Drain, Wide Hollow Creek, and the Naches River had intermediate concentrations of DDT+DDE+DDD. Although the composite fish sample from Moxee Drain had an intermediate concentration of DDT+DDE+DDD, bed-sediment samples from that tributary had high concentrations of DDT+DDE+DDD. No large fish, however, could be captured for the sample from Moxee Drain. The average fish length at Moxee Drain was 199 mm (millimeters), whereas the average length for all samples approached 400 mm. These smaller and, therefore, younger fish in Moxee Drain may not have had sufficient time to accumulate contaminants when compared to the older fish in the other samples.

Lipid-normalized concentrations of DDT+DDE+DDD in nine samples of mountain whitefish were similar to concentrations in largescale suckers. For example, concentrations of DDT+DDE+DDD in largescale

Table 49. Occurrence of carcinogenic polycyclic aromatic hydrocarbons (PAHs) in biological samples, Yakima River Basin, Washington, 1989-90

[Minimum reporting level is 0.01 micrograms per gram, wet weight; --, not detected; yes, detected]

Station name	Were PAHs detected in aquatic plants, mollusks, fish, or crayfish?					
	1,2-Benzanthracene	Benzo(b)fluoranthene	Benzo(k)fluoranthene	Benzo(a)pyrene	Chrysene	1,2,5,6-Dibenzanthracene
Tributary stations						
Cherry Creek at Thrall	--	--	--	--	--	--
Umtanum Creek near mouth at Umtanum	--	--	--	--	--	--
Little Naches River at mouth near Cliffdell	--	--	--	--	--	--
Moxee Drain at Thorp Road near Union Gap	--	--	--	--	--	--
Wide Hollow Creek at old sewage treatment plant at Union Gap	--	--	--	--	--	--
Wide Hollow Creek at West Valley Middle School near Ahtanum	--	--	--	--	--	--
Toppenish Creek at Indian Church Road near Granger	--	--	--	--	--	--
Satus Creek at gage at Satus	--	--	--	--	yes	--
Spring Creek at mouth at Whitstran	--	--	--	--	--	--
Main-stem stations						
Yakima River at Parker	--	--	--	--	--	--
Yakima River at Umtanum	--	--	--	--	--	--
Yakima River at river mile 72 above Satus Creek near Sunnyside	yes	--	--	--	--	--
Yakima River at Euclid Bridge at river mile 55 near Grandview	--	--	--	--	--	--
Yakima River at Kiona	--	--	--	--	--	--

suckers from the Naches River near North Yakima were almost identical to concentrations in mountain white fish collected from the Naches River station. Samples from the Yakima River at Kiona had the highest concentrations of DDT+DDE+DDD, whereas samples from the Yakima River at RM 72, Moxee Drain, and the Naches River had intermediate concentrations. Samples from stations upstream from the city of Yakima had low concentrations (fig. 35).

Samples of Asiatic clams were collected from three main-stem stations and one tributary, all in the lower parts of the basin. Samples from Spring Creek had the highest lipid-normalized concentrations of DDT+DDE+DDD, followed by the Yakima River near Grandview, the Yakima River at Kiona, and the Yakima River at RM 72. No fish samples were collected from the Spring Creek station for comparison, but duplicate samples of Asiatic clams from this station also had high concentrations of DDT. Relatively high concentrations of DDT+DDE+DDD also were detected in water samples from Spring Creek in July 1988 and June 1989.

By examining the lipid-normalized concentrations, an overall pattern emerges that shows most elevated DDT+DDE+DDD concentrations in biota occur downstream from the city of Yakima, where agriculture is the dominant land use. The highest concentrations in biota occur in a section of the Yakima River near Grandview, downstream from several agricultural return flows.

The NAS-NAE (1973) have recommended that DDT+DDE+DDD concentrations in whole fish not exceed 1.0 μg/g, wet weight, for the protection of fish-eating predators. The concentration in the sample of largescale suckers from Sulphur Creek Wasteway is almost five times this threshold concentration. In addition, 17 fish samples and 2 of 14 mollusk samples from the Yakima River Basin exceeded the criterion. These samples were distributed among the four main-stem sampling stations located downstream from the town of Granger and the six tributaries downstream from the city of Yakima. No concentrations of DDT+DDE+DDD exceeding the NAS-NAE recommendation were found in samples upstream from the city of Yakima, in either tributaries or the main stem.

Table 50. Summary of concentrations of organochlorine compounds in whole-fish samples, Yakima River Basin, Washington, 1989–90

[Concentrations reported in micrograms per gram, wet weight; data include analyses of bridgelip suckers, carp, chiselmouth, largescale suckers, mountain whitefish, smallmouth bass, and rainbow trout; sculpin data are not included; NCBP, National Contaminant Biomonitoring Program; NSCRF, National Study of Chemical Residues in Fish; DDT, dichlorodiphenyltrichloroethane; DDE, dichlorodiphenyldichloroethylene; DDD, dichlorodiphenyldichloroethane; DDT+DDE+DDD, the sum of 2,4'-DDT, 4,4'-DDT, 2,4'-DDE, 4,4'-DDE, 2,4'-DDD, and 4,4'-DDD; chlordane-related compounds, the sum of cis-chlordane, trans-chlordane, cis-nonachlor, trans-nonachlor, oxychlordane, and heptachlor epoxide; <, less than; NR, not reported]

Compound	Maximum	Arithmetic mean	Median	NCBP geometric mean[1]	NSCRF arithmetic mean[2]	NAS-NAE guideline for protection of fish-eating predators[3]	EPA fish-tissue concentration resulting in cancer risk of [4]1:1,000,000
2,4'-DDT	0.04	0.01	<0.01	NR	NR	1.0	NR
4,4'-DDT	.96	.19	.07	0.03	NR	1.0	0.0316
2,4'-DDE	.03	.01	< .01	NR	NR	1.0	NR
4,4'-DDE	3.4	.85	.51	.19	0.30	1.0	.0316
2,4'-DDD	.07	.02	.02	NR	NR	1.0	NR
4,4'-DDD	.42	.11	.07	.06	NR	1.0	.0449
DDT+DDE+DDD	4.8	1.12	.67	.26	NR	1.0	NR
Dicofol	.3	.02	< .01	NR	.001	NR	NR
Total polychlorinated biphenyls (PCBs)	.90	.15	< .05	.39	7.6	NR	.0014
Hexachlorobenzene (HCB)	< .01	< .01	< .01	NR	.006	NR	NR
α-hexachlorocyclohexane (HCH)	< .01	< .01	< .01	< .01	.002	NR	.0017
β-HCH	< .01	< .01	< .01	NR	NR	NR	.006
δ-HCH	< .01	< .01	< .01	NR	NR	NR	NR
Lindane (γ-HCH)	< .01	< .01	< .01	< .01	.003	.1	.0081
cis-Chlordane	.04	.01	.01	.03	.021	.1	.0083
trans-Chlordane	.01	.01	< .01	.02	.017	.1	.0083
cis-Nonachlor	< .01	< .01	< .01	.02	.009	NR	NR
trans-Nonachlor	.04	.01	.01	.03	.017	NR	NR
Oxychlordane	.02	.01	< .01	.01	.005	NR	NR
Heptachlor epoxide	.02	.01	< .01	.01	.002	.1	.0012
Chlordane-related compounds	.12	.05	.04	.12	NR	NR	NR
Toxaphene	1.2	.19	< .05	.14	NR	.1	.0098
Dieldrin	.17	.03	.02	.04	.028	.1	.00067
Endrin	< .01	< .01	< .01	< .01	.002	.1	3.23
Mirex	< .01	< .01	< .01	< .01	.004	NR	NR
Kepone	< .01	< .01	< .01	NR	NR	NR	NR
Octachlorostyrene	< .01	< .01	< .01	NR	.002	NR	NR

[1]Schmitt and others, 1990.
[2]U.S. Environmental Protection Agency, 1992c.
[3]National Academy of Sciences-National Academy of Engineers Committee on Water Quality Criteria (NAS-NAE), 1973.
[4]Nowell and Resek, 1994.

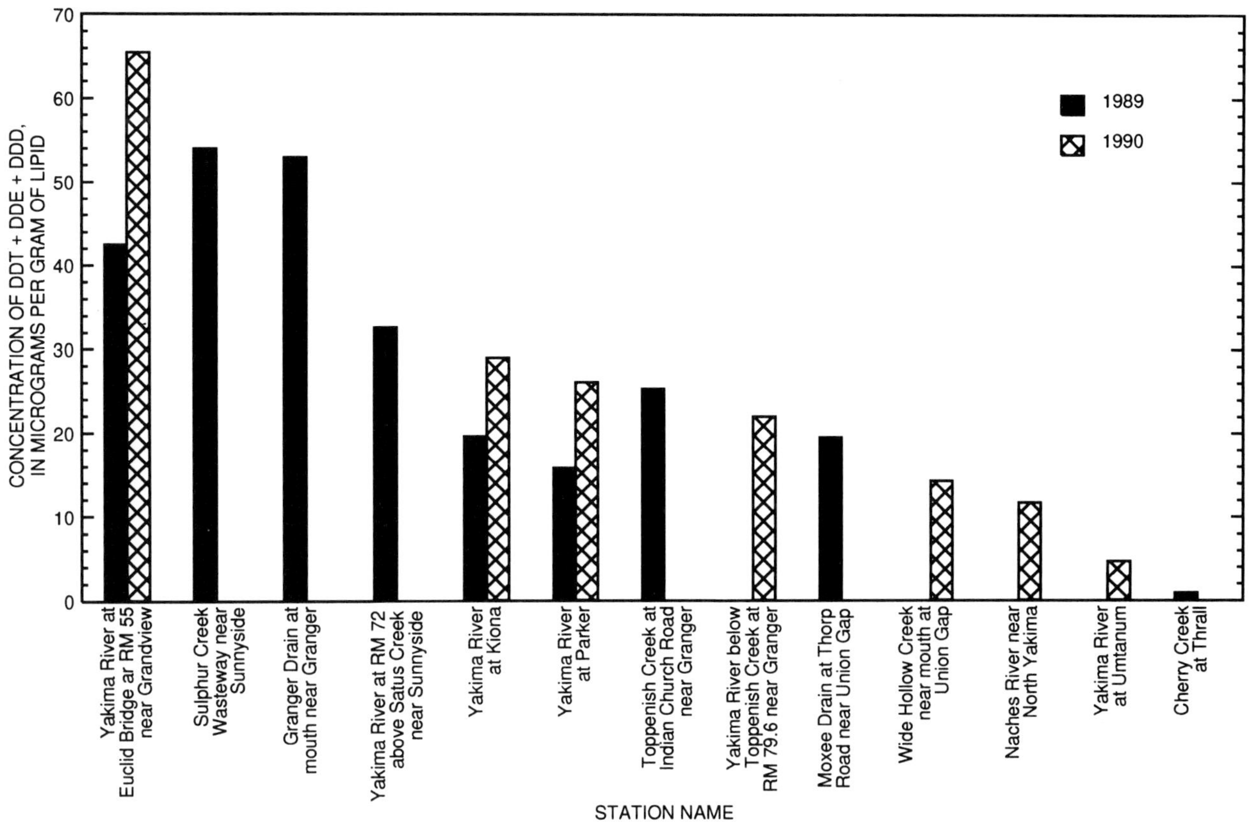

Figure 34. Lipid-normalized concentrations of dichlorodiphenyltrichloroethane (DDT) plus dichlorodiphenyldichloroethylene (DDE) plus dichlorodiphenyldichloroethane (DDD) in whole largescale suckers, Yakima River Basin, Washington, 1989–90. (DDT+DDE+DDD, the sum of 2,4'-DDT, 4,4'-DDT, 2,4'-DDE, 4,4'-DDE, 2,4'-DDD, and 4,4'-DDD; RM, river mile.)

The U.S. Food and Drug Administration (FDA) (1992) has established an action level of 5.0 µg/g for DDT in the edible portions of fish. No concentrations in samples of fish from the Yakima River Basin study were as high as this, even though whole fish were analyzed rather than edible portions of fish. Because the lipid content of fish is normally lower in muscle tissues than in whole fish, DDT+DDE+DDD concentrations would be expected to be lower in the commonly eaten fish tissues (muscle tissues) than in whole fish.

The U.S. Environmental Protection Agency has calculated concentrations of various contaminants in fish, which, if the fish were consumed at an average rate of 6.5 g/d by a human weighing 70 kg, would result in an increased lifetime (70 years) cancer risk of 1:1,000,000. For DDT, DDE, and DDD, the concentrations are 0.0316, 0.0316, and 0.0449 µg/g, respectively (table 50; Nowell and Resek, 1994). DDE concentrations in resident whole fish from the Yakima River Basin, excluding sculpins, averaged more than 25 times this criterion level. Even though contaminant concentrations in whole fish are expected to be higher that those in fish fillets, these data suggest a concern. Consequently, fish-fillet samples were collected in 1991 from the Yakima River at Kiona, where DDT+DDE+DDD concentrations in whole fish were high. These samples were analyzed for several contaminants, including DDT+DDE+DDD. Fillet samples from largescale suckers and mountain whitefish were analyzed in the USGS's National Water Quality Laboratory. Concentrations in these samples consistently exceeded EPA's calculated concentration that would result in an increased lifetime cancer risk of 1:1,000,000. The highest concentration of DDE detected in fish-fillet samples from the Yakima River at Kiona was about 2.2 µg/g, which is about 70 times larger than the EPA fish-tissue concentration associated with an increase in cancer risk of 1:1,000,000.

The calculated risk associated with consumption of contaminated fish is highly dependent on the consumption rate of the contaminated fish by the human

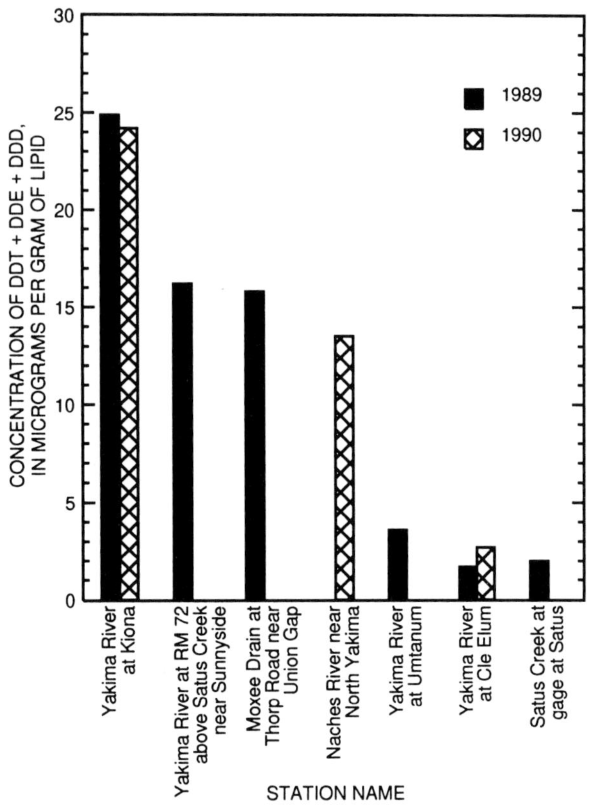

Figure 35. Lipid-normalized concentrations of dichlorodiphenyltrichloroethane (DDT) plus dichlorodiphenyldichloroethylene (DDE) plus dichlorodiphenyldichloroethane (DDD) in whole mountain whitefish, Yakima River Basin, Washington, 1989–90. (DDT+DDE+DDD, the sum of 2,4'-DDT, 4,4'-DDT, 2,4'-DDE, 4,4'-DDE, 2,4'-DDD, and 4,4'-DDD; RM, river mile.)

consumer. For risk calculations in this report, the consumption rate was assumed to be 6.5 g/d, which is the average consumption rate of fish and shellfish from estuarine and fresh waters by the general population in the United States. An increase in consumption rate is directly proportional to an increase in the calculated risk. For example, the average concentration of DDE in fish-fillet samples from resident fish from the Yakima River at Kiona is 0.82 μg/g. If these fish fillets were consumed at a rate of 6.5 g/d by a human weighing 70 kg, the increased lifetime cancer risk associated with the DDE contamination would be 26:1,000,000. The average consumption rate by adults in the Columbia River Basin Tribes (Yakama, Umatilla, Nez Perce, and Warm Springs Tribes), who eat suckers and whitefish, is 11.7 g/d (Columbia River Inter-Tribal Fish Commission, 1994). This consumption rate is 1.8 times larger than the average consumption rate of fish and shellfish for the general population in the United States, and would result in an increased lifetime cancer risk for the Columbia River Basin Tribes of about 47:1,000,000. Subsistence fisherman in the United States consume 140 g/d (U.S. Environmental Protection Agency, 1993), which is about 22 times larger than the average consumption rate for the general population. If subsistence fishermen primarily consumed suckers and whitefish from the Yakima River at Kiona, their increased lifetime cancer risk would be about 570:1,000,000. In addition to consumption rate, these risk calculations are highly dependent on the other estimates and assumptions used in the equations. For example, the above risk calculations only consider carcinogenic effects from DDE contamination. Additional contaminants, including DDT, DDD, PCB, dieldrin, and toxaphene, also were detected in the fish from the Yakima River, and their added carcinogenic effects are not considered in the above risk calculations.

PCBs

The highest concentration of total PCBs (0.90 μg/g, wet weight) occurred in a composite sample of largescale suckers from the Yakima River at Kiona (table 50). For the 1984 sampling cycle, the NCBP determined a nationwide geometric mean of 0.39 μg/g, wet weight, for total PCBs (Schmitt and others, 1990). Only four samples from the current study had higher concentrations than this nationwide mean. Samples from EPA's NSCRF had a mean concentration of 7.6 μg/g, wet weight, for total PCBs (table 50). None of the samples from the Yakima River Basin approached this mean. In the WDOE study, Johnson and others (1986) reported samples of Yakima River fish having PCB concentrations ranging from 0.11 to 0.30 μg/g, wet weight, which is similar to results from the current study.

The lipid-normalized data show that a composite largescale-sucker sample from the Yakima River at Kiona had the highest concentration of total PCBs (14.7 μg/g-lipid). Samples from the Yakima River near Grandview and Sulphur Creek Wasteway were moderately elevated compared to the sample from Kiona. No detectable concentrations of total PCBs were found in any sample upstream from the city of Yakima.

EPA calculated that a total-PCB concentration of 0.0014 μg/g, wet weight, in fish, which if consumed at an average rate of 6.5 g/d by a human weighing 70 kg, would result in an increased lifetime (70 years) cancer risk of 1:1,000,000 (Nowell and Resek, 1994). PCB concentrations in resident whole fish from the Yakima River Basin averaged more than 100 times this criterion level (table 50). PCBs were detected in 11 whole-fish samples collected from the Sulphur Creek Wasteway and Yakima River downstream from the city of Yakima at RMs 103.7, 79.6, 72, 55, and 29.9. Concentrations in mountain whitefish, chiselmouth (*Acrocheilus alutaceus*), bridgelip suckers (*Catostomus columbianus*), and largescale suckers ranged from 0.10 to 0.90 µg/g, wet weight, with a median PCB concentration of 0.37 µg/g. The increased lifetime cancer risk associated with consuming fish containing the maximum concentration of 0.90 µg/g would be more than 640:1,000,000.

In 1991, more than 90 fillet samples from largescale suckers and mountain whitefish were collected from the Yakima River at Umtanum (RM 140.4) and the Yakima River at Parker (RM 103.7). The median total-PCB concentration (0.025 µg/g) was about 18 times the EPA-criterion concentration of 0.0014 µg/g. In this study, concentrations of specific PCB congeners were not determined. To provide better risk estimates in future studies, concentrations of PCB congeners should be determined because some congeners, which may be included in the total-PCB concentration reported in this study, do not have carcinogenic effects. The FDA (1992) action level for PCBs is 2.0 µg/g, wet weight. No sample from the Yakima River Basin exceeded this level.

Chlordane-Related Compounds

Chlordane-related compounds analyzed in the Yakima River study include cis- and trans-chlordane, cis- and trans-nonachlor, oxychlordane, and heptachlor epoxide (many other metabolites and isomers of chlordane exist that were not analyzed). In this report, concentrations of the six compounds listed above are summed to yield a "total" concentration for interpretation of chlordane-related compounds.

In the Yakima River Basin, samples of whole fish had concentrations of chlordane-related compounds ranging from less than the minimum reporting level of 0.01 µg/g to 0.12 µg/g, wet weight (table 50). Of the 58 fish samples collected, 26 (45 percent) had detectable concentrations of chlordane-related compounds. The highest concentration was identical to the national geometric mean of 0.12 µg/g detected in the 1984 NCBP (Schmitt and others, 1990). In the NSCRF, the EPA (1992c) did not report a total for chlordane-related compounds, but if arithmetic means for the six chlordane-related compounds are summed, the total chlordane-related concentration is 0.071 µg/g, wet weight. This concentration is about 60 percent of the highest concentration detected in the present study, and it is approximately equal to the arithmetic mean concentration detected in the Yakima River Basin (table 50). Johnson and others (1986), using a minimum reporting level of 0.1 µg/g, did not detect chlordane-related compounds in their 1985 sampling of the Yakima River Basin.

Data for whole fish show that, similar to DDT+DDE+DDD, concentrations of chlordane-related compounds are highest in the lower basin, with the highest concentrations in the main stem occurring in samples from the Yakima River at Kiona. Concentrations of chlordane-related compounds in the Yakima River Basin biota, however, are not elevated relative to the geometric mean concentration detected in the NCBP study.

The FDA (1992) has established a chlordane concentration of 0.3 µg/g as the action level for edible fish. In every case, whole-fish samples from the Yakima River Basin had concentrations lower than this level. EPA calculated that a chlordane concentration of 0.0083 µg/g in fish, which if consumed at an average rate of 6.5 g/d by a human weighing 70 kg, would result in an increased lifetime cancer risk of 1:1,000,000 (table 50 in this report; Nowell and Resek, 1994). The corresponding criterion for heptachlor epoxide is 0.0012 µg/g. Concentrations of cis-chlordane exceeded EPA's criterion in 25 whole-fish samples of species that included mountain whitefish, chiselmouth, largescale suckers, bridgelip suckers, and carp. All of these samples had concentrations at or slightly greater than the minimum reporting level of 0.01 µg/g, indicating an increased lifetime cancer risk slightly greater than 1:1,000,000. Cis-chlordane generally was detected in agricultural return flows and at sampling stations downstream from the city of Yakima. Trans-chlordane was detected in only a few samples. Chlordane and heptachlor epoxide were not detected in rainbow trout from the upper basin. Concentrations of heptachlor epoxide exceeded EPA's 1:1,000,000 risk level of 0.0012 µg/g in 10 samples of mountain whitefish, sculpin, chiselmouth, largescale suckers, and bridgelip suckers, with concentrations generally ranging near the minimum reporting level, also indicating increased risk levels slightly greater than 1:1,000,000.

Dieldrin

Dieldrin was detected in 35 of the 58 fish samples (60 percent) collected in the Yakima River Basin. The highest concentration was 0.17 µg/g, wet weight (table 50), from a sample of largescale suckers collected from Sulphur Creek Wasteway. This concentration is more than four times the national geometric mean concentration of 0.04 µg/g, wet weight, for the 1984 collection period of the NCBP (Schmitt and others, 1990). The arithmetic mean dieldrin concentration from the NSCRF was 0.028 µg/g, similar to the arithmetic mean of 0.03 µg/g, wet weight, detected in the Yakima River Basin study. The maximum concentration detected in the 1985 WDOE study was 0.24 µg/g, wet weight, in whole largescale suckers from the Yakima River at Kiona. Largescale-sucker samples from Kiona averaged 0.08 µg/g, wet weight, during this study.

Similar to other organic contaminants, dieldrin was detected most often in fish collected at sampling stations downstream from the city of Yakima. Fish also had low concentrations of dieldrin at five stations upstream from Yakima: Yakima River at Cle Elum, Cherry Creek, Yakima River at Umtanum, Umtanum Creek, and Naches River.

Lipid-normalized concentrations of dieldrin in largescale suckers were highest in two tributaries, Granger Drain and Sulphur Creek Wasteway (fig. 36). Concentrations in samples from those locations were almost twice as high as in any other sample. Other sampling stations where fish have relatively high concentrations were the Yakima River near Grandview (RM 55), the Yakima River at Kiona (RM 29.9), and the Yakima River at RM 72. Of the seven locations where mountain whitefish were sampled, samples from the Yakima River at Kiona and the Yakima River at RM 72 had the highest lipid-normalized concentrations of dieldrin (fig. 37). Once again, the highest dieldrin concentrations were detected in fish samples from the lower basin.

The NAS-NAE (1973) have recommended that, for the protection of fish-eating predators, concentrations of dieldrin in whole fish should not exceed 0.1 µg/g, wet weight. Only one fish sample in the present study had concentrations that exceeded that guideline, the composite sample of largescale suckers from Sulphur Creek Wasteway.

The FDA (1992) action level for dieldrin in edible fish tissue is 0.3 µg/g. No sample from the Yakima study had a concentration of dieldrin this high. EPA calculated that a dieldrin concentration of 0.00067 µg/g in fish (table 50 in this report; Nowell and Resek, 1994), which if consumed at an average rate of 6.5 g/d by a human weighing 70 kg, would result in an increased lifetime cancer risk of 1:1,000,000. Dieldrin concentrations in 35 samples of resident whole fish from the Yakima River Basin exceeded this risk level. Most of the exceedances were detected in samples of largescale suckers, mountain whitefish, bridgelip suckers, and chiselmouth. The mean concentration (0.03 µg/g, wet weight) was more than 40 times the 1:1,000,000 risk level. The increased lifetime cancer risk associated with consumption of fish with the maximum concentration detected in the basin (0.17 µg/g, wet weight) is 250:1,000,000. The analysis of 50 fillet samples of resident fish from the Yakima River at Kiona indicate that the increased lifetime cancer risk associated with consumption of those fish with an mean dieldrin concentration of 0.058 µg/g is greater than 85:1,000,000.

Toxaphene

Toxaphene was detected in 12 of the 58 whole-fish samples (21 percent). The highest concentration, 1.2 µg/g, wet weight (table 50), was in a composite sample of largescale suckers from Sulphur Creek Wasteway, a value 12 times the NAS-NAE (1973) recommended level (0.1 µg/g) for the protection of fish-eating predators. Detectable toxaphene concentrations in the other 11 samples also exceeded this recommended level.

The geometric mean concentration for toxaphene in all samples collected for the 1984 NCBP was 0.14 µg/g, wet weight (table 50). Toxaphene concentrations in 12 samples of whole fish from the Yakima River Basin were larger than this nationwide mean. Toxaphene was not detected in the 1985 WDOE study of contaminants in fish from the Yakima River Basin (Johnson and others, 1986), probably because the minimum reporting level of 2.0 µg/g was too high. As a comparison, the maximum concentration in the present study was 1.2 µg/g.

Lipid-normalized concentrations of toxaphene were high in largescale-sucker samples from the Yakima River near Grandview (6.87 and 17.5 µg/g-lipids) and from Sulphur Creek Wasteway (13.4 µg/g) when compared with samples from other locations. A sample of chiselmouth from Sulphur Creek Wasteway was similarly higher (10.3 µg/g). Sampling locations where concentrations of toxaphene in largescale

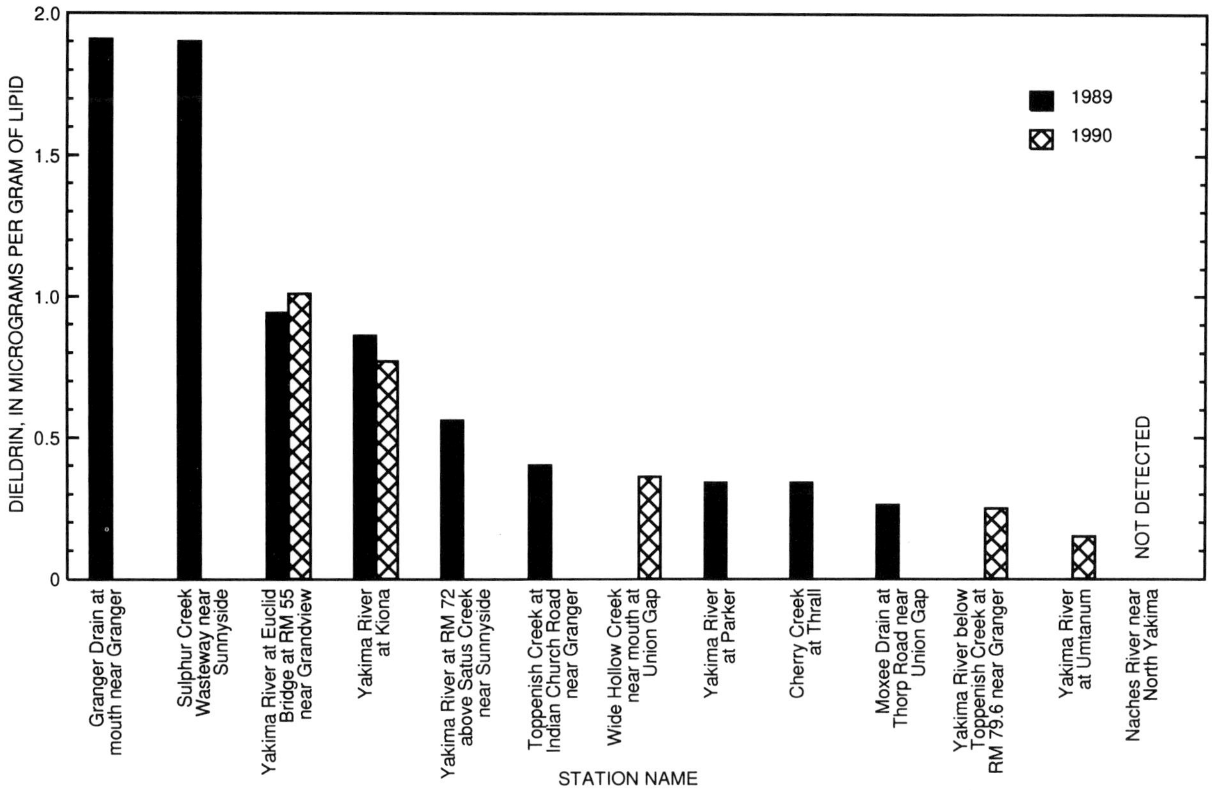

Figure 36. Lipid-normalized concentrations of dieldrin in whole largescale suckers, Yakima River Basin, Washington, 1989–90. (RM, river mile.)

suckers were lower were the Yakima River at Kiona (2.67 to 8.89 µg/g) and the Yakima River at RM 79.6 (3.69 µg/g) and RM 72 (4.92 µg/g).

FDA (1992) has established an action level of 5 µg/g for toxaphene in edible fish tissues. No sample from the Yakima study had a concentration this high. EPA calculated that a toxaphene concentration of 0.0098 µg/g in fish (table 50 in this report; Nowell and Resek, 1994), which if consumed at an average rate of 6.5 g/d by a human weighing 70 kg, would result in an increased lifetime cancer risk of 1:1,000,000. Toxaphene in 12 samples of resident whole fish from the lower Yakima River Basin exceeded this risk level. These exceedances were detected in samples of largescale suckers, mountain whitefish, chiselmouth, bridgelip suckers, and carp from Granger Drain, Sulphur Creek Wasteway, and Yakima River at RMs 79.6, 72, 55, and 29.9. These concentrations averaged 0.19 µg/g, wet weight, almost 20 times the concentration associated with the 1:1,000,000 risk level. The increased lifetime cancer risk associated with consumption of fish with the maximum concentration detected in the basin (1.2 µg/g, wet weight) is about 120:1,000,000.

Dicofol

Of the 45 samples of whole fish from the Yakima River Basin analyzed, dicofol was detected in 15 samples (33 percent of the samples) at concentrations greater than 0.01 µg/g, wet weight. The largest concentration, 0.30 µg/g, wet weight (table 50), was detected in a composite chiselmouth sample from Moxee Drain. This concentration is five times higher than the next highest concentration detected in this study. Dicofol was not a target compound in the NCBP so comparisons can not be made against this national data base. The EPA detected an arithmetic mean concentration of dicofol of 0.001 µg/g, wet weight, in their NSCRF (U.S. Environmental Protection Agency, 1992c). Dicofol was not analyzed in the WDOE study of the Yakima River in 1985 (Johnson and others, 1986). The NAS-NAE (1973) have not established recommended dicofol concentration limits for the protection of fish predators.

When normalized for lipid content, the chiselmouth sample from Moxee Drain contained a dicofol concentration of 4.5 µg/g-lipid. By comparison, a composite sample of largescale suckers from Moxee Drain had a lipid-normalized dicofol concentration of 0.77

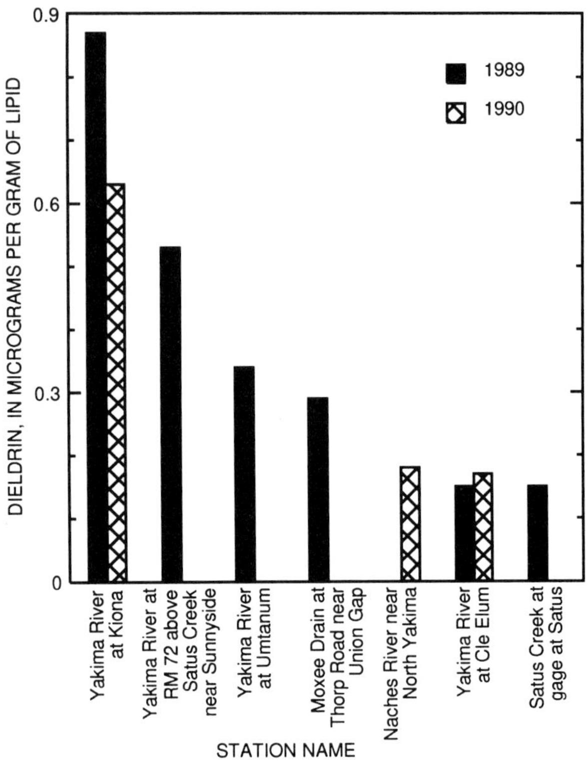

Figure 37. Lipid-normalized concentrations of dieldrin in whole mountain whitefish, Yakima River Basin, Washington, 1989–90. (RM, river mile.)

µg/g-lipid and a composite sample of mountain whitefish from this station had a concentration of 0.59 µg/g-lipid. Other sampling locations having large concentrations of dicofol in fish samples were Sulphur Creek Wasteway, Granger Drain, Yakima River near Grandview (RM 55), and Yakima River below Toppenish Creek (RM 79.6). At the farthest downstream station, the Yakima River at Kiona (RM 29.9), three samples of largescale suckers, two samples of mountain whitefish, and three samples of Asiatic clams had concentrations below the minimum reporting level of 0.01 µg/g, wet weight.

According to EPA risk calculations, a dicofol concentration of 0.024 µg/g in fish would result in an increased lifetime cancer risk of 1:1,000,000, if the fish were consumed at an average rate of 6.5 g/d by a human weighing 70 kg (Nowell and Resek, 1994; Smith, 1995). Dicofol concentrations in 12 samples of resident whole fish from major agricultural return flows (Moxee Drain, Granger Drain, Toppenish Creek, Satus Creek, Sulphur Creek Wasteway, and Spring Creek) and from the lower Yakima River (RMs 79.6, 72, and 55) exceeded this risk level. With the exception of a chiselmouth sample from Moxee Drain, which was more than 10 times larger than the 1:1,000,000 risk level, the samples had dicofol concentrations that were less than or equal to 0.060 µg/g with an associated risk less than 3:1,000,000.

Polycyclic Aromatic Hydrocarbons

Most detections of PAHs occurred in samples from the main stem of the Yakima River. Five different PAH compounds were detected in samples from the Yakima River at RM 72 and from the Yakima River near Grandview (tables 48 and 49). Four different PAHs were found in samples from the Yakima River at Kiona. Except for Spring Creek, where six PAH compounds were detected, two or fewer PAH compounds were detected in biological samples from the tributaries.

In general, PAH detections were more prevalent in mollusks than in rooted aquatic plants, probably because of the higher lipid concentrations in mollusks. In addition, mollusks actively filter water and suspended particles, which may account for the higher incidence of PAH detections in biological samples from main-stem stations and from Spring Creek, where Asiatic clams were available for sampling.

Concentrations of PAHs were typically at or slightly greater than the minimum reporting level of 0.01 µg/g, wet weight. Four of the five main-stem sampling stations and six of the nine tributary stations had PAH concentrations at or greater than the minimum reporting level (tables 48 and 49). The highest PAH concentration found was a phenanthrene concentration of 0.09 µg/g, wet weight, in a sample of Asiatic clams from the Yakima River at Kiona; the duplicate field sample had a phenanthrene concentration of 0.04 µg/g. No other sample had a concentration of any of the PAH compounds that exceeded 0.05 µg/g.

Neither the USFWS nor the EPA analyzed PAH compounds in their nationwide monitoring studies (Schmitt and others, 1990; U.S. Environmental Protection Agency, 1992c), so comparable national data do not exist. Similarly, the WDOE did not analyze PAHs in their study of contaminants in fish from the Yakima River Basin (Johnson and others, 1996). Furthermore, FDA (1992) has not established action levels for PAH compounds in aquatic organisms for the protection of human health.

EPA has calculated fish/shellfish tissue concentrations for total carcinogenic PAHs of 0.000933 µg/g (Nowell and Resek, 1994), which, if consumed at an average rate of 6.5 g/d by humans weighing 70 kg, would result in an increased cancer risk of 1:1,000,000. Of the PAH compounds analyzed in this study, only those six listed in table 49 are carcinogenic. Only two of these compounds were detected in Yakima River biological samples. Chrysene was detected in one sample of the rooted aquatic plant, waterweed (*Elodea* sp.), from Satus Creek and 1,2-benzanthracene was detected in one sample of Asiatic clams from the Yakima River at RM 72. Both of these concentrations were at the minimum reporting level of 0.01 μg/g, wet weight. On the basis of rare occurrence of carcinogenic PAHs in the Yakima River Basin samples, PAHs probably are not a widespread problem in the basin. Using the EPA-criterion fish-tissue concentration as a guide for human consumption, however, any carcinogenic PAH found at the minimum reporting level of 0.01 μg/g, would indicate an increased cancer risk of approximately 11:1,000,000. Consideration of these risk factors is presented here only as a point of reference. These risk factors are applicable to fish and shellfish tissues that are consumed by humans. The samples collected for the Yakima River NAWQA study for PAH analysis were primarily aquatic plants and mollusks (shellfish), which are not typically consumed by humans. The data, however, indicate a potential for cancer risk on the basis of PAH concentrations in aquatic biota from the Yakima River. As such, additional study is warranted and should be focused on species consumed by humans; collections should be made at stations where commercial and (or) sport fishing are prevalent.

The EPA also lists fish-tissue-concentration criteria for noncarcinogenic PAHs (Nowell and Resek, 1994). All biological samples from the Yakima River Basin had concentrations of noncarcinogenic PAHs that were substantially lower than the EPA criteria.

RELATIONS OF COMPOUND CONCENTRATIONS AMONG SAMPLING MEDIA

Concentrations in Aquatic Biota and Water

Bioaccumulation is the uptake of organic compounds from either food or water by aquatic biota. The extent to which aquatic biota accumulate organic compounds is quantified by calculating bioconcentration factors (BCFs). A BCF is the ratio of the steady-state concentration of a compound in aquatic tissue to its concentration in water, thereby, describing the relation between concentrations in biota and water (Smith and others, 1988).

To determine whether aquatic biota or water is the more sensitive sampling medium for detecting nonionic hydrophobic organic compounds in streams, concentrations in biota and water are compared in this section of the report. Prior to examining actual concentrations of hydrophobic organic compounds in samples from the Yakima River Basin, BCFs from EPA (U.S. Environmental Protection Agency, Region IV, Toxic Substance Spreadsheet, Atlanta, Georgia, written commun., January 25, 1993) were used to calculate concentrations in unfiltered-water samples when concentrations in aquatic biota were at minimum reporting levels. These BCFs were used by EPA to develop national water-quality criteria for human health (U.S. Environmental Protection Agency, 1986). If the concentration of a hydrophobic organic compound in water is calculated to be lower than the minimum reporting level for the water analysis, then aquatic biota is considered to be the more suitable sampling medium for detecting low levels of organic compound.

Many of the laboratory studies used by EPA to develop aquatic-life criteria were made by adding hydrophobic organic compounds to water that contained small concentrations of suspended sediment; therefore, it is uncertain whether bioaccumulation responses resulted from partitioning of the dissolved or suspended compound (Charles Delos, U.S. Environmental Protection Agency, oral commun., 1989). In addition, some of EPA's calculations for BCFs were based on field data. Consequently, EPA's BCFs were used in this report to estimate hydrophobic organic-compound concentrations in unfiltered water.

In this study, BCFs were used to estimate hydrophobic organic-compound concentrations in unfiltered water when the concentrations of these compounds in fish tissue were at the minimum reporting level. The compounds, listed in table 51, are the most frequently detected compounds in biota, bed sediment, filtered water, suspended sediment, and (or) unfiltered water in the Yakima River Basin. If a tissue sample of aquatic plants (typically 0.2-percent lipids in the Yakima River Basin) had a PCB concentration of 0.05 μg/g (minimum reporting level), then the corresponding unfiltered-water sample would have an estimated concentration of 24 ηg/L (table 51). If a tissue sample of largescale sucker (typically 6-percent lipids in the Yakima River Basin) had a PCB concentration of

Table 51. Estimated concentrations of selected organic compounds in unfiltered-water samples assuming concentrations of organic compounds at minimum reporting levels in biological-tissue samples

[Concentrations in unfiltered-water samples were calculated as follows:

$$\text{concentration in unfiltered water} = \frac{\text{concentration in tissue} \times 3{,}000{,}000}{\text{BCF} \times \text{lipid content}},$$

where

concentration in unfiltered water reported in nanograms per liter,
concentration in tissue = minimum reporting level = 0.01 micrograms per gram (μg/g), except for PCB and toxaphene, which equal 0.05 μg/g,
BCF, bioconcentration factor in liters per kilogram of tissue normalized for 3-percent lipids, is from the U.S. Environmental Protection Agency (Region IV Toxic Substance Spreadsheet, Atlanta, Georgia, written commun., October 29, 1991; 1992e),
and lipid content is in percent.

Because of variability in a BCF from station to station and from species to species, an estimated concentration in unfiltered water should be used only as a relative indicator of the actual concentration; DDT+DDE+DDD, dichlorodiphenyltrichloroethane (DDT) plus dichlorodiphenyldichloroethylene (DDE) plus dichlorodiphenyldichloroethane (DDD); PCB, polychlorinated biphenyls; --, not analyzed]

Compound	BCF	Estimated compound concentration in unfiltered water				Minimum reporting level for 1-liter samples of unfiltered water, in nanograms per liter
		Aquatic plant	Sculpin, clams, and fish fillets	Rainbow trout	Largescale sucker, mountain whitefish, and chiselmouth	
		0.2-percent lipids	1.5-percent lipids	3-percent lipids	6-percent lipids	
Organochlorine compounds						
Chlordane (technical)	14,100	11	1.4	0.71	0.36	100
DDT+DDE+DDD	53,600	2.8	.37	.19	.09	1
Dieldrin	4,670	32	4.3	2.1	1.0	1
Endosulfan I	270	560	74	37	18	1
Endrin	3,970	38	5.0	2.5	1.2	1
Heptachlor	11,200	13	1.8	.89	.44	1
Heptachlor epoxide	11,200	13	1.8	.89	.44	1
Hexachlorobenzene	8,690	17	2.3	1.2	.58	--
PCB	31,200	24	3.2	1.6	.78	100
Toxaphene (technical)	13,100	57	7.6	3.8	1.9	1,000
Polycyclic aromatic hydrocarbons						
Fluoranthene	1,150	130	17	8.7	4.3	5,000
Naphthalene	10.5	14,000	1,900	950	480	5,000
Phenanthrene	30	5,000	670	330	170	5,000

0.05 μg/g, the corresponding estimated concentration in unfiltered water would be 0.78 ηg/L. With a PCB concentration of 0.78 ηg/L in unfiltered water, the aquatic-plant sample would have an estimated PCB concentration less than the minimum reporting level of 0.05 μg/g. According to equilibrium-partitioning theory, biological-tissue samples with the highest lipid concentrations tend to bioconcentrate the highest concentrations of nonionic hydrophobic organic compounds in the whole organism (Chiou, 1985). As a result, samples with larger lipid fractions should facilitate the detection of lower levels of nonionic hydrophobic organic compounds in the aquatic environment. Additional factors that control bioaccumulation include diet, physiology, metabolism, age, and trophic level (Steven Goodbred, U.S. Geological Survey, Sacramento, California, written commun., January 1993). These effects of these factors have not been quantified and, therefore, were not included in calculations made in table 51.

The relation between lipid concentration and concentrations of DDT+DDE+DDD and dieldrin in fish tissues is apparent in data collected from the Yakima River Basin in 1989–90 (table 52). Aquatic plants, sculpins, rainbow trout, largescale sucker, and mountain whitefish from the Yakima River Basin have median lipid concentrations of 0.2, 1.5, 3.6, 5.9 and 6.6 percent, respectively (generalized values are used in table 51). As expected, increasing concentrations of DDT+DDE+DDD and dieldrin in whole-fish samples from a station coincide with increasing concentrations of lipids in the samples. For example, at the Yakima River at Cle Elum, DDT+DDE+DDD and dieldrin were detectable in mountain whitefish and not detectable in sculpins; at Cherry Creek, they were detectable in largescale sucker and not detectable in aquatic plants. The ability to detect hydrophobic nonionic organic compounds in species containing low concentrations of lipids, however, can be improved by increasing the mass of sample, thereby increasing the amount of lipids that are extracted for analysis—assuming that (1) lipid content is one of the major controlling factors of bioaccumulation, and (2) the higher lipid concentrations do not make sample clean-up and analysis more difficult.

Minimum reporting levels for organochlorine compounds in water using conventional analyses (1-L samples) are 1 ηg/L for DDT, DDE, DDD, dieldrin, endosulfan I, endrin, heptachlor, and heptachlor epoxide; and 100 ηg/L for technical chlordane and PCB; 1,000 ηg/L for technical toxaphene. Except for endosulfan I, the calculated concentrations of compounds in unfiltered water, based on fish samples containing compound concentrations equal to the minimum reporting level and 6-percent lipids, are less than or equal to the minimum reporting levels for the unfiltered-water analysis (table 51). The relation suggests that analyses of fish samples are more suitable than conventional analyses of water samples for detecting the low-level occurrence of many organochlorine compounds. Additionally, when compared to instantaneous measurements of compound concentrations in water, compound concentrations in fish provide a relatively long-term integration of contaminants in the aquatic environment, including contaminants associated with food source, water, and sediment. With current minimum reporting levels of 100 to 1,000 ηg/L for chlordane, PCB, and toxaphene in unfiltered water, analyses of fish containing 6-percent lipids are more than 100 times more sensitive than analyses of water for the detection of nonionic hydrophobic organic compounds. Furthermore, in future NAWQA studies, minimum reporting levels for tissue samples will be reduced from 0.01 μg/g to 0.0025 μg/g, making fish analyses even more sensitive indicators (Tom Leiker, Chemist, U.S. Geological Survey, Arvada, Colorado, written commun., December 1992). Moreover, tissue analyses provide critical information for directly determining bioavailability to aquatic biota and for determining critical routes of contaminant exposure to fish predators, including humans.

Unfiltered-water samples were collected throughout the basin for conventional analyses (1-L samples) of PCBs and toxaphene, and these compounds were not detected in any of the water samples. In fish samples, however, both PCB and toxaphene were detected in Sulphur Creek Wasteway, Granger Drain, and the Yakima River at RM 79.6 (downstream from Toppenish Creek), RM 72 (upstream from Satus Creek), RM 55 (near Grandview), and RM 29.9 (at Kiona). These detectable concentrations in largescale suckers ranged from 0.10 to 0.90 μg/g, wet weight, of PCBs and from 0.30 to 1.2 μg/g, wet weight, of toxaphene (table 50).

In addition to conventional unfiltered-water analyses in this study, large volumes of water (up to 224 liters) were filtered, extracted, and analyzed in order to reduce minimum reporting levels by one or more orders of magnitude (Rinella, McKenzie, Crawford, and others, 1992). The filtrate was analyzed for organic compounds, and the large mass of suspended sediment was analyzed to determine suspended-

Table 52. Concentrations of dichlorodiphenyltrichloroethane (DDT) plus dichlorodiphenyldichloroethylene (DDE) plus dichlorodiphenyldichloroethane (DDD) and dieldrin in aquatic biota, bed sediment, filtered water, and suspended sediment, Yakima River Basin, Washington, 1988–90

[Aquatic biota collected from 1989–90, bed sediment collected from 1988–90, and filtered water and suspended sediment collected June 25–30, 1989; concentrations in fish and bed sediment reported in micrograms per kilogram of tissue (wet weight) and sediment (dry weight), respectively; concentrations in filtered water and suspended phase reported in nanograms per liter; DDT+DDE+DDD, 4,4'-DDT plus 4,4'-DDE plus 4,4'-DDD; for aquatic-biota analyses, DDT+DDE+DDD also includes 2,4'-DDT, 2,4'-DDE, and 2,4'-DDD; RT, rainbow trout; MW, mountain whitefish; LS, largescale sucker; BS, bridgelip sucker; CM, chiselmouth; C, finer than 2,000-micrometer diameter; F, finer than 62- or 180-micrometer diameter; Trace, compound was detected but not quantified; <, less than; --, not analyzed]

Station number	Station name	Whole fish	Whole sculpins	Plants	Bed sediment	Filtered water	Suspended phase
		DDT+DDE+DDD					
12479720	Jungle Creek	10 RT	< 10	--	< 0.3 C	--	--
12479500	Yakima River at Cle Elum	135 MW	< 10	--	--	0.5	< 0.13
12479750	North Fork Teanaway River	--	< 10	--	.5 F	--	--
12480000	Teanaway River	10 RT	--	--	< .3 C	--	--
12481900	Taneum Creek	--	< 10	--	.4 F	--	--
12483190	South Fork Manastash Creek	--	< 10	--	.6 F	--	--
12483750	Naneum Creek	--	< 10	--	.7 F	--	--
12484480	Cherry Creek	50 LS	--	< 10	54.3 C	Trace	8.0
12484500	Yakima River at Umtanum	310 LS	--	< 10	5 C	Trace	.43
12484550	Umtanum Creek	40 RT	10	--	.4 C	.5	< .26
12487200	Little Naches River	20 RT	--	--	.5 F	--	--
12488250	American River	--	< 10	--	.2 C	--	--
12489100	Rattlesnake Creek	10 RT	< 10	--	.1 C	--	--
12499000	Naches River	540 LS	--	--	--	.7	.28
12500430	Moxee Drain	760 LS	--	20	129 C	15	21
12500437	Wide Hollow Creek	1,050 BS	--	--	209 C	--	--
12500442	Wide Hollow Creek	400 LS	--	10	80.9 C	1.4	1.3
12500450	Yakima River above Ahtanum Creek	--	--	--	< .3 F	2.8	.95
12500900	South Fork Ahtanum Creek	--	< 10	--	1.4 F	--	--
12502500	Ahtanum Creek at mouth	--	110	--	--	1.1	.31
12503950	Yakima River at Parker	690 LS	--	< 10	--	--	--
12505460	Granger Drain	1,940 LS	--	--	97 C	2.6	80
12507508	Toppenish Creek	1,280 LS	--	< 10	18.8 C	1.7	2.6
12507585	Yakima river mile 72	2,320 LS	--	10	--	3.7	6.6
12507594	Satus Creek at Wilson-Charley Canyon	10 RT	< 10	--	--	.6	.008
12508620	Satus Creek at Satus	130 MW	45	10	--	.9	1.6
12508850	Sulphur Creek	4,840 LS	--	--	13.8 C	38	19
12509050	Yakima river mile 55	3,340 LS	--	30	--	6.2	5.9
12509710	Spring Creek	480 CM	--	40	--	9.0	41
12510500	Yakima River at Kiona	2,570 LS	--	20	4.3 C	4.2	11
		Dieldrin					
12479720	Jungle Creek	< 10 RT	< 10	--	< .1 C	--	--
12479500	Yakima River at Cle Elum	10 MW	< 10	--	--	Trace	< .02
12479750	North Fork Teanaway River	--	< 10	--	< .1 F	--	--
12480000	Teanaway River	< 10 RT	--	--	< .1 C	--	--
12481900	Taneum Creek	--	< 10	--	< .1 F	--	--
12483190	South Fork Manastash Creek	--	< 10	--	< .1 F	--	--
12483750	Naneum Creek	--	< 10	--	< .1 F	--	--
12484480	Cherry Creek	20 LS	--	< 10	35 C	3.7	.96
12484500	Yakima River at Umtanum	10 LS	--	< 10	1 C	.93	.04
12484550	Umtanum Creek	10 RT	< 10	--	< .1 C	< .10	< .02
12487200	Little Naches River	< 10 RT	--	--	< .1 F	--	--
12488250	American River	--	< 10	--	< .1 C	--	--
12489100	Rattlesnake Creek	< 10 RT	< 10	--	< .1 C	--	--
12499000	Naches River	< 10 LS	--	--	--	< .33	.44
12500430	Moxee Drain	10 LS	--	< 10	1.7 C	Trace	.86

Table 52. Concentrations of dichlorodiphenyltrichloroethane (DDT) plus dichlorodiphenyldichloroethylene (DDE) plus dichlorodiphenyldichloroethane (DDD) and dieldrin in aquatic biota, bed sediment, filtered water, and suspended sediment, Yakima River Basin, Washington, 1988–90—Continued

Station number	Station name	Whole fish	Whole sculpins	Plants	Bed sediment	Filtered water	Suspended phase
		Dieldrin—Continued					
12500437	Wide Hollow Creek	30 BS	--	--	1.0 C	--	--
12500442	Wide Hollow Creek	10 LS	--	< 10	3.2 C	1.0	0.019
12500450	Yakima River above Ahtanum Creek	--	--	--	< .1 F	.77	Trace
12500900	South Fork Ahtanum Creek	--	< 10	--	< .1 F	--	--
12502500	Ahtanum Creek at mouth	--	< 10	--	--	<1.1	.003
12503950	Yakima River at Parker	< 10 LS	--	< 10	--	--	--
12505460	Granger Drain	70 LS	--	--	1.8 C	Trace	3.6
12507508	Toppenish Creek	20 LS	--	< 10	.5 C	Trace	.055
12507585	Yakima river mile 72	40 LS	--	< 10	--	.98	.12
12507594	Satus Creek at Wilson-Charley Canyon	< 10 RT	< 10	--	--	< .34	< .05
12508620	Satus Creek at Satus	10 MW	5	< 10	--	1.6	.13
12508850	Sulphur Creek Wasteway	170 LS	--	--	4.5 C	35	2.1
12509050	Yakima river mile 55	65 LS	--	< 10	--	5.7	Trace
12509710	Spring Creek	100 CM	--	< 10	--	7.2	.7
12510500	Yakima River at Kiona	83 LS	--	< 10	.2 C	3.6	.08

compound concentrations. Data listed in table 52 show concentrations of DDT+DDE+DDD and dieldrin in aquatic-biota samples collected during 1989–90 and in water samples (dissolved and suspended phases) collected during peak irrigation in June 1989. Results from the comparison of these data are not conclusive because the tissue data and the water-column data were collected on different sampling dates. These data, however, suggest that analyses of tissue samples and (or) large-volume water samples with minimum reporting levels lower than 1 ηg/L can be used to determine the occurrence of DDT+DDE+DDD and dieldrin in the Yakima River Basin. When DDT+ DDE+DDD and dieldrin were detected in fish tissues, the compounds also were detected in either the dissolved or suspended phases in the water column, except in Umtanum Creek where the concentrations in the dissolved and suspended phases were less than the minimum reporting level (table 52).

Increasing concentrations of dissolved DDT+DDE+DDD and dieldrin in Yakima River Basin streams are associated with increasing lipid-normalized concentrations in aquatic biota (fig. 38). This association is not conclusive, because fish samples generally were collected in November 1989 and November 1990 and water samples were collected in June 1989. As a result, these relations assume reasonably uniform concentrations of dissolved compounds during the summer and fall time periods (data collected from major drains and from the Yakima River at Kiona from May-September 1991 show some uniformity, with dissolved concentrations of DDT+DDE+DDD and dieldrin varying within about one order of magnitude [Rinella, McKenzie, Crawford, and others, 1992]). In addition, these relations assume that fish primarily reside in the reach where the water was collected. The relations appear to be consistent with the theory that accumulation of nonionic organic compounds is controlled, in part, by solute partitioning between water and lipid reservoirs of organisms (Smith and others, 1988). In this study, data were not collected to determine the importance of food-chain effects and solute partitioning relative to bioaccumulation. Regardless of whether bioaccumulation is caused primarily by food uptake or solute partitioning, the dissolved phase is a large environmental compartment in streams that appears to be associated with concentrations of nonionic organic compounds in lipids.

BCFs for DDT+DDE+DDD and dieldrin data from the Yakima River Basin are listed in table 53. For comparison, the factors were calculated from compound concentrations measured in unfiltered-water and filtered-water samples collected in June 1989. As expected, BCFs calculated from compound concentrations in filtered water are higher than those calculated from concentrations in unfiltered water. Differences between the unfiltered- and filtered-water BCFs are larger for DDT+DDE+DDD than for dieldrin because a larger portion of the DDT, DDE, and DDD compounds are associated with the suspended sediment (K_{oc} values for DDT, DDE, and DDD are higher

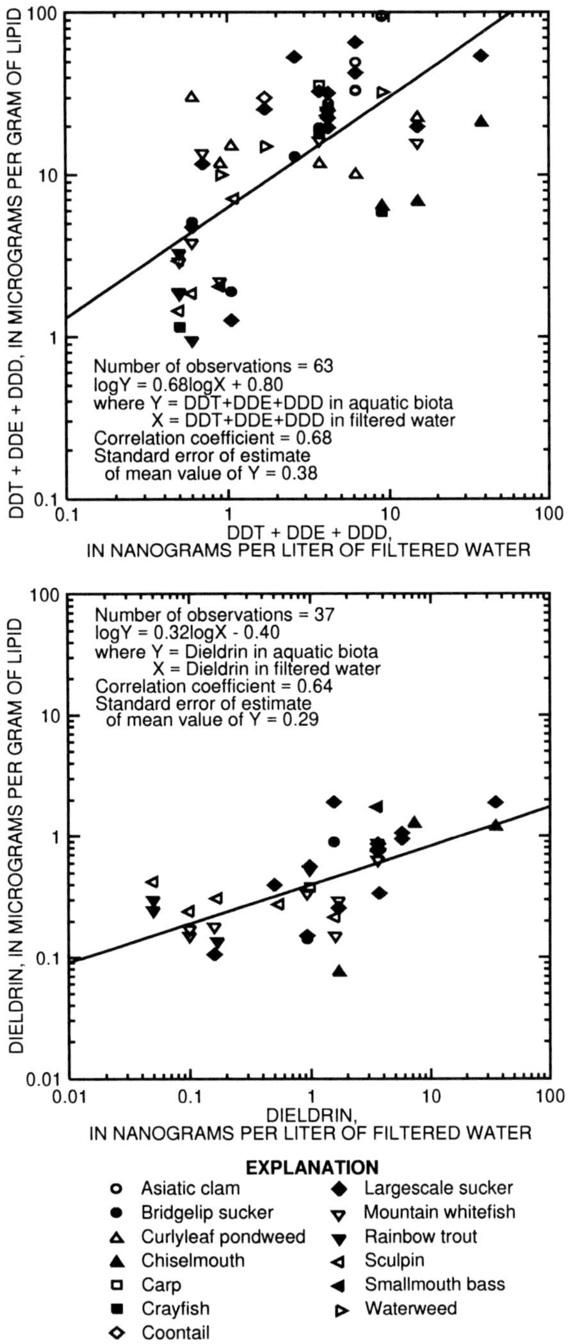

Figure 38. Relation of concentrations of 4,4'-dichlorodiphenyltrichloroethane (DDT) plus 4,4'-dichlorodiphenyldichloroethylene (DDE) plus 4,4'-dichlorodiphenyldichloroethane (DDD) and dieldrin in aquatic-biota and filtered-water samples at selected stations, Yakima River Basin, Washington, 1989–90. (Aquatic-biota samples were collected from 1989–90; filtered-water samples were collected from June 25–30, 1989. Aquatic-biota concentrations are lipid normalized. Analyses of aquatic biota for DDT+DDE+DDD also include 2,4'-DDT, 2,4'-DDE, and 2,4'-DDD. Values less than the minimum reporting level are graphically represented as one-half the minimum reporting level.)

than the K_{oc} value for dieldrin [table 61, at back of report]).

BCFs listed in table 53 were normalized to 3-percent lipid content for comparison to BCFs values listed in EPA's toxic-substance spreadsheet, which also were normalized to 3-percent lipids (U.S. Environmental Protection Agency, Region IV, Atlanta, Georgia, Toxic Substance Spreadsheet, written commun., October 29, 1991). Even after normalizing, BCFs based on concentrations in unfiltered-water samples are variable and range from 2,200 to 410,000 liters per kilogram (L/kg) for DDT+DDE+DDD (median 62,000 L/kg), and from 960 to greater than 61,000 L/kg for dieldrin (median 6,300 L/kg). Similarly, the values for filtered water are variable (table 53). This variability indicates that either equilibrium has not been reached or that factors other than chemical partitioning between water and lipids are controlling bioaccumulation. Some differences in BCFs among species are discernible; for example, BCFs for chiselmouth, crayfish, and aquatic plants appear to be lower than others, but owing to the small number of samples, these differences are not conclusive. The median BCFs for the unfiltered-water samples from the Yakima River Basin are similar to the BCFs used by EPA (table 51) for developing water-quality criteria (53,600 L/kg for DDT, DDE, and DDD; 4,670 L/kg for dieldrin).

Fish samples from sampling stations that had among the highest concentrations of DDT+DDE+DDD and dieldrin in water had among the lowest BCFs (fig. 39). The Yakima River at Kiona, for example, had among the lowest concentrations of DDT+DDE+DDD in unfiltered water (mean concentration of 4.4 ηg/L during May-September 1991) in the lower agricultural valley, and Granger Drain had the highest concentration (60 ηg/L). Accordingly, the highest BCFs for DDT+DDE+DDD were measured in fish samples from the Yakima River at Kiona (greater than 140,000 L/kg); in contrast, the lowest BCFs were measured in fish samples from Granger Drain (6,900 to 28,000 L/kg). Other factors in addition to concentrations of lipids in fish and concentrations of organic compounds in unfiltered water appear to be controlling bioaccumulation. For example, Granger Drain had an average concentration of DDT+DDE+DDD in unfiltered water (60 ηg/L) that was more than 13 times higher than the concentration in the Yakima River at Kiona (4.4 ηg/L). Even after normalizing tissue samples for lipid content, however, the concentration of DDT+DDE+DDD in largescale suckers at Granger

Table 53. Bioconcentration factors for dichlorodiphenyltrichloroethane (DDT) plus dichlorodiphenyldichloroethylene (DDE) plus dichlorodiphenyldichloroethane (DDD) and dieldrin based on aquatic-biota and water samples, Yakima River Basin, Washington, 1989–90

[Aquatic-biota samples collected from 1989–90 and water samples collected from June 25–30, 1989; bioconcentration factor (BCF) reported in liters per kilogram and arbitrarily normalized to 3-percent lipids; when concentrations in filtered or unfiltered water are reported as less-than values, BCFs are reported as greater-than (>) values; --, compound not detected in aquatic-biota samples]

Station number	Station name	Species	BCF for DDT+DDE+DDD, based on analysis of: Filtered water	BCF for DDT+DDE+DDD, based on analysis of: Unfiltered water	BCF for dieldrin, based on analysis of: Filtered water	BCF for dieldrin, based on analysis of: Unfiltered water
12507585	Yakima River at river mile 72 above Satus Creek near Sunnyside	Asiatic clam	160,000	47,000	13,000	11,000
		Asiatic clam	140,000	43,000	13,000	12,000
		Asiatic clam	150,000	45,000	12,000	10,000
		Asiatic clam	140,000	42,000	16,000	14,000
12509050	Yakima River at Euclid Bridge at river mile 55 near Grandview	Asiatic clam	160,000	82,000	4,800	>4,700
		Asiatic clam	160,000	80,000	6,400	>6,300
		Asiatic clam	240,000	120,000	3,000	>3,000
12509710	Spring Creek at mouth at Whitstran	Asiatic clam	320,000	57,000	8,300	7,600
		Asiatic clam	320,000	58,000	6,300	5,800
12510500	Yakima River at Kiona	Asiatic clam	180,000	71,000	3,800	3,700
		Asiatic clam	170,000	66,000	4,300	4,200
		Asiatic clam	200,000	76,000	4,100	4,000
12484440	Cherry Creek above Wipple Wasteway at Thrall	Bridgelip sucker	>27,000	>5,600	3,100	2,400
12484500	Yakima River at Umtanum	Bridgelip sucker	>120,000	>92,000	4,600	4,400
12500442	Wide Hollow Creek at old sewage treatment plant at Union Gap	Bridgelip sucker	150,000	80,000	24,000	23,000
12505460	Granger Drain at mouth near Granger	Bridgelip sucker	150,000	4,700	>8,600	>4,000
12507585	Yakima River at river mile 72 above Satus Creek near Sunnyside	Carp	290,000	87,000	12,000	10,000
12500430	Moxee Drain at Thorp Road near Union Gap	Chiselmouth	13,000	5,600	--	--
12508850	Sulphur Creek Wasteway near Sunnyside	Chiselmouth	16,000	11,000	1,000	960
12509710	Spring Creek at mouth at Whitstran	Chiselmouth	20,000	3,600	5,300	4,800
		Crayfish	12,000	2,200	--	--
12484480	Cherry Creek at Thrall	Largescale sucker	>12,000	>2,500	2,700	2,200
12484500	Yakima River at Umtanum	Largescale sucker	>120,000	>86,000	4,900	4,600
12499000	Naches River near North Yakima	Largescale sucker	490,000	350,000	--	--
		Largescale sucker	510,000	370,000	--	--
12500430	Moxee Drain at Thorp Road near Union Gap	Largescale sucker	39,000	16,000	>2,300	>1,800
12500442	Wide Hollow Creek at old sewage treatment plant at Union Gap	Largescale sucker	310,000	160,000	11,000	10,000
12503950	Yakima River at Parker	Largescale sucker	170,000	120,000	11,000	>8,600
		Largescale sucker	280,000	200,000	--	--
12505460	Granger Drain at mouth near Granger	Largescale sucker	610,000	19,000	>18,000	>8,600
12507508	Toppenish Creek at Indian Church Road near Granger	Largescale sucker	450,000	180,000	>12,000	>11,000
12507585	Yakima River at river mile 72 above Satus Creek near Sunnyside	Largescale sucker	260,000	79,000	17,000	15,000
12508850	Sulphur Creek Wasteway near Sunnyside	Largescale sucker	43,000	28,000	1,600	1,500

Table 53. Bioconcentration factors for dichlorodiphenyltrichloroethane (DDT) plus dichlorodiphenyldichloroethylene (DDE) plus dichlorodiphenyldichloroethane (DDD) and dieldrin based on aquatic-biota and water samples, Yakima River Basin, Washington, 1989–90—Continued

Station number	Station name	Species	BCF for DDT+DDE+DDD, based on analysis of: Filtered water	BCF for DDT+DDE+DDD, based on analysis of: Unfiltered water	BCF for dieldrin, based on analysis of: Filtered water	BCF for dieldrin, based on analysis of: Unfiltered water
12509050	Yakima River at Euclid Bridge at river mile 55 near Grandview	Largescale sucker	210,000	110,000	5,600	>5,500
		Largescale sucker	210,000	110,000	5,000	>5,000
		Largescale sucker	320,000	160,000	5,000	>4,900
12510500	Yakima River at Kiona	Largescale sucker	140,000	54,000	7,200	7,000
		Largescale sucker	230,000	88,000	6,200	6,100
		Largescale sucker	240,000	94,000	6,400	6,300
		Largescale sucker	180,000	69,000	6,500	6,300
12479500	Yakima River at Cle Elum	Mountain whitefish	99,000	99,000	>16,000	>15,000
		Mountain whitefish	160,000	160,000	>18,000	>17,000
12484500	Yakima River at Umtanum	Mountain whitefish	>89,000	>65,000	11,000	10,000
12499000	Naches River near North Yakima	Mountain whitefish	580,000	410,000	>16,000	>7,000
12500430	Moxee Drain at Thorp Road near Union Gap	Mountain whitefish	31,000	13,000	>2,600	>2,000
12507585	Yakima River at river mile 72 above Satus Creek near Sunnyside	Mountain whitefish	130,000	39,000	16,200	14,000
12508620	Satus Creek at gage at Satus	Mountain whitefish	65,000	23,000	2,800	2,600
12510500	Yakima River at Kiona	Mountain whitefish	170,000	67,000	7,200	7,000
		Mountain whitefish	180,000	69,000	5,300	5,200
12484550	Umtanum Creek near mouth at Umtanum	Rainbow trout	110,000	110,000	--	--
		Rainbow trout	73,000	73,000	>73,000	>61,000
12507594	Satus Creek above Wilson-Charley Canyon near Toppenish	Rainbow trout	14,000	13,000	--	--
12484550	Umtanum Creek near mouth at Umtanum	Sculpin	50,000	50,000	--	--
12508620	Satus Creek at gage at Satus	Sculpin	57,000	20,000	4,000	3,700
		Sculpin	14,000	5,200	--	--
12510500	Yakima River at Kiona	Smallmouth bass	160,000	62,000	14,000	14,000
		Coontail	71,000	27,000	--	--
12500430	Moxee Drain at Thorp Road near Union Gap	Curlyleaf	20,000	8,400	--	--
12500442	Wide Hollow Creek at old sewage treatment plant at Union Gap	Curlyleaf	110,000	56,000	--	--
12507585	Yakima River at river mile 72 above Satus Creek near Sunnyside	Curlyleaf	27,000	8,200	--	--
12508620	Satus Creek at gage at Satus	Curlyleaf	110,000	40,000	--	--
12509050	Yakima River at Euclid Bridge at river mile 55 near Grandview	Curlyleaf	48,000	25,000	--	--
12500442	Wide Hollow Creek at old sewage treatment plant at Union Gap	Waterweed	210,000	110,000	--	--
		Waterweed	110,000	56,000	--	--
12509710	Spring Creek at mouth at Whitstran	Waterweed	67,000	12,000	--	--

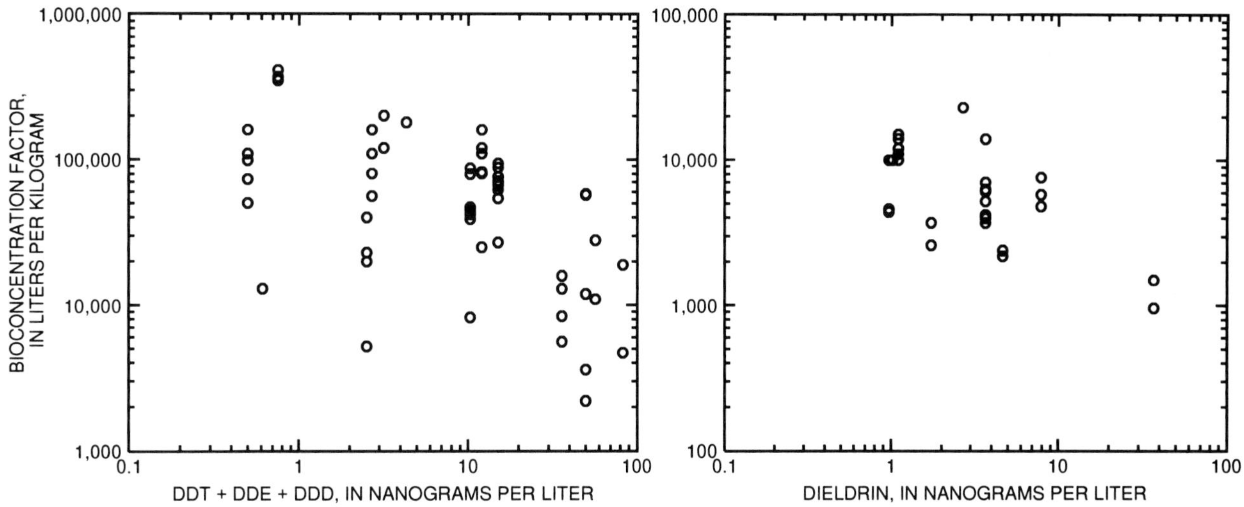

Figure 39. Relation of concentrations of 4,4'-dichlorodiphenyltrichloroethane (DDT) plus 4,4'-dichlorodiphenyldichloroethylene (DDE) plus 4,4'-dichlorodiphenyldichloroethane (DDD) and dieldrin in unfiltered-water samples and bioconcentration factors for selected stations, Yakima River Basin, Washington, 1989–90. (Data reported in table 53. Aquatic-biota samples were collected from 1989–90; unfiltered-water samples were collected from June 25–30, 1989. Bioconcentration factors are normalized to 3-percent lipids and are based on compound concentrations in unfiltered water. Greater-than values for bioconcentration factors are excluded.)

Drain (53 μg/g-lipid) was only twice the mean concentration in largescale suckers at Kiona (26 μg/g-lipid). One possibility is that as concentrations of suspended sediment and organic compounds increase in unfiltered water, concentrations of colloidal particles also increase. Consequently, colloidal particles along with the suspended sediment may sorb organic compounds rendering them less available for uptake in aquatic biota (Larsson and others, 1992). Another possibility is that fish obtain DDT through ingestion of benthic invertebrates that live in and consume bed sediment. Considering that fish samples were collected near the mouths of agricultural return flows, it also is possible that the fish may have migrated in and out of the drains prior to collection. This migration may have resulted in nonequilibrium conditions, which may account for BCF differences observed among stations.

Concentrations in Aquatic Biota and Bed Sediment

Organochlorine compounds detected in aquatic biota were DDT+DDE+DDD, dicofol, dieldrin, cis- and trans-chlordane, oxychlordane, trans-nonachlor, heptachlor epoxide, PCB, and toxaphene (table 47). These organochlorine compounds also were detected in bed sediment, except for PCB and toxaphene, which were not detected, and for trans-nonachlor, oxychlordane, and dicofol, which were not analyzed. Most frequently occurring compounds in biota were DDT+DDE+DDD, dieldrin, cis-chlordane, and trans-nonachlor; most frequently occurring compounds in bed sediment were DDT+DDE+DDD, dieldrin, and endosulfan I (endosulfan I was not analyzed in biota).

Increasing concentrations of DDT+DDE+DDD and dieldrin in bed sediment generally were associated with increasing concentrations of these compounds in aquatic biota (fig. 40). Both of these compounds are hydrophobic and tend to sorb to organic carbon and lipids. As a result of these sorption characteristics, their concentrations in bed sediment and tissues were normalized to organic-carbon and lipid concentrations, respectively. Data for largescale suckers from Cherry Creek do not appear to fit the positive relation, probably because the sampling station at Cherry Creek is near Wipple Creek Wasteway, Wilson Creek, and the Yakima River. Consequently, fish collected at Cherry Creek probably move to different reaches for periods of time and are affected by water quality from several sources. The X = Y line is shown in figure 40 as a reference to aid in determining whether compound concentrations are higher in the organic carbon of the sediment or the lipid phase of the biota. The data show that concentrations of DDT+DDE+DDD were about 10 to 15 times higher in the tissue samples when concentrations in the bed sediment were less than 3 μg/g-organic carbon. Concentrations of dieldrin were about 5 to 10 times higher in the tissue samples when concentrations in the bed sediment were less than 0.1 μg/g-organic carbon. In this study, minimum reporting levels for DDT, DDE, DDD, and dieldrin were 0.01 μg/g for tissue samples and 0.0001 μg/g for bed sediment. In the Yakima River Basin, largescale suckers, mountain whitefish, and chiselmouth had median lipid concentrations of about 6 percent, and bed sediment had organic-carbon concentrations of about 1 percent. After normalizing the minimum reporting levels of these compounds for concentrations of lipids and organic carbon, the minimum reporting levels were 0.01 μg/g-organic carbon for bed sediment and 0.17 μg/g-lipid for tissue samples. Even though lipids in fish tissue accumulate about 10 times higher concentrations of dieldrin and DDT+DDE+DDD than organic carbon in bed sediment (thereby effectively lowering the minimum reporting level by a factor of 10 to 0.017 μg/g), analyses of tissue or sediment should be equally sensitive for detecting the occurrence of these compounds (0.017 μg/g for tissues as compared to 0.01 μg/g for bed sediment). Because future-NAWQA tissue samples will have minimum reporting levels of 0.0025 μg/g or less, analyses of tissue samples for DDT+DDE+DDD and dieldrin may be slightly more sensitive than analyses of bed sediment. Data from the Yakima River Basin generally show that whenever DDT+DDE+DDD or dieldrin were detected in bed sediment, they also were detected in tissue (table 52). Data were not adequate in this study to determine the analytical sensitivity of using fish tissue or bed sediment to detect the occurrence of PCBs and toxaphene; however, in a few samples, these compounds were detected in fish tissue and not in bed sediment.

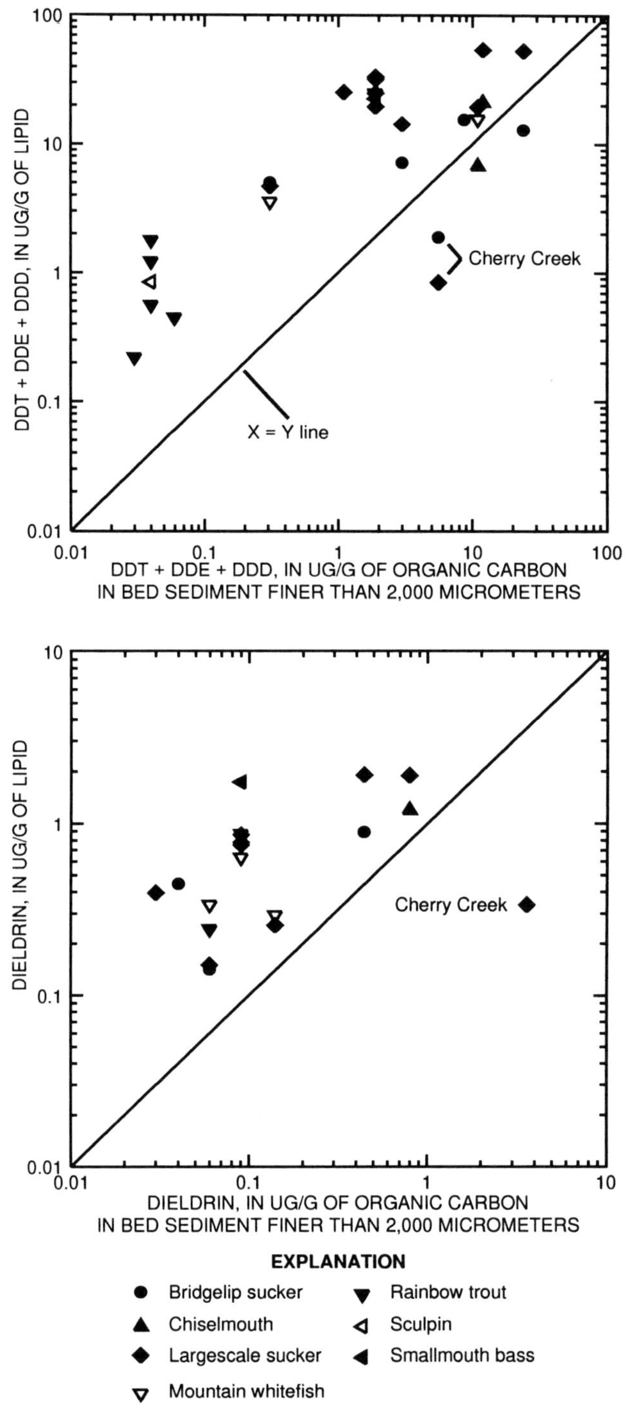

Figure 40. Relation of concentrations of 4,4'-dichlorodiphenyltrichloroethane (DDT) plus 4,4'-dichlorodiphenyldichloroethylene (DDE) plus 4,4'-dichlorodiphenyldichloroethane (DDD) and dieldrin in aquatic-biota and bed-sediment samples from selected stations, Yakima River Basin, Washington, 1988–90. (Aquatic-biota samples were collected from 1989–90; bed-sediment samples were collected from 1988–90. Analyses of aquatic biota for DDT+DDE+DDD also include 2,4'-DDT, 2,4'-DDE, and 2,4'-DDD.)

Concentrations in Bed Sediment and Suspended Sediment

DDT+DDE+DDD and dieldrin were the two compounds most frequently detected in bed- and suspended-sediment samples in the Yakima River Basin. Concentrations of DDT+DDE+DDD and dieldrin in bed and suspended sediment are shown in figure 41 for Cherry Creek, Yakima River at Umtanum, Umtanum Creek, Wide Hollow Creek, Toppenish Creek, Sulphur Creek, and Yakima River at Kiona. These sampling stations were selected, because data were available for normalizing DDT+DDE+DDD and dieldrin to organic-carbon concentrations (data with organic-carbon concentrations reported as "greater

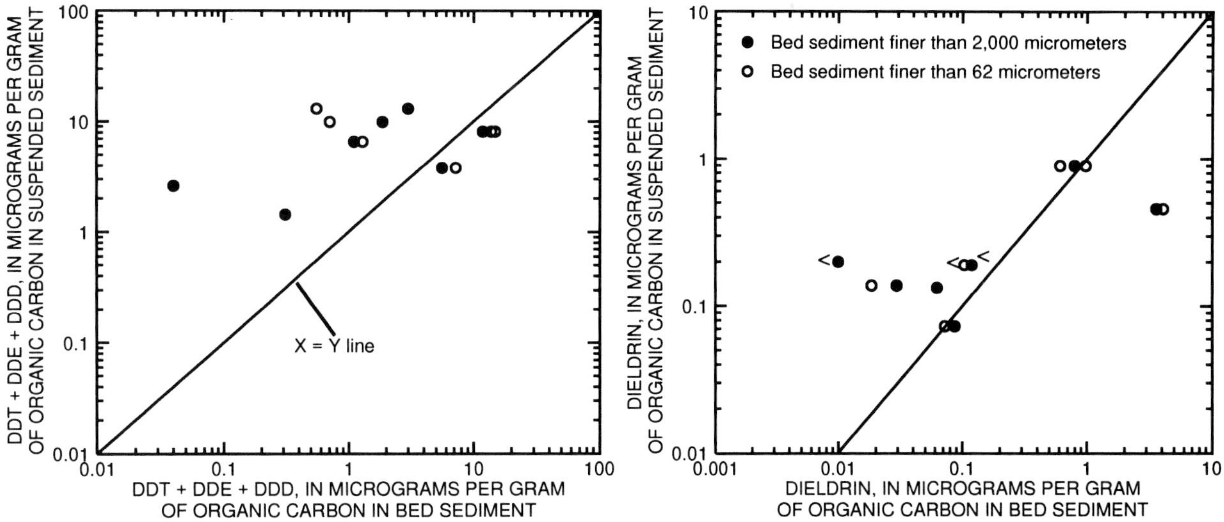

Figure 41. Relation of concentrations of 4,4'-dichlorodiphenyltrichloroethane (DDT) plus 4,4'-dichlorodiphenyldichloroethylene (DDE) plus 4,4'-dichlorodiphenyldichloroethane (DDD) and dieldrin in suspended-sediment and bed-sediment samples at selected stations, Yakima River Basin, Washington, 1988–90. (Suspended-sediment samples were collected from June 25–30, 1989; bed-sediment samples were collected from 1988–90. <, data point represents a value less than the minimum reporting level.)

than" values were excluded from figure 41). Bed-sediment data were collected during 1988–90 and compared to suspended-sediment data collected during peak irrigation in June 1989. Surficial bed-sediment samples (upper 1–2 centimeters) were collected from August through November in order to reflect recent deposition of suspended sediment during irrigation season. Normalized concentrations of DDT+DDE+DDD and dieldrin in the bed and suspended sediment generally were within an order of magnitude of one another, and many of these suspended-contaminant concentrations were higher than those concentrations associated with bed sediment. This pattern is consistent with data collected by Gilliom and Clifton (1990) in the San Joaquin River Basin. The data suggest that normalized contaminant concentrations in bed sediment may be used in conjunction with concentrations of suspended sediment and suspended organic carbon to estimate a lower bounds for concentrations of suspended DDT+DDE+DDD and dieldrin in the water column. These estimates could be used to identify stream locations where water-column data should be collected in a study area to characterize contaminant transport.

RELATIONS BETWEEN WATER-QUALITY CONDITIONS AND PESTICIDE USE

Most of the modern-day pesticides that were applied in the basin and analyzed in this study were detected in Yakima River Basin streams. In addition, several organochlorine compounds that were used historically also were detected in agricultural soils, stream water, suspended sediment, bed sediment, and aquatic biota throughout the basin downstream from agricultural areas. The flushing of pesticides from soil-pore water, the erosion of soil with sorbed pesticides, and the dissolution of pesticides from the soil into the water are major pathways for pesticides to be conveyed from agricultural fields to streams. Controlling excessive applications of pesticides and irrigation water, however, will help to reduce overland runoff and the subsequent erosion, dissolution, and transport of pesticides to agricultural return flows and streams.

Compounds that most frequently exceeded EPA (1986) water-quality criteria or NAS-NAE (1973) guidelines for the protection of freshwater aquatic life include DDT+DDE+DDD, dieldrin, diazinon, and parathion (tables 21 and 30). On the basis of a few analyses, azinphos-methyl and PCB also may be of concern. Except for diazinon, these nonionic compounds have sediment-water partition coefficients (K_{oc}) greater than 1,000 mL/g (table 61). Assuming equilibrium partitioning between soil and soil-pore water in the agricultural fields, more than 90 percent (by weight) of each of these compounds, except diazinon, is partitioned to soils, and the remainder is dissolved in pore water. For compounds that have partition coefficients less than 1,000 mL/g, a larger portion is expected to be dissolved in the soil-pore water in the agricultural field. For example, diazinon has a K_{oc} value of 85 mL/g (table 61); assuming equilibrium partitioning, about 36 percent of the diazinon is dissolved in soil-pore water and about 64 percent is partitioned to soils. Many pesticides used today have K_{oc} values greater than 85 mL/g; therefore, more than 64 percent of the pesticide mass is expected to be partitioned to soils in the agricultural fields. Once the soil becomes suspended in water at suspended-sediment concentrations less than 1,000 mg/L, dissolution is favored. For example, if a pesticide has a K_{oc} value less than 250,000 mL/g (the K_{oc} value for DDT is about 240,000 mL/g), more than 50 percent of the pesticide mass will undergo dissolution from the suspended sediment to the surface water. Therefore, keeping the soil on the fields and out of the streams is an effective method for controlling pesticide transport.

If pesticides are persistent in agricultural soil (for example, with half-lives longer than 3 weeks), then the soil will serve as a contaminant reservoir. During periods of irrigation or storms, these reservoirs gradually will be flushed into ground and surface water by dissolution and (or) erosion. For lipophilic compounds, including DDT+DDE+DDD, gradual flushing of these contaminants from soils may result in bioaccumulations in fish and other aquatic organisms at levels of concern for many years. To determine the importance of pesticide contributions from agricultural sources, research should be done to better define (1) the persistence of pesticides in soils and water, (2) transport processes of pesticides from source areas (agricultural fields, urban areas, and forests) to soil-pore water, ground water, and streams, and (3) the mechanism for uptake by aquatic organisms.

Data Limitations

Although this study represents the most extensive sampling of organic compounds in the Yakima River Basin to date (including hundreds of analyses of unfiltered-water, filtered-water, suspended-sediment, bed-sediment, and aquatic-biota samples), many questions and data needs still exist:

- For currently applied pesticides that were not analyzed in this study, are their concentrations at levels of ecological concern?

  In 1989, more than 100 compounds that were applied were not analyzed. Potential concern for these compounds has not been addressed in this study; however, many of the pesticides that were applied most abundantly in 1989 were analyzed (table 3).

- Considering that water-quality criteria or standards are not available for many pesticides, are any of these pesticides at levels of ecological concern?

  An extensive literature search and (or) additional research could be done to address this question. This issue would best be addressed in terms of acute and chronic toxicity and ecological effects, including the structure and function of aquatic communities.

- Considering that many pesticides are being detected at low concentrations in streams in the Yakima River Basin, are aquatic biota being affected by additive or synergistic effects?

  As of 1995, water-quality criteria had not been established for the occurrence of multiple compounds in water, probably because the array of possible mixtures is nearly infinite. To help address the potential for additive or synergistic effects, (1) bioassay tests could be done using native water and sensitive test organisms that represent those present in the basin or (2) multivariate analysis of contaminants in relation to aquatic community structure and function could be done.

- What is the temporal variability of organic-compound concentrations in water, suspended sediment, bed sediment, and aquatic biota?

  In this study, water samples were collected at eight stations to determine instantaneous pesticide concentrations during (1) irrigation season in May, June, July, August, and September 1988, and June 1989, (2) low flows in November 1988, (3) low or high flows in March 1989, and (4) high flows at Yakima River at Kiona in December 1989.

  Water-column data indicate that high pesticide concentrations occur during peak irrigation following pesticide applications and during periods of storm runoff from the agricultural fields. Although the data were collected monthly during the intensive irrigation period, the temporal data coverage is inadequate for determining (1) maximum pesticide concentrations that occurred in 1988 and 1989, and (2) the duration of the maximum pesticide concentrations. For example, assume that the highest pesticide concentrations in a stream occurred from July 15–31, 1988, during a period of intensive irrigation immediately following pesticide application. If instantaneous samples were collected prior to July 15, 1988, and then sometime later in August 1988, the measured instantaneous pesticide concentrations may represent only a small fraction of the actual peak concentrations. As a consequence, the pesticide concentrations reported in this study do not necessarily represent worst-case conditions because the data are not continuous over time.

Additionally, water-quality criteria are based on average compound concentrations over a specified time period. For example, the chronic-toxicity criterion for the protection of freshwater aquatic life indicates that concentrations of DDT+DDE+DDD should not exceed an average concentration of 1 ηg/L over a 24-hour period. Most water-quality data from the Yakima River Basin, however, are instantaneous data collected over a 15-to-60-minute period, so that the 24-hour period variability is unknown. Consequently, instantaneous data are not suitable for determining whether violations of criteria have occurred.

In this study, temporal variability of contaminant concentrations in bed sediment and aquatic biota was not determined.

- How do analytical recoveries for determining concentrations of sediment-sorbed pesticides affect results?

  Currently, analytical recoveries are unknown for the multitude of sediment types and environmental settings. Sediment samples can be spiked with known analyte concentrations; however, analytical recoveries of these spiked analytes are not necessarily representative of recoveries of pesticides that have gradually sorbed deep into the sediment matrix over time.

- What is the spatial distribution of sources of DDT+DDE+DDD in the Yakima River Basin?

  Additional soil samples will have to be collected and analyzed to answer this question. With this information, control efforts could focus on areas of highest contamination.

- How well do whole-fish samples represent contaminant concentrations detected in the edible fish portions?

  Most fish samples collected for this study were whole fish. These data are appropriate for evaluating general water-quality questions concerning the occurrence of hydrophobic organic compounds in the Yakima River Basin; however, they may not be representative of contaminant concentrations in edible fish portions. Concentrations of some compounds in resident whole-fish samples suggest increased cancer risk levels for human consumption in excess of 1 in 1,000,000 people. Additional data are needed to properly evaluate whether edible fish portions also pose a significant risk.

Thus, although this study has answered many questions concerning organic compounds in the Yakima River Basin, many questions are still unresolved.

Exceedances of Water-Quality Criteria and Guidelines

To meet water-quality criteria and recommendations for the protection of freshwater aquatic life in streams in the Yakima River Basin, concentrations of 15 pesticides (chlordane, DDT+DDE+DDD, dieldrin, endosulfan I, endrin, PCB, toxaphene, azinphos-methyl, diazinon, disulfoton, ethion, malathion, parathion, phosphamidon, and 2,4-D) would have to be reduced at one or more stations. In this section of the report, the occurrence of DDT+DDE+DDD, dieldrin, diazinon, and parathion are discussed, because these pesticides most often exceeded water-quality criteria and guidelines during 1988–91 (table 62, at back of report).

Using synoptic data from June 1989, concentrations of selected pesticides were compared to water-quality criteria and guidelines for the protection of freshwater aquatic life (table 54). Exceedances for dieldrin and parathion were observed at relatively few stations. For diazinon, 14 stations in the lower basin, downstream from the city of Yakima, had exceedances. To meet the guideline for the protection of freshwater aquatic life, the median reduction in the diazinon concentration at these stations would have to be 76 percent, with reductions ranging from 31 to 98 percent. Concentrations of DDT+DDE+DDD at Cherry Creek and all major tributary and main-stem stations in the lower basin exceeded the water-quality criterion of 1 ηg/L for the protection of freshwater aquatic life. To meet the water-quality criterion, concentrations of DDT+DDE+DDD would have to be reduced by 95 percent or more at stations that had the highest concentrations of suspended sediment (Eastside tributaries and South Drain; table 60, at back of report). Concentrations of DDT+DDE+DDD at main-stem stations downstream from Union Gap would have to be reduced by 90 percent or more.

The importance of controlling soil erosion in order to control the transport of hydrophobic pesticides can be seen in the relation between concentrations of DDT+DDE+DDD and suspended sediment at Moxee Drain (fig. 12). To meet the water-quality criterion in Moxee Drain, concentrations of suspended sediment would have to be reduced to less than 50 mg/L.

To determine the likelihood of reducing concentrations of DDT+DDE+DDD, dieldrin, parathion, and diazinon during irrigation season, data collected in November 1988 were examined from three major agricultural return flows and the Yakima River at Kiona (table 55). November data were selected because they represent conditions when soil erosion and overland runoff are minimal; thus, simulating the effects of controlling erosion and overland runoff during irrigation season. If water-management decisions are made to reduce overland runoff, ground-water quality probably should be monitored to ensure that pesticide concentrations are not increasing over time. Assuming that instream pesticide concentrations during irrigation season can be reduced to the concentrations measured in Nov-ember 1988, parathion and diazinon would meet the water-quality guidelines (table 55). Dieldrin would meet the criterion in the main stem and Moxee Drain, and DDT+DDE+DDD would meet the criteria in the main stem. In the other agricultural return flows, however, the criteria would be exceeded by as much as 3 ηg/L for dieldrin and 18 ηg/L for DDT+DDE+DDD.

Minimizing Pesticide Transport in Streams

Pesticide use and transport depend on basin characteristics that include hydrology, climate, basin slope, soil type, and water and land use (crop types, farming practices, and irrigation methods). The fact that these characteristics vary widely from basin to basin makes it difficult to establish a set of guidelines for controlling pesticide contamination of streams in all basins. Therefore, if the adverse effects of pesticide use on streams are to be minimized, pesticide users will need to develop a general understanding of how pesticides are transported to streams. For this purpose, a water-quality training program has been established by the U.S. Department of Agriculture for controlling erosion and sediment pollution from irrigation (Hedlund, 1992).

Pesticide use

Data from the Yakima River Basin indicate that increases in pesticide use generally coincide with increases in the number of pesticide detections in streams; therefore, reducing or discontinuing (for example, organic farming) pesticide use would be expected to reduce pesticide occurrence in streams. In an integrated pest-management program, pesticides are not applied automatically on a fixed time schedule. Instead, applications are based on extensive field observations relative to crop needs. A sufficient number of pests must be observed to warrant pesticide application. Pesticide use can be minimized by carefully coordinating pesticide applications with crop needs, pest cycles, and expected climatological conditions—for example, rainstorms (U.S. Environmental Protection Agency, 1988a). Crop rotation, as well as selection of pest-resistant varieties of crops, also can be used to minimize pesticide use. Insect populations that attack specific crops can be reduced in a field by planting non-host crops in a crop-rotation program. Mechanical cultivation can be used to control weeds and to help reduce herbicide use; however, mechanical disturbance of soil aggregates also enhances erosion and thereby promotes transport of soil-sorbed pesticides.

Table 54. Reductions that are needed in selected pesticide concentrations to meet water-quality criteria or guidelines for the protection of freshwater aquatic life, Yakima River Basin, Washington, June 25–30,1989

[Two well stations, with all concentrations below the criteria or guidelines, were excluded from this listing; DDD+DDE+DDD, dichlorodiphenyltrichloroethane (DDT) plus dichlorodiphenyldichloroethylene (DDE) plus dichlorodiphenyldichloroethane (DDD); B, concentration is lower than water-quality criteria or guideline; --, not applicable; >, greater than; see table 60 for listing of instantaneous concentrations and loads]

Yakima River mile (RM)	Station number	Station name	Streamflow, in cubic feet per second	Reduction needed in the pesticide concentration to meet water-quality criteria or guidelines, in percent			
				DDT+DDE+ DDD[1]	Dieldrin[1]	Diazinon[2]	Parathion[1]
185.6	12478200	Cooper River at Salmon LaSac	229	B	B	B	B
183.1	12479500	Yakima River at Cle Elum	3,780	B	B	B	B
147.0	12484100	Wilson Creek above Cherry Creek	144	B	B	B	B
147.0	12484480	Cherry Creek at Thrall	126	88	60	B	B
140.4	12484500	Yakima River at Umtanum	3,630	B	B	B	B
139.8	12484550	Umtanum Creek near mouth	1	B	B	B	B
116.3	12496510	Pacific Power and Light Co. Wasteway	--	44	B	B	B
116.3	12496511	City of Yakima—Finish water	--	B	B	B	B
116.3	12499000	Naches River near North Yakima	1,190	B	B	B	B
107.4	12500445	Wide Hollow Creek near mouth	33	63	B	B	B
107.3	12500430	Moxee Drain at Thorp Road	65	97	B	98	B
107.3	12500450	Yakima River above Ahtanum Creek	3,630	69	B	B	B
106.9	12502500	Ahtanum Creek at Union Gap	23	29	B	B	B
86.0	12505350	East Toppenish Drain at Wilson Road	53	92	B	68	B
83.2	12505410	Sub 35 Drain at Parton Road	56	90	B	64	B
82.8	12505460	Granger Drain at mouth	37	99	47	31	B
82.6	12505510	Marion Drain at Indian Church Road	71	78	B	84	B
80.4	12507508	Toppenish Creek at Indian Church Road	39	77	B	71	19
72.0	12507585	Yakima River at river mile 72	1,070	90	B	74	B
69.6	12507594	Satus Creek above Wilson-Charley Canyon	12	B	B	B	B
69.6	12508620	Satus Creek at gage at Satus	124	60	B	59	93
69.3	12508630	South Drain near Satus	82	95	B	94	43
61.0	12508850	Sulphur Creek Wasteway	259	98	95	86	B
55.0	12509050	Yakima River at Euclid Bridge at river mile 55	1,560	92	>67	92	57
41.8	12509710	Spring Creek at mouth	47	98	76	88	B
41.8	12509829	Snipes Creek at mouth	50	95	69	72	B
29.9	12510500	Yakima River at Kiona	1,500	93	49	77	B

[1]Relative to chronic-toxicity water-quality criteria (U.S. Environmental Protection Agency, 1986).

[2]Relative to recommended maximum concentrations (National Academy of Sciences-National Academy of Engineering Committee on Water Quality Criteria, 1973).

Pesticide Characteristics

Use of pesticides that include one or more of the following characteristics would minimize transport: (1) short half-lives, ideally less than 3 weeks; (2) water solubilities lower than 30 mg/L, to minimize flushing of dissolved pesticides from soils; and (3) K_{oc} values higher than 300 to 500 mL/g, to increase the propensity of the pesticide to remain sorbed to agricultural soils (see table 61, at back of report, to determine which pesticides have these characteristics). Use of pesticides with short half-lives in both soils and water likely would minimize the occurrence of pesticides in ground water and streams. Pesticides with soil half-lives longer than 3 weeks generally will be transported into ground water or surface runoff before degrading (Becker and others, 1989). Additionally, degradation products should have low or no toxicity. If large quantities of a pesticide are used in a subbasin, the pesticide might still be transported and detected in streams regardless of the compound's half-life. For example, a large quantity of 2,4-D is used in the Yakima River Basin. Even though its half-life in soils is about 1 week, the quantity that reaches the agricultural return flows and streams after degradation results in concentrations well above minimum reporting levels. In the Yakima River Basin, however, 2,4-D generally

Table 55. Concentrations of suspended sediment and selected pesticides in unfiltered-water samples from three major agricultural-return flows and the Yakima River at Kiona, Yakima River Basin, Washington, 1988

[Concentrations and criteria reported in nanograms per liter, except for suspended sediment which is in milligrams per liter; DDT+DDE+DDD, 4,4'-dichlorodiphenyltrichloroethane (DDT) plus 4,4'-dichlorodiphenyldichloroethylene (DDE) plus 4,4'-dichlorodiphenyldichloroethane (DDD)]

Constituent	Chronic-toxicity water-quality criteria for freshwater aquatic life[1]	Concentrations in unfiltered-water samples				
		May	June	July	August	November
Moxee Drain						
Suspended sediment	--	134–143	296	443–607	157	58
DDT+DDE+DDD	1	8–9	21	31–76	14	3
Dieldrin	1.9	1	4	6–8	2	<1
Diazinon	[2]9	<10	10	130–630	30	<10
Parathion	13	<10	<10	10–100	<10	<10
Granger Drain						
Suspended sediment	--	205	526	421–432	282	62
DDT+DDE+DDD	1	41	40	96–122	36	19
Dieldrin	1.9	6	9	17–30	8	5
Diazinon	[2]9	<10	<10	10–30	10	<10
Parathion	13	<10	<10	<10	<10	<10
Sulphur Creek Wasteway						
Suspended sediment	--	70	204	99–128	67–83	19
DDT+DDE+DDD	1	12	27	35	13–14	10
Dieldrin	1.9	2	5	14–15	6	4
Diazinon	[2]9	<10	<10	10–20	<10–10	<10
Parathion	13	<10	10	<10	<10	<10
Yakima River at Kiona						
Suspended sediment	--	28	35	22	35	10
DDT+DDE+DDD	1	5	3	1	2	1
Dieldrin	1.9	1	2	3	2	<1
Diazinon	[2]9	<10	<10	250	10	10
Parathion	13	<10	<10	60	<10	<10

[1]24-hour average values, except parathion which is a 4-day average value (U.S. Environmental Protection Agency, 1986).

[2]Guideline (at any time or any place) for the protection of freshwater aquatic life (National Academy of Science-National Academy of Engineering Committee on Water Quality Criteria, 1973).

does not exceed NAS-NAE (1973) guidelines for the protection of freshwater aquatic life.

Pesticides with water solubilities higher than 30 mg/L have a high propensity to be flushed from soils and to be transported in solution to ground water and streams (U.S. Environmental Protection Agency, 1988a). Use of pesticides with solubilities lower than 30 mg/L is expected to minimize the leaching of pesticides to ground water and streams.

Nonionic pesticides with K_{oc} values higher than 300–500 mL/g tend to be partitioned mostly to soils in the agricultural fields. Application of these pesticides will help to reduce leaching to ground water. Their transport to streams in the Yakima River Basin also could be reduced by controlling soil erosion from fields. The field also serves as a terrestrial habitat for many species; controlled application of these pesticides can help to minimize bioaccumulation in terrestrial species. Another method to insure that pesticides will not be washed from plant foliage is to use a "sticker" agent during application. (Steven L. Goodbred, U.S. Fish and Wildlife Service, Sacramento, California, written commun., February 1993). Sorption of nonionic pesticides to soil particles may be enhanced by increasing the content of organic carbon in the soils. It is advisable, however, to keep the amount of active pesticide and the amount of pesticide sorbed to soil in balance, because organically enriched soils may bind the active ingredients, thus requiring increased applications of certain pesticides.

Irrigation

Pesticide transport from fields to streams could be reduced by minimizing the amount of irrigation immediately following pesticide applications. This reduction in irrigation water will reduce the potential for (1) soluble pesticides to leach from the soils, (2) soluble pesticides to be washed from the crop foliage, and (3) soil-sorbed pesticides to erode to the streams. Soil moisture could be monitored to determine the minimum volume of irrigation water required for crop success. Reduced irrigation will help to minimize pesticide wash-off from the fields, thus increasing the length of time for the pesticides to degrade in the soils. Coincident crop needs for pesticides and irrigation during the warm summer months places water quality and agricultural production in a delicate balance.

Methods of irrigation also can affect pesticide transport. Four types of irrigation methods are used commonly in the Yakima River Basin: rill (also known as furrow), flood, sprinkle, and drip irrigation (also known as trickle). In rill irrigation, water flows between crop rows in rills (small channels or furrows); in flood irrigation, water is retained in the fields within dikes. In these methods, irrigation water often is overapplied, and excess water infiltrates to the ground water, carrying soluble pesticides with it (U.S. Environmental Protection Agency, 1988a). In addition, flowing water in the rills or furrows promotes erosion of freshly cultivated soils and soil-sorbed pesticides. To minimize erosion and reduce pesticide transport of soil-sorbed compounds, surge systems can be installed to control water flow in the rills. Initially, water surges into a rill for a period of time, so that the water is permitted to flow and saturate soils part way down the row. After the second or third surge, irrigation water reaches the end of the row and then the flow of water is stopped to conserve water use and minimize overland runoff and erosion.

In sprinkler irrigation, water is sprayed into the air through nozzles. Sprinkler irrigation promotes pesticide leaching when the irrigation water washes pesticides off the crop foliage and weed surfaces (U.S. Environmental Protection Agency, 1988a). To minimize this washing effect, sprinkler irrigation could be delayed until after adequate time has elapsed for the pesticide to perform its function and degrade.

In drip irrigation, water is applied slowly through spray or drop emitters that are located along a water-delivery pipe. These pipes usually are positioned near the base of the plants, above or below the land surface. Buried drip systems are designed to wet a portion of soil near the roots of the plants in order to minimize the overapplication of irrigation water. In addition to irrigation water, drip and sprinkle systems can be used to distribute nutrients and some pesticides to the plants. The use of buried drip systems minimizes erosion, overland runoff, and weed growth.

Conservation of water helps to reduce the leaching of pesticides into the ground water; however, it is necessary to apply a minimum leaching fraction to prevent accumulation of salts in the root zone of the soil. In an ideal design, the drip system would (1) efficiently irrigate crops by conserving water, (2) minimize pesticide transport by reducing tail-water flow to surface water and leaching to ground water, and (3) maintain an adequate downward seepage of irrigation water into soils to minimize future salinity problems. Drip systems require irrigation water with low concentrations of suspended sediment to avoid clogging the small openings in the drop emitters. Larger openings, however, are used in pulse drip systems. To reduce concentrations of suspended sediment, irrigation water may require expensive sand filtration prior to distribution.

Pesticide Transport

Before pesticides are applied, it is important of the user to understand how pesticides are routed from the fields to the receiving streams. Relevant questions or concerns include

- Are streamflows sufficiently large and are pesticide concentrations sufficiently low in the receiving streams to dilute pesticide concentrations below levels of ecological concern?
- Is there a depositional reach (low-stream-velocity reach), wetland area, or sediment-detention pond between the field and receiving stream which may help reduce loadings of sediment-sorbed pesticides to streams?

Further research is needed to determine relations between pesticide concentrations and particle sizes of suspended sediment. For example, when pesticides are associated primarily with the finer-sized sediment, wetland areas or sediment-detention ponds may not be effective for controlling pesticide transport. Even it wetlands could serve to reduce pesticide contamination in streams, the fact that a variety of animal species attracted to wetlands might bioaccumulate contaminants and be affected could mitigate against using wetland for that purpose. The use of a wetland to control pesticide transport is an important research topic for future study (Steven L. Goodbred, U.S. Fish and Wildlife Service, Sacramento, California, written commun., February 1993).

Location of sediment-detention ponds in areas of impervious soils would reduce the opportunity for ground-water contamination. Periodic dredging of the ponds and disposal of dredged material in accordance with local, State, Tribal and Federal laws also would reduce the potential for contamination.

- Is irrigation water reused between the field and the receiving stream? As a result of reuse, will pesticide concentrations and loads be larger or smaller?

  For example, if suspended-sediment concentrations increase, concentrations of sediment-sorbed pesticides would likely increase. In contrast, water reuse may increase travel time to the stream, so the pesticide has greater opportunity to degrade or the sediment-sorbed pesticides have greater opportunity to settle out in the fields.

- Does the soil in the field have a high infiltration rate so that soluble pesticides would likely leach to the ground water?

- Are soil types and field slopes conducive to erosion?

  For example, rill irrigation of freshly tilled soils promotes erosion of sediment-sorbed pesticides, especially when slopes exceed 3 percent (U.S. Environmental Protection Agency, 1988a). The practice of mulching furrows with straw has been shown to reduce soil erosion from the field (North Yakima Conservation District, 1991). In areas of sprinkler irrigation, vegetative cover may be planted between rows of crops to control erosion.

Pesticide-Application Methods

Common types of pesticide application include aerial spraying, ground application, and chemigation (pesticide application through irrigation systems). Aerial applications are made by plane or helicopter, ground applications are made using land vehicles, and chemigation entails mixing the pesticide with irrigation water.

Aerial applications might not be uniform because some of the pesticide might drift away from target fields by wind or pilot error (U.S. Environmental Protection Agency, 1988a), and might deposit directly into the streams. Sprays deposited on foliage make pesticides vulnerable to wash off during sprinkler irrigation or precipitation. Problems associated with drift can be minimized by applying the pesticides when the winds are calm, applying the pesticide at low altitudes, and adjusting spray drop sizes to account for air turbulence (U.S. Environmental Protection Agency, 1988a). In ground applications and chemigation, pesticides can be applied to the foliage or soil surfaces. Chemigation systems could promote leaching of dissolved pesticides into ground water and streams when the fields are overirrigated. For example, when using rill irrigation, more than 50 percent of the applied water runs off as tail water. This problem could be alleviated by using chemigation in sprinkler or subsurface drip systems to conserve water and minimize pesticide use.

SUMMARY

The Yakima River Basin in Washington was one of four surface-water study areas selected by the U.S. Geological Survey in 1986 to test and refine concepts for implementing a National Water-Quality Assessment program. As part of this study, water, suspended sediment, streambed sediment, aquatic biota (fish, crayfish, mollusks, and plants), and agricultural soils were collected and analyzed for a variety of pesticides and other organic compounds that have been and (or) continue to be used or produced in the basin.

This report presents monitoring results on the occurrence, distribution, transport, fate, and potential effects of organic compounds at about 100 stations in the Yakima River Basin during 1987–91. More specifically, this report provides information about the major natural and human factors that affect (1) the spatial distribution of organic-compound concentrations in water, suspended sediment, bed sediment, aquatic biota, and soils, (2) the seasonal variations of organic-compound concentrations and loads in streams, (3) the suitability of surface water for the protection of aquatic biota and human health, based on Federal and State water-quality guidelines, criteria, and regulations, and (4) the patterns of organic-compound concentrations which are common among water, sediment, and aquatic biota. The implications of major findings as they relate to water-resource monitoring and management also are discussed in this report. Pesticide and other organic-compound data that are interpreted in this report have been published in Rinella, McKenzie, Crawford, and others (1992).

The Yakima River Basin is one of the most intensively irrigated areas in the United States. As a result, the occurrence and distribution of agricultural pesticides in surface water have been emphasized in this study. Estimates for 1989 indicate that about 3 million kilograms (3,300 tons) of active pesticide ingredients were applied in the basin to increase the production and quality of agricultural crops. Although pesticide use in the basin has been and continues to be extensive, relatively few historical water-quality samples have been collected to determine the spatial distribution and temporal variation of pesticide concentrations in the aquatic environment. the small amount of data available from prior to 1987 shows that concentrations of many pesticides in water were less than the conventional minimum reporting levels. These minimum reporting levels, however, were too high for water managers to adequately assess stream-quality conditions because the reporting levels sometimes exceeded levels of environmental concern. Consequently, research procedures were developed in this study to reduce minimum reporting levels by an order of magnitude or greater by analyzing large sample volumes, up to 120 L for filtered water and 224 L for the suspended phase. The use of these new techniques provides water managers with important information concerning the occurrence, transport, and fate of these compounds in the Yakima River Basin.

In 1989, about 180 pesticides were applied in the Yakima River Basin. Fifty-four of these compounds were analyzed in this study, and 43 of the 54 compounds (80 percent) were detected in soil, bed sediment, suspended sediment, water, and (or) aquatic biota at one or more sampling stations at trace or quantifiable concentrations. Including other organic compounds associated with industrial and urban activities, as well as persistent pesticides that were used historically, more than 110 compounds were detected in Yakima River Basin streams during 1987–91.

Temporal variation of pesticide concentrations in streams—Early in the study, seasonal fixed-station data were collected to determine the temporal variation of pesticide concentrations in Yakima River Basin streams. These data were used to design and conduct a synoptic study to determine the spatial distribution, sources, and transport of pesticides during a period when compound concentrations and loads were expected to be high. The seasonal data were collected from eight stations (one forested, pristine station; five agricultural return flows; one urban station; and one main-stem station) in May, June, July, August, and November 1988, and March 1989.

In 1988, concentrations of soluble and relatively insoluble pesticides generally began to increase in agricultural runoff in June in response to increased irrigation following springtime pesticide applications. This pattern of occurrence in the Yakima River Basin is similar to the runoff pattern of herbicides in the Mississippi River and other streams in the midwestern United States, where the highest concentrations occurred in response to flushing by late-spring and early summer rainfall immediately following pesticide applications. Unlike the midwestern streams, rainfall in the Yakima River Basin is low during the spring and summer periods following pesticide applications. As a result, irrigation at or near peak water use flushes relatively large pesticide loads to streams in June and July. In addition, large pesticide loads are flushed to the streams in the Yakima River Basin by storm runoff from agricultural fields, as observed in March 1989.

The highest concentrations of suspended sediment also occurred in June and July 1988 and during storm runoff in March 1989; thereby suggesting that eroding soil is associated with the transport of sorbed pesticides from fields to streams during periods of overland flow. Theoretical equilibrium-partitioning calculations indicate that hydrophilic and hydrophobic compounds with sediment-water partition coefficients (K_{oc}) larger than about 85 mL/g (milliliters per gram) are sorbed mostly to soils prior to soil erosion (for perspective, (2,4-dichlorophenoxy) acetic acid [2,4-D] is a hydrophilic compound with a K_{oc} of 20 mL/g, and 4,4'-dichlorodiphenyltrichloroethane [DDT] is a hydrophobic compound with a K_{oc} of about 240,000 mL/g). For example, diazinon is highly soluble in water and has a K_{oc} of 85 mL/g. For soils in the Yakima River Basin that contain 1-percent organic carbon, about 64 percent of the diazinon is theoretically sorbed to the soil and 36 percent is associated with soil-pore water. As soils and the soil-pore water are flushed and (or) eroded by the irrigation water or precipitation during overland flow, both the sorbed and soluble phases of diazinon are transported to the streams. Once the soil is suspended in surface water and the mass of soil decreases from the high concentrations in the fields (about 2,000,000 milligrams of dry soil per liter of moist soil) to lower concentrations in agricultural return flows (typically 500 milligrams per liter or less), equilibrium partitioning favors desorption of hydrophilic compounds into the dissolved phase. For example, more than 99 percent of the total diazinon mass will equilibrate into the dissolved phase in the water column.

In the Yakima River Basin, pesticide concentrations decreased by the end of irrigation season because of decreases in pesticide application, erosion, and overland runoff. Also, pesticides that were applied early in the growing season have had time to undergo chemical and biological degradation and volatilization throughout the summer. Nonetheless, compounds that are persistent in soils will continue to be transported throughout the year, especially during storm runoff from agricultural areas; however, annual precipitation in the agricultural areas is typically small (less than 10 inches).

Spatial distribution of pesticide concentrations in streams—Pesticide concentrations in the Yakima River are controlled primarily by dilution from streamflow from the upper basin and pesticide contributions from agricultural areas. During irrigation season, streamflow throughout the basin is regulated extensively by storage reservoirs and irrigation diversions. Upstream from the Roza Canal diversion (river mile [RM] 127.9), streamflow in the Yakima River is augmented by releases from three large storage reservoirs, resulting in main-stem flows that typically range from 3,000 to 4,000 ft^3/s (cubic feet per second). In this upper reach, agricultural return flows from the Kittitas Valley convey pesticides to the Yakima River, where flow augmentation substantially dilutes the pesticide concentrations in the main stem. Downstream from the Wapato and Sunnyside Canal diversions (RMs 106.7 and 103.8, respectively) streamflow in the Yakima River is reduced to only about 200 ft^3/s during irrigation season. Many agricultural return flows then drain into the lower Yakima River and typically account for as much as 80 percent of the main-stem flow near the downstream terminus of the basin. As a result, concentrations of many pesticides increase substantially in the lower main stem as the proportion of agricultural return flow increases.

In June 1989, when pesticide concentrations were expected to be high, water samples were collected from 29 stations in the Yakima River Basin to determine the spatial distribution of hydrophilic and hydrophobic pesticide concentrations and loads. Twenty of the most frequently detected compounds in this synoptic sampling are listed in table 56. With the exception of chlordane, DDT+DDE+DDD, dieldrin, and prometon, these compounds were also among the most abundantly used compounds in the basin in 1989, which probably accounts for their frequent detections. Although the use or sale of DDT, dieldrin, and chlordane was banned or restricted in 1972, 1974, and 1987, respectively, these persistent compounds were detected frequently at trace or quantifiable concentrations (28 stations for DDT+DDE+DDD, 20 stations for dieldrin, and 11 stations for chlordane). The widespread occurrence of these pesticides suggests broad historical usage throughout the basin. Use of prometon in the basin was minimal; however, it has a relatively long half-life in soils (up to 500 days), which probably accounts for its increased number of detections.

The predominant source area for many pesticides in the basin is the East-side area (area east of the Yakima River from Moxee Valley downstream to Benton City near Kiona including the Moxee, Granger, Sunnyside, and Whitstran areas) (table 57). This area has the largest acreage of irrigated land and generally receives the largest application of pesticides. Owing, in part, to the highly erosive soil characteristics of the Warden-Esquatzel association, steep subbasin slopes, and rill irrigation of tilled crops, the suspended-sediment load in June 1989 to the Yakima River from the East-side area was five or more times larger than loads from the other areas. Similarly, several of the more hydrophobic-compound (DDT+DDE+DDD, dieldrin, endosulfan I, phorate, and propargite) loads were four or more times larger than their corresponding loads from other areas. Concentrations of DDT+DDE+DDD, dieldrin, diazinon, malathion, phosphamidon, propargite, atrazine, and simazine significantly increased ($\rho \leq 0.02$) in Yakima River Basin streams as concentrations of suspended sediment increased. These correlations are consistent with the equilibrium-partitioning calculations, which indicate that hydrophobic and hydrophilic compounds (K_{oc} > 85 mL/g) are primarily mobilized and transported, at least initially, in the suspended phase from the agricultural fields. Tributaries with the highest suspended-sediment concentrations in June 1989 were agricultural return flows, which also had among the highest concentrations of many pesticides (table 60, at back of report). As expected, background stations upstream from agricultural activities (for example, Cooper River at Salmon LaSac, Umtanum Creek near

Table 56. Streamflow and concentrations of suspended sediment and the most frequently detected pesticides in water samples from selected stations, Yakima River Basin, Washington, June 25–30, 1989

[Concentrations reported in nanograms per liter, except for streamflow which is in cubic feet per second and suspended sediment which is in milligrams per liter; DDT+DDE+DDD, 4,4'-dichlorodiphenyltrichloroethane (DDT) plus 4,4'-dichlorodiphenyldichloroethylene (DDE) plus 4,4'-dichlorodiphenyldichloroethane (DDD); EPTC, *S*-ethyl dipropylthiocarbamate; 2,4-D, (2,4-dichlorophenoxy) acetic acid; <, less than; --, not analyzed; T, present but below quantifiable concentration; >, greater than—the compound was quantified in one phase and detected but not quantified in the other phase; river mile is shown in parentheses]

	Yakima River main-stem stations						Agricultural return-flows near mouth of Yakima River (confluence)					
Constituent	Cle Elum (183.1)	Umtanum (140.4)	Union Gap (107.3)	Near Sunnyside (72.0)	Near Grandview (55.0)	Kiona (29.9)	Cherry Creek (147.0)	Moxee Drain (107.3)	Granger Drain (82.8)	Satus Creek (69.6)	Sulphur Creek Wasteway (61.0)	Spring Creek (41.8)
Streamflow	3,780	3,630	3,630	1,070	1,560	1,500	126	65	37	124	259	47
Suspended sediment	6	25	24	44	47	30	120	440	640	38	240	200
Organochlorine compounds in unfiltered water (suspended plus dissolved phases)												
Chlordane, cis- and trans-	<.46	T <1.4	<1.4	<2.5	--	<1.4	>4.0	T <3.5	.14	<4.0	T <8.5	1.4
DDT+DDE+DDD	.50	.43	3.2	10	12	15	>8.0	36	83	2.5	57	50
Dieldrin	T <.23	.97	.77	1.1	>5.7	3.7	4.7	>.86	>3.6	1.7	37	7.9
Endosulfan I	T <.17	.82	.92	<.10	<.80	1.2	<.30	12	T <1.4	<.80	6.8	<.81
Organophosphorus compounds in unfiltered water (suspended plus dissolved phases)												
Chlorpyrifos	<.79	T <2.5	<2.3	T <3.2	T <2.3	>9.6	>.3	>.48	T <6.1	>.20	.16	3.0
Diazinon	.67	T <2.1	>7.9	35	>120	39	T <2.8	410	13	22	66	72
Dimethoate[1]	<1.2	<3.7	<3.5	<3.5	7.3	<3.7	<3.5	<13	T<12	7.7	330	<4.6
Malathion	1.5	T <4.0	25	14	8.6	70	T <2.8	T<11	T<10	14	T <44	48
Parathion	<1.0	T <3.3	T <3.2	11	30	T <3.3	<3.1	.14	<12	180	.10	12
Phorate	<.92	<2.9	3.6	T <3.5	T <3.0	T <2.9	5.5	130	<8.1	<4.9	T <40	<5.0
Phosphamidon[1]	T <2.1	T <6.5	T <6.1	T <2.4	22	5.3	<2.4	43	56	<4.5	T <37	6.7
Thiocarbamate and sulfite compounds in filtered water												
EPTC	<.35	1.6	1.1	7.3	3.8	3.3	37	T <2.7	3.1	2.9	T <5.0	3.3
Propargite	T <.68	T <2.1	3.0	17	44	7.2	T <4.6	210	40	14	260	3.0
Acetamide and triazine compounds in filtered water												
Alachlor	<.52	<1.6	<1.5	11	19	12	<1.2	<4.0	33	13	33	<2.0
Atrazine	.27	10	5.6	26	61	32	71	8.5	48	46	49	13
Metolachlor	<.42	T <1.3	T <1.2	.93	1.3	1.9	<.70	<3.1	T <2.8	4.1	T <15	<1.8
Prometon	<.27	1.1	1.4	2.4	5.2	3.0	1.4	3.8	32	<2.5	26	<2.5
Simazine	.25	2.2	3.9	16	33	18	11	<3.6	130	27	81	10
Chlorophenoxy-acetic acid and benzoic compounds in filtered water												
2,4-D	<10	<10	70	<10	70	90	290	120	140	<10	80	60
Dicamba	<10	10	<10	<10	<10	<10	100	<10	50	<10	10	10

[1]Dimethoate and phosphamidon were not analyzed for in suspended sediment.

Table 57. Instantaneous loads of suspended sediment and selected pesticides in water samples, Yakima River Basin, Washington, June 25–30, 1989

[Loads reported in grams per day, except suspended sediment which is in kilograms per day; East side, the area east of the Yakima River from Moxee Valley downstream to Benton City near Kiona including the Moxee, Granger, Sunnyside, and Whitstran areas; West side, the area west of the Yakima River downstream from Ahtanum Ridge to the city of Mabton including the Wapato, Toppenish, and Mabton areas; if all tributaries in an area had compound concentrations less than minimum reporting levels then the load value is reported as less than (<); quantifiable loads exclude compound concentrations less than the minimum reporting level; DDT+DDE+DDD, 4,4'-dichlorodiphenyltrichloroethane (DDT) plus 4,4'-dichlorodiphenyldichloroethylene (DDE) plus 4,4'-dichlorodiphenyldichloroethane (DDD); EPTC, *S*-ethyl dipropylthiocarbamate; 2,4-D, (2,4-dichlorophenoxy) acetic acid]

Constituent	Kittitas area	Tieton area	East side	West side	Yakima River at Kiona
Suspended sediment	41,600	1,150	317,000	59,500	110,000
Organochlorine compounds in unfiltered water (suspended plus dissolved phases)					
Chlordane	1.2	.08	.21	.76	<5.2
DDT+DDE+DDD	2.5	.30	58	9.0	55
Dieldrin	1.4	.08	26	.94	14
Endosulfan I	<.38	.34	6.2	.87	4.4
Organophosphorus compounds in unfiltered water (suspended plus dissolved phases)					
Chlorpyrifos	.09	<.39	.54	.55	35
Diazinon	<1.9	.66	120	56	140
Dimethoate[1]	<2.7	.46	210	3.1	<14
Malathion	<3.0	1.4	8.8	15	260
Parathion	<3.0	<.68	1.5	64.8	<12
Phorate	1.7	<.57	21	4.6	<10
Phosphamidon[1]	<2.3	.95	16	11	19
Thiocarbamate and sulfite compounds in filtered water					
EPTC	12	<.10	.84	3.5	12
Propargite	<2.7	<.41	200	47	26
Acetamide and triazine compounds in filtered water					
Alachlor	<1.1	<.23	24	26	44
Atrazine	32	.55	40	81	120
Metolachlor	<.89	<.20	<11	3.2	7.0
Prometon	3.2	2.0	20	.26	11
Simazine	4.9	.95	65	20	66
Chlorophenoxy-acid and benzoic compounds in filtered water					
2,4-D	89	<.81	90	35	330
Dicamba	49	<.81	13	<10	<37

[1]Dimethoate and phosphamidon were not analyzed for in suspended sediment.

mouth, and Satus Creek above Wilson-Charley Canyon) had among the lowest concentrations.

Ratios of pesticide loads in runoff to annual pesticide applications in 1989 in the Yakima River Basin generally were less than 0.6 percent. For these calculations, instantaneous pesticide loads, measured near peak irrigation in June 1989, were computed in grams per day and divided by the 1989 pesticide applications, computed in grams per day. Ratios were computed for the most frequently quantified compounds (listed in table 56) except DDT+DDE+DDD, dieldrin, and chlordane, which were not applied in 1989. Instantaneous runoff loads near or during peak irrigation are expected to be larger than those instantaneous loads during non-irrigation season; consequently, the ratios computed in this study may be biased high when compared to ratios between annual loads in streams and annual applications. If, however, major storm-runoff had occurred, the ratios may be biased low. These calculations indicate that less than 0.6 percent of the applied pesticide loads were detected in instantaneous loads in runoff. For comparison, ratios of annual herbicide loads in runoff to annual applications for midwestern streams in the United States range from less than 0.1 to 3.7 percent and are similar to those in Yakima River Basin streams.

Organic compounds that exceeded chronic-toxicity water-quality criteria or guidelines for the protection of freshwater aquatic life are listed in table 58. Most of the exceedances of the chronic-toxicity criteria occurred in agricultural return flows and main-stem stations down-

Table 58. Number of stations and samples with organic-compound concentrations that equaled or exceeded chronic-toxicity water-quality criteria for the protection of freshwater aquatic life, Yakima River Basin, Washington, 1988–91

[Criteria reported in nanograms per liter; DDT+DDE+DDD, 4,4'-dichlorodiphenyltrichloroethane (DDT) plus 4,4'-dichlorodiphenyldichloroethylene (DDE) plus 4,4'-dichlorodiphenyldichloroethane (DDD); PCB, polychlorinated biphenyls; 2,4-D, (2,4-dichlorophenoxy) acetic acid]

Compound	Chronic-toxicity water-quality criteria for freshwater aquatic life[1]	Stations: Number sampled	Stations: Number of exceedances of criteria or guidelines	Samples: Number sampled	Samples: Number of exceedances of criteria or guidelines	Samples: Frequency of exceedance
Organochlorine compounds						
Chlordane (technical)	4.3	37	1	133	1	0.75
DDT+DDE+DDD	1.0	37	21	133	105	79
Dieldrin	1.9	37	12	133	65	49
Endosulfan I	56	37	3	133	8	6.0
Endrin	2.3	37	3	133	2	1.5
PCB	14	25	5	133	9	6.8
Toxaphene	.2	25	1	133	1	.75
Organophosphorus compounds						
Azinphos-methyl	10	18	[2]2	18	2	11
Diazinon	[3]9	37	16	133	55	41
Disulfoton	[3]50	33	1	57	2	3.5
Ethion	[3]20	37	1	133	7	5.3
Malathion	100	37	1	133	1	.75
Parathion	13	37	11	133	11	8.3
Phosphamidon	[3]30	29	4	29	3	10
Chlorophenoxy-acetic acid compound						
2,4-D	[3]3,000	31	1	106	1	.94

[1]U.S. Environmental Protection Agency, 1986.

[2]Two stations sampled by North Yakima Conservation District.

[3]Recommended maximum concentration sampled at any time and any place (National Academy of Science-National Academy of Engineering Committee of Water Quality Criteria, 1973).

stream from the city of Yakima. Criteria for acute toxicity were not exceeded in the basin. Water-quality criteria established by the U.S. Environmental Protection Agency (EPA) and (or) recommended guidelines by the National Academy of Sciences-National Academy of Engineering (NAS-NAE) Committee on Water Quality Criteria for the protection of freshwater aquatic life are not available for all of the pesticides that were applied in the basin in 1989. In addition, additive or synergistic effects that may result from the low-level occurrence of several compounds in the water were not evaluated in this study.

In June 1989, the city of Yakima's finished drinking-water supply was analyzed for 67 pesticides, using research procedures that make possible minimum reporting levels near 1 nanograms per liter (ηg/L). The only pesticide detected was DDE, at a concentration of 0.36 ηg/L. Assuming that a person ingests 2 liters of this water each day, the increased lifetime (70 years) cancer risk is relatively small—4 in 1,000,000,000 people using EPA's risk-assessment methodology calculations. For this calculation, drinking water was assumed to be the sole source of DDE ingested by the person.

Spatial distribution of organic-compound concentrations in bed sediment—Results from the sampling of 59 stations in the basin show that several semivolatile and organochlorine compounds were detected in streambed sediment. Highest concentrations were detected in agricultural return flows, many of which also receive point-source discharges and urban runoff. DDT+DDE+DDD and dieldrin are organochlorine insecticides that were detected in more than 50 percent of the samples; phenol, di(2-ethylhexyl) phthalate, naphthalene, and phenanthrene are semivolatile organic compounds that were detected in more than 30 percent of the samples. Semivolatile compound concentrations did not exceed EPA's interim sediment-quality criteria for the protection of benthic fauna. Concentrations of several organochlorine compounds generally exceeded the criteria in the agricultural areas, whereas relatively few exceedances were detected in the nonagricultural areas. DDT+ DDE+DDD and dieldrin exceeded criteria at most of the agricultural stations, and to a lessor extent, endosulfan I, endrin, and chlordane also exceeded the criteria. The detection of DDT+DDE+DDD and dieldrin in the bed sediment in agricultural return flows generally coincided with the detection of these compounds in the water column. Concentrations of DDT+DDE+ DDD in bed-sediment samples collected upstream from nonagricultural areas in the Yakima River Basin ranged from less than 0.3 to 1.4 µg/kg. These lower concentrations suggest that the primary source of DDT+DDE+ DDD in the nonagricultural areas is atmospheric deposition.

Spatial distribution of organic-compound concentrations in aquatic biota—In 1989–90, samples of fish, mollusks, and aquatic plants were collected from 33 stations for analyses of organic compounds. About two-thirds of these stations receive agricultural return flow, point-source discharges, and (or) urban runoff. The other stations are influenced minimally by human activity. Samples were analyzed for 25 organochlorine compounds and 14 polycyclic aromatic hydrocarbons.

One DDT metabolite, 4, 4'-DDE, was the most widely occurring organic compound detected in aquatic biota. This organochlorine compound was detected in fish samples at all stations sampled in 1989 and 68 percent of the stations in 1990. Other organochlorine compounds including DDT, DDD, cis-chlordane, trans-nonachlor, and dieldrin were each detected at 32 percent or more of the stations. Dicofol, PCBs (polychlorinated biphenyls), toxaphene, and other chlordane-related compounds were detected less frequently. Highest organochlorine-compound concentrations generally were detected at main-stem and tributary stations downstream from the city of Yakima, where agriculture is the primary land use.

Mean concentrations of DDT+DDE+DDD (1.12 micrograms per gram [µg/g], wet weight) and dicofol (0.02 µg/g, wet weight) in resident fish from the Yakima River Basin were about 4 and 20 times higher, respectively, than concentrations detected in national studies (U.S. Fish and Wildlife's National Contaminant Biomonitoring Program and EPA's National Study of Chemical Residues in Fish). The other 18 organochlorine compounds had mean concentrations that generally were less than or equal to those concentrations detected in the national studies.

In this study, wet-weight concentrations of most organochlorine compounds in whole fish correlated significantly with lipid concentrations in the fish, suggesting that concentrations of these compounds accumulate in or are controlled by fish lipids. This relation was evaluated further by normalizing organochlorine-compound concentrations for lipid concentrations in samples of smallmouth bass, mountain whitefish, largescale sucker, and Asiatic clams collected from a Yakima River station near the terminus of the basin. Prior to normalization the concentrations ranged from 0.47 to 2.27 µg/g, wet weight. After normalization, DDT+DDE+DDD concentrations were remarkably similar among the four species, ranging from 19.6 to 25.9 µg/g-lipid.

In the Yakima River Basin, most samples of whole resident fish collected downstream from the city of Yakima had concentrations of DDT+DDE+DDD, PCB's, chlordane-related compounds, dieldrin, toxaphene, and dicofol that were higher than those concentrations expected to result in an increased lifetime cancer risk for an individual compound of 1:1,000,000. The highest increased cancer risk was computed to be 640:1,000,000 and was based on the detection of a high PCB concentration in a composited sample of largescale suckers from the Yakima River at Kiona. While these results indicate possible human-health effects from eating certain species of resident fish in the Yakima River Basin, several caveats must be considered: (1) in this study, whole fish generally were analyzed, not edible portions, so the data may not be representative of what people are consuming, (2) different species may accumulate different amounts of contaminants, so the fish and associated contaminants that people are consuming may not be represented in this study, (3) the higher contaminant concentrations were detected primarily in the lower basin downstream from the city of Yakima, and, (4) seasonal variations in contaminant concentrations in biota were not determined in this study.

In the lower Yakima River Basin, concentrations of DDT+DDE+DDD, toxaphene, and dieldrin in whole fish exceeded guidelines recommended by the National Academy of Sciences-National Academy of Engineering Committee on Water Quality for the protection of fish-eating predators. Seventeen, 12, and 1 whole-fish sample(s) had concentrations of DDT+ DDE+DDD, toxaphene, and dieldrin, respectively, that exceeded their guidelines of 1.0, 0.1, and 0.1 µg/g, wet weight, respectively. The highest concentrations of DDT+DDE+DDD (4.8 µg/g, wet weight), toxaphene (1.2 µg/g, wet weight), and dieldrin (0.17 µg/g, wet weight) were detected in largescale suckers from Sulphur Creek Wasteway, a major agricultural return flow near Sunnyside, Washington. Upstream from the city of Yakima, the guidelines were not exceeded in whole-fish samples.

Naphthalene, fluoranthene, phenanthrene, and benzo(e)pyrene were the most frequently detected polycyclic aromatic hydrocarbons (PAHs) found in the biological samples collected from the Yakima River downstream from the city of Yakima. Only two carcinogenic PAHs were detected in this study, chrysene in a waterweed sample from Satus Creek and 1,2-benzanthracene in an Asiatic clam sample from the Yakima River at RM 72, both at the minimum reporting level of 0.01 µg/g, wet weight.

Relations of organic-compound concentrations among sampling media—In aquatic environments, organic compounds are partitioned among water (dissolved phase), dissolved organic carbon (dissolved and colloidal phases), particulate organic carbon (soils, suspended sediment, and bed sediment), and the lipids of aquatic biota. In this study, organochlorine compounds were detected in all environmental compartments that were sampled including water, suspended sediment, bed sediment, aquatic biota, and soils; PAHs were only analyzed and detected in bed sediment and aquatic biota. Carbamate compounds, chlorophenoxy-acid herbicides, organophosphorus compounds, and triazine herbicides have relatively high water solubilities and low K_{oc} values, so these compounds were analyzed primarily in filtered water, where many were detected.

To explore the relations among sampling media, concentrations of DDT+DDE+DDD and dieldrin were chosen, because they were detected frequently throughout the Yakima River Basin. For these relations, contaminant concentrations in sediment were normalized to organic carbon and concentrations in aquatic biota were normalized to lipid content. Results indicated that increased concentrations of DDT+DDE+DDD and dieldrin in filtered water, unfiltered water, or bed sediment also were associated with increased concentrations of these compounds in aquatic biota. Generally, the contaminant concentrations associated with suspended sediment were higher than those in bed sediment. Overall, consistent results among the sampling media for DDT+DDE+DDD and dieldrin provide multiple lines of evidence of pesticide occurrence. Results also indicated that contaminant concentrations in one sampling medium could be used to estimate contaminant concentrations in another medium.

Sampling media were also evaluated to determine which medium would be best suited for detecting the low-level occurrence of organochlorine compounds. The sampling media evaluated in this study included aquatic biota with varying concentrations of lipids, bed sediment, 1-L volumes of unfiltered water (conventional analyses), large volumes of filtered water (up to 120 L), and large amounts of suspended sediment (from sample volumes up to 224 L). Overall, analyses of bed sediment and aquatic biota were the most sensitive sampling media for detecting hydrophobic compounds. At a particular sampling station, a contaminant generally was detected in both the bed sediment and the biota if the biota sample had a lipid concentration of about 6 percent or more (typical lipid concentrations in largescale sucker, mountain whitefish, and chiselmouth from the Yakima River Basin). The least sensitive medium for detecting low-level concentrations of organochlorine compounds was aquatic plants. Aquatic plants have low lipid concentrations (about 0.2 percent) and tend to bioaccumulate lower concentrations of these hydrophobic compounds. At three of the stations (Granger Drain, Sulphur Creek Wasteway, and Yakima River at Kiona), PCB and toxaphene were detected in fish-tissue samples, but were not detected in bed sediment. At these stations, PCB concentrations in the whole fish exceeded EPA health advisory estimates for an increased lifetime cancer risk of more than 1 in 1,000,000 people. For assessment purposes, it may be advantageous to analyze tissue samples rather than bed-sediment samples, because tissue analyses provide a direct measure of bioavailability and may be used for determining critical routes of contaminant exposure to fish predators, including humans. Fish, however, are mobile, which makes it difficult to use tissue data to identify whether the contaminant source is upstream or downstream from the sampling station. In contrast, the source of bed-sediment contamination is generally at or upstream from the sampling location.

Implications concerning the use of pesticides, water, and land—Results of this study showed that there is a clear seasonal increase in stream concentrations of hydrophobic and hydrophilic pesticides at or near peak irrigation periods after pesticide applications in the spring, when pesticides were flushed from the soils and (or) eroded with the contaminated soils from the agricultural fields. Pesticide concentrations also increased during stormwater runoff from agricultural areas in the winter and summer.

The flushing of compounds from soil-pore water, the eroding of soil-sorbed compounds, and the dissolving of compounds from soil and sediment into surface water are major pathways for pesticides to travel from agricultural fields to streams. Controlling excessive applications of irrigation water will help to reduce overland runoff, and, therefore, the subsequent dissolution, erosion, and (or) transport of pesticides to streams. If management decisions are made to reduce overland runoff, increased monitoring of ground-water quality would be advisable because pesticide concentrations could increase in the aquifers. Further, it would be prudent to monitor agricultural return flows because less irrigation water would be available to dilute hydrophilic pesticide concentrations in the surface-water runoff.

Data from the Yakima River Basin indicate that increases in pesticide use generally coincide with increases in the number of pesticide detections in streams. This relation seems obvious; nonetheless, newly developed analytical techniques were required in this study to enable detection of many of the currently used pesticides at concentrations below 100 ηg/L.

To minimize the likelihood of pesticide transport from agricultural fields to streams, the use of pesticides with one or more of the following characteristics would be desirable:

(1) half-lives in soils and water of less than 3 weeks, so compounds have more opportunity to degrade in the fields prior to stream transport;

(2) water solubilities smaller than 30 mg/L, to minimize flushing of dissolved pesticides from soils; and

(3) K_{oc} values larger than 300 to 500 mL/g, to increase the propensity of the pesticide to remain sorbed to agricultural soils.

These characteristics, in concert with reductions in overland runoff and erosion, are key factors that could result in reduced pesticide concentrations in streams.

Methods of irrigation can greatly effect overland runoff and erosion, which subsequently affect pesticide transport to streams. Of the four types of irrigation methods commonly used in the Yakima River Basin (rill, flood, sprinkler, and drip), drip irrigation can be used effectively to reduce erosion and overland runoff, because a minimum amount of water is applied to subsurface soils in the root zone. Other factors that affect pesticide transport to streams include (1) timing of irrigation relative to pesticide application and the resulting likelihood for increased overland runoff, (2) location of pesticide application relative to the potential for stream contamination, and (3) pesticide application methods (ground-vehicle sprays, aerial sprays, and chemigation).

Reductions in pesticide concentrations needed to meet water-quality criteria—Pesticides that most frequently exceeded EPA chronic-toxicity water-quality criteria or NAS-NAE guidelines for the protection of freshwater aquatic life in June 1989 included DDT+DDE+DDD, dieldrin, diazinon, and parathion. Reductions in pesticide concentrations that were needed to meet criteria or guidelines for DDT+DDE+DDD, dieldrin, diazinon, and parathion ranged from 29 to 99 percent at 19 of 29 stations, 47 to 95 percent at 7 stations, 31 to 98 percent at 14 stations, and 19 to 93 percent at 4 stations, respectively. Assuming that instream pesticide concentrations during the irrigation season can be reduced by controlling erosion and overland runoff, parathion and diazinon concentrations would then meet guidelines. Dieldrin would meet the criteria in the main stem and Moxee Drain, and DDT+DDE+ DDD would meet the criteria in the main stem. In the other agricultural return flows, however, the criteria would still be exceeded by as much as 3 ηg/L for dieldrin and 18 ηg/L for DDT+DDE+DDD.

Future studies—To better understand the occurrence, distribution, sources, transport, fate, and effects of organic compounds in the Yakima River Basin, the following tasks could be performed:

- Analyze more pesticides and their breakdown products at lower detection limits—In this study, about 70 percent of the 180 compounds applied in 1989 were not analyzed. Therefore, ecological consequences for more than 180 of these compounds in water, sediment, and aquatic biota in the Yakima River Basin have not been identified.

- Increase temporal coverage for organic-compound concentrations in water, sediment, and aquatic biota—To better understand major sources and potential effects on aquatic biota, the variations (duration and range) in seasonal concentrations must first be determined. This assessment may require weekly, daily, and (or) hourly sampling during selected time periods when concentrations are expected to be high (for example, irrigation season or storm runoff following pesticide application).

- Increase spatial and media coverage for determining the occurrence of organic-compound concentrations in water, bed sediment, aquatic biota, and soils and the processing of these compounds within and among these media. Identify major sources of organic compounds in order to model and assess cause and effect processes and pesticide transport in the basin.
- Relate pesticide loads to contributing factors, including crop types, soil characteristics, basin characteristics, and farming practices (pesticide application, irrigation, and cultivation methods). Better quantification of these factors would provide managers with valuable information for improving and (or) maintaining good water-quality conditions in the basin.
- Conduct research to determine whether erosion controls could effectively reduce concentrations of dissolved and suspended organic compounds in streams.
- Conduct bioassays using native water, native sediment, and appropriate sensitive test organisms and ecological studies of the structure and function of aquatic communities to help address the potential for additive or synergistic effects.
- Conduct research to determine the synergistic effects of multiple toxic compounds and elements at high water temperatures on aquatic life.
- Obtain accurate information about the quantity of pesticides applied in the Yakima River Basin.

REFERENCES

Aiken, G.R., and Gillam, A.H., 1989, Determination of molecular weights of humic substances by colligative property measurements, *in* Hayes, M.H.B., MacCarthy, P., Malcolm, R.L., and Swift, R.S., eds., Humic substances II—In search of structure: New York, John Wiley & Sons, p. 515–544.

Aiken, G.R., and Malcolm, R.L., 1987, Molecular weight of aquatic fulvic acids by vapor pressure osmometry: Geochimica et Cosmochimica Acta, v. 51, p. 2177–2184.

Aiken, G.R., McKnight, D.M., Thorn, K.A., and Thurman, E.M., 1992, Isolation of hydrophilic organic acids from water utilizing macroporous resins: Organic Geochemistry, v. 18, p. 567–573.

Bain R.C., Jr., 1990, Kittitas, Nonpoint pollution control project—Summary: Ellensburg, Washington, Kittitas County Conservation District, 18 p.

Becker, R.L., Herzfeld, D., Ostlie, K.R., and Stamm-Katovich, E.J., 1989, Pesticides—Surface runoff, leaching, and exposure concerns: Minnesota Extension Service, AG–BU–3911, 32 p.

Beers, E.H., and Brunner, J.F., 1991, Washington State apple and pear pesticide use survey, 1989–90: Wenatchee, Washington, Washington State University Tree Fruit Research and Extension Center, unpublished report on file in Cooperative Extension Office, 135 p.

Biggar, J.W., and Seiber, J.N., eds., 1987, Fate of pesticides in the environment—Proceedings of a technical seminar: Agricultural Experiment Station, Division of Agriculture and Natural Resources University of California, Publication 3320, 157 p.

Bowles, E.C., Antweiler, R.C., and MacCarthy, P., 1989, Acid-base titration and hydrolysis of fulvic acid from the Suwannee River, *in* Averett, R.C., Leenheer, J.A., McKnight, D.M., and Thorn, K.A., eds., Humic substances in the Suwannee River, Georgia—Interactions, properties, and proposed structures: U.S. Geological Survey Open-File Report 87–557, p. 99–118.

Callahan, M.A., Slimak, M.W., Gabel, N.W., May, I.P., Fowler, C.F., Freed, J.R., Jennings, P., Durfee, R.L., Whitmore, F.C., Maestri, B., Mabey, W.R., Holt, B.R., and Gould, C., 1979, Water-related environmental fate of 129 priority pollutants, v. 1 and 2: Washington, D.C., U.S. Environmental Protection Agency, EPA–440/4–79–029 [Available from National Technical Information Service, Springfield, VA 22161 as NTIS Report PB80–204381].

Carter, C.W., and Suffet, I.H., 1982, Binding of DDT to dissolved humic materials: Environmental Science and Technology, v. 16, p. 735–740.

Chapman, P.M., Romberg, G.P., and Vigers, G.A., 1982, Design of monitoring studies for priority pollutants: Journal of the Water Pollution Control Federation, v. 54, no. 3, p. 292–297.

Chen, R.F., and Bada, J.L., 1989, Seawater and porewater fluorescence in the Santa Barbara Basin: Geophysical Research Letters, v. 16, p. 687–690.

Chiou, C.T., 1985, Partition coefficients of organic compounds in lipid-water systems and correlations with fish bioconcentration factors: Environmental Science and Technology, v. 19, no. 1, p. 57.

Chiou, C.T., Freed, V.H., Schmedding, D.W., and Kohnert, R.L., 1977, Partition coefficient and bioaccumulation of selected organic chemicals: Environmental Science and Technology, v. 11, no. 5, p. 475–478.

Chiou, C.T., Malcolm, R.L., Brinton, T.I., and Kile, D.E., 1986, Water solubility enhancement of some organic pollutants and pesticides by dissolved humic and fulvic acids: Environmental Science and Technology, v. 20, p. 502–508.

Columbia River Inter-Tribal Fish Commission, 1994, A fish consumption survey of the Umatilla, Nez Perce, Yakama, and Warm Springs Tribes of the Columbia River Basin: Portland, Oregon, Columbia River Inter-Tribal Fish Commission, Technical Report 94–3, 172 p.

Columbia Basin Inter-Agency Committee, 1964, River-mile index, Yakima River: Columbia River Basin Hydrology Subcommittee Report 15, 31 p.

Confederated Tribes and Bands of the Yakama Indian Nation, Washington Department of Fisheries, and Washington Department of Wildlife, 1990, Yakima River Subbasin—Salmon and steelhead production plan [draft]: Toppenish, Washington, Confederated Tribes and Bands of the Yakama Indian Nation [unpaginated].

Crawford, J.K., and Luoma, S.N., 1993, Guidelines for studies of contaminants in biological tissues for the National Water-Quality Assessment program: U.S. Geological Survey Open-File Report 92–494, 69 p.

Custom Chemicides, [n.d], Quick guide to buffering: Fresno, California, Custom Chemicides, 43 p.

Dearth, M.A., and Hites, R.A., 1991, Complete analysis of technical chlordane using negative ionization mass spectrometry: Environmental Science and Technology, v. 25, no. 2, p. 245–254.

Eisler, Ronald, 1987, Polycyclic aromatic hydrocarbon hazards to fish, wildlife, and invertebrates—A synoptic review: U.S. Fish and Wildlife Service, Contamination Hazard Reviews Report no. 11, 81 p.

Foster, G.D., and Rogerson, P.F., 1990, Enhanced preconcentration of pesticides from water using the Goulden large-samples extractor: International Journal of Environmental Analytical Chemistry, v. 41, p. 105–117.

Foster, G.D., Foreman, W.T., and Gates, P.M., 1991, Performance of the Goulden large-sample extractor in multiclass pesticide isolation and preconcentration from stream water: Journal of Agricultural and Food Chemistry, v. 39, p. 1618–1622.

Foster, G.D., Gates, P.M., Foreman, W.T., McKenzie, S.W., and Rinella, F.A., 1993, Determination of dissolved-phase pesticides in surface water from the Yakima River Basin, Washington, using the Goulden large-sample extractor and gas chromatography/mass spectrometry: Environmental Science and Technology, v. 27, p. 1911–1917.

Gilliom, R.J., and Clifton, D.G., 1990, Organochlorine pesticide residues in bed sediments of the San Joaquin River, California: Water Resources Bulletin, v. 26, p. 11–24.

Gilliom, R.J., Alexander, R.B., and Smith, R.A., 1985, Pesticides in the Nation's rivers, 1975–1980, and implications for future monitoring: U.S. Geological Survey Water-Supply Paper 2271, 26 p.

Goolsby, D.A., Coupe, R.C., and Markovchick, D.J., 1991, Distribution of selected herbicides and nitrate in the Mississippi River and its major tributaries—April through June 1991: U.S. Geological Survey Water-Resources Investigations Report 91–4163, 35 p.

Goulden. P.D., and Anthony, D.H.J., 1985, Design of a large-scale extractor for the determination of organics in water: Burlington, Ontario, Environment Canada, National Water Research Institute, Report 85–121.

Guy, H.P., 1978, Fluvial sediment concepts: U.S. Geological Survey Techniques of Water-Resources Investigations, Book 3, Chapter C1, 55 p.

Hale, R.C., 1988, Disposition of polycyclic aromatic compounds in blue crabs, *Callinectes sapidus*, from the Southern Chesapeake Bay: Estuaries, v. 11, p. 255–263.

Hayes, M.H.B., and Swift R.S., 1978, The chemistry of soil organic colloids, *in* Greenland, D.J.,and Hayes, M.H.B.,eds., The chemistry of soil constituents: New York, John Wiley and Sons, p. 179–230.

Hedlund, J.D., 1992, Controlling erosion and sediment pollution from irrigation: Portland, Oregon, U.S. Soil Conservation Service, West National Technical Center, 109 p.

Hirsch, R.M., Alley, W.M., and Wilber, W.G., 1988, Concepts for a National Water-Quality Assessment program: U.S. Geological Survey Circular 1021, 42 p.

Hopkins, B.S., Clark, D.K., Schlender, M., and Stinson, M., 1985, Basic water monitoring program—Fish tissue and sediment sampling for 1984: Olympia, Washington State Department of Ecology, 43 p.

Huckins, J.N., Schwartz, T.R., Petty, J.D., and Smith, L.M., 1988, Determination, fate, and potential significance of PCBs in fish and sediment samples with emphasis on selected AAH-inducing congeners: Chemosphere, v. 17, no. 10, p. 1995–2016.

Huling, S.G., 1989, Facilitated transport: U.S. Environmental Protection Agency Superfund Ground Water Issue, EPA/540/4–89/003, 5 p.

Hurle, K.B., and Freed, V.H., 1972, Effect of electrolytes on the solubility of some 1,3,5-triazine and substituted ureas and their adsorption on soil: Weed Res., v. 12, p. 1–10.

Johnson, A., Norton, D., and Yake, W., 1986, Occurrence and significance of DDT compounds and other contaminants in fish, water, and sediment from the Yakima River Basin: Olympia, Washington Department of Ecology, 89 p.

Kenaga, E.E., 1980, Predicted bioconcentration factors and soil sorption coefficients of pesticides and other chemicals: Ecotoxicology and Environmental Safety, v. 4, p. 26–38.

Kidd, Hamish, and James, D.R., eds., 1991, The agrochemicals handbook (3rd ed.): Cambridge, England, The Royal Society of Chemistry, 1470 p.

Larsson, Per, Collvin, Lars, Lennart, Okla, and Meyer, Gabrielle, 1992, Lake productivity and water chemistry as governors of the uptake of persistent pollutants in fish: Environmental Science and Technology, v. 26, no. 2, p. 346–351.

Leahy, P.P., Rosenshein, J.S., and Knopman, D.S., 1990, Implementation plan for the National Water-Quality Assessment program: U.S. Geological Survey Open-File Report 90–174, 10 p.

Lenfesty, C.D., and Reedy, T.E., 1985, Soil survey of Yakima County area, Washington: U.S. Department of Agriculture, Soil Conservation Service Report 0–389–957 QL 3, 345 p.

Malins, D.C., and Hodgins, H.O., 1981, Petroleum and marine fishes—A review of uptake, disposition and effects: Environmental Science and Technology, v. 17, p. 1273–1280.

McCarthy, J.F., and Shugart, L.R., 1990, Biological markers of environmental contamination, *in* McCarthy, J.F., and Shugart, L.R., eds., Biomarkers of environmental contamination: Boca Raton, Florida, Lewis Publishers, p. 3–14.

McCarthy, J.F., and Zachara, J.M., 1989, Subsurface transport of contaminants, Environmental Science and Technology, v. 23, p. 496–502.

McKnight, D.M., Aiken, G.R., and Smith, R.L., 1991, Aquatic fulvic acids in microbially based ecosystems—Results from two lakes in Antarctica: Limnology and Oceanography, v. 36, p. 998–1006.

McKnight, D.M., Bencala, K.E., Zellweger, G.W., Aiken, G.R., Feder, G.L., and Thorn, K.A., 1992, Sorption of dissolved organic carbon by hydrous aluminum and iron oxides occurring at the confluence of Deer Creek with the Snake River, Summit County, Colorado: Environmental Science and Technology, v. 26, p. 1388–1396.

McKnight, D.M., Thorn, K.A., Wershaw, R.L., Bracewell, J.M., and Robertson, G.W., 1988, Rapid changes in dissolved humic substances in Spirit Lake and South Fork Castle Lake, Washington: Limnology and Oceanography, v. 33, 1527–1541.

Mercer, J.W., Skipp, D.C., and Giffin, Daniel, 1990, Basics of pump-and-treat ground-water remediation technology: Ada, Oklahoma, U.S. Environmental Protection Agency, EPA/600/8–90/003, 31 p.

Moore, D.G., and Loper, B.R., 1980, Soils—DDT residues in forest floors and soils of Western Oregon, September-November 1966: Corvallis, Oregon, U.S. Forest Service, Forestry Sciences Laboratory, p. 77–85.

Mopper, K., Zhou, X., Kieber, R.J., Kieber, D.J., Sikorski, R.J., and Jones, R.D., 1991, Photochemical degradation of dissolved organic carbon and its impact on the oceanic carbon cycle: Nature, v. 353, p. 60–62.

National Academy of Sciences-National Academy of Engineering Committee on Water Quality Criteria (NAS-NAE), 1973, Water quality criteria 1972: Washington, D.C., U.S. Environmental Protection Agency Ecological Research Series, EPA–R3–73–033, 594 p.

Nowell, L.H., and Resek, E.A., 1994, Summary of national standards and guidelines for pesticides in water, bed sediment, and aquatic organisms and their application to water-quality assessments: U.S. Geological Survey Open-File Report 94–44, 115 p.

North Yakima Conservation District, 1991, Final report Moxee/Naches water quality project 1989–1991: Yakima, Washington, 32 p.

Oregon State University, Washington State University, and the University of Idaho Extension Services, 1992a, Pacific Northwest 1992 insect control handbook: Corvallis, Oregon, Oregon State University, Agricultural Communications, 308 p.

———1992b, Pacific Northwest 1992 weed control handbook: Corvallis, Oregon, Oregon State University, Agricultural Communications, 326 p.

———1992c, Pacific Northwest 1992 plant disease control handbook: Corvallis, Oregon, Oregon State University, Agricultural Communications, 335 p.

Pankow, J.F., and McKenzie, S.W., 1991, Parameterizing the equilibrium distribution of chemicals between the dissolved, solid particulate matter, and colloidal matter compartments in aqueous systems: Environmental Science & Technology, v. 25, p. 2046–2053.

Pearson, W.H., Sugarman, P.C., Woodruff, D.L., Blaylock, J.W., and Olla, B.L., 1980, Detection of petroleum hydrocarbons by the dungeness crab, *Cancer magister*: Fishery Bulletin, v. 78, p. 821–826.

Pereira, W.E., Rostad, C.E., Chiou, C.T., Brinton, T.I., and Barber, L.B.,II, 1988, Contamination of estuarine water, biota, and sediment by halogenated organic compounds—A field study: Environmental Science and Technology, v. 22, p. 772–778.

Phillips, D.J., 1980, Quantitative aquatic biological indicators: Barking, Essex, England, Applied Science Publishers Ltd., 488 p.

Poplyk, J., ed., 1989, Farm chemicals handbook: Willoughby, Ohio, Meister Publishing Company, v. 75, 676 p.

Preston, C.M., 1987, Review of solution NMR of humic substances, *in* Wershaw, R.L., and Mikita, M.A., eds., NMR of humic substances and coal: Chelsea, Michigan, Lewis Publishers, Inc., p. 3–32.

Rasmussen, J.J., 1976, Soil survey of Yakima Indian Reservation irrigated area, Washington, part of Yakima County: U.S. Soil Conservation Service Report 1976–596–665/26, 51 p.

Rinella, J.F., McKenzie, S.W., Crawford, J.K., Foreman, W.T., Gates, P.M., Fuhrer, G.J., Janet, M.L., 1992, Surface-water-quality assessment of the Yakima River Basin, Washington—Pesticides and other trace-organic-compound data for water, sediment, soil, and aquatic biota, 1987–91: U.S. Geological Survey Open-File Report 92–644, 154 p.

Rinella, J.F., McKenzie, S.W., and Fuhrer, G.J., 1992, Surface-water-quality assessment of the Yakima River Basin, Washington—Analysis of available water-quality data through 1985 water year: U.S. Geological Survey Open-File Report 91–453, 244 p.

Sawyer, C.N., McCarty, P.L., 1967, Chemistry for sanitary engineers: New York, McGraw-Hill Book Company, 518 p.

Schmitt, C.J., Zajicek, J.L., and Peterman, P.H., 1990, National contaminant biomonitoring program—Residues of organochlorine chemicals in United States freshwater fish, 1976–1984: Archives of Environmental Contamination and Toxicology, v. 19, p. 748–781.

Sittig, Marshall, 1981, Handbook of toxic and hazardous chemicals: Park Ridge, New Jersey, Noyes Publications, 729 p.

Smith, J.A., Witkowski, P.J., and Fusillo, T.V., 1988, Man-made organic compounds in the surface waters of the United States—A review of current understanding: U.S. Geological Survey Circular 1007, 92 p.

Smith, R.L., 1995, EPA Region III risk-based concentration table, background information: U.S. Environmental Protection Agency, February 9, 1995, 22 p.

Squillace, P.J., and Thurman, E.M., 1992, Herbicide transport in rivers: Importance of hydrology and geochemistry in nonpoint-source contamination: Environmental Science and Technology, v. 26, no. 3, p. 538–545.

Steelink, C., 1985, Implications of elemental characteristics of humic substances, *in* Aiken, G.R., McKnight, D.M., Wershaw, R.L., and MacCarthy, P., eds., Humic substances in soil, sediment and water: New York, John Wiley & Sons, p. 457–476.

Steelink, C., Wershaw, R.L., Thorn, K.A., and Wilson, M.A., 1990, Application of liquid state NMR spectroscopy to humic substances, *in* Hayes, M.H.B., MacCarthy, P., Malcolm,R.L., and Swift, R.S., eds., Humic substances II—In search of structure: New York, John Wiley & Sons, p. 281–308.

Steinberg, C. and Muenster, U., 1985, Geochemistry and ecological role of humic substances in lake water, *in* Aiken, G.R., McKnight, D.M., Wershaw, R.L., and MacCarthy, P., eds., Humic substances in soil, sediment and water: New York, John Wiley & Sons, p. 433–455

Tetra Tech, Inc., 1988, Pesticides of concern in the Puget Sound Basin—A review of contemporary pesticide usage: Prepared for U.S. Environmental Protection Agency by Tetra Tech, Inc., Bellevue, Washington, Final Report TC–3338–32, 200 p.

Thurman, E.M., 1985, Organic geochemistry of natural water: Hingham, Massachussets, Kluwer Academic Publishers, 497 p.

Thurman, E.M., and Malcolm, R.L., 1981, Preparative isolation of aquatic humic substances: Environmental Science and Technology, v. 15, p. 463–466.

Thurman, E.M., Goolsby, D.A., Meyer, M.T., and Kolpin, D.W., 1991, Herbicides in surface waters of the Midwestern United States—The effect of spring flush: Environmental Science and Technology, v. 25, no. 10, p. 1794–1796.

Thurman, E.M., Goolsby, D.A., Meyer, M.T., Mills, M.S., Pomes, M.L., and Kolpin, D.W., 1992, A reconnaissance study of herbicides and their metabolites in surface water of the Midwestern United States using immunoassay and gas chromatography/mass spectrometry: Environmental Science and Technology, v. 26, no. 12, p. 2440–2447.

U.S. Environmental Protection Agency, 1982, Aquatic fate process data for organic priority pollutants: Washington, D.C., EPA–440/4–81–014, 434 p.

———1986, Quality criteria for water, update no. 1 and update no. 2 (May 1, 1987): Washington, D.C., EPA–440/5–86–001, 13 p.

———1988a, Protecting ground water—Pesticides and agricultural practices: Washington, D.C., EPA–440/6–88–001, 53 p.

———1988b, Interim sediment criteria values for nonpolar hydrophobic organic contaminants: Washington, D.C., Office of Water Regulations and Standards, Criteria and Standards Division, May 1988, SCD#17, 34 p.

———1990, Ambient water quality criteria—Notice of request for comments on proposed ambient water quality criteria for protection of aquatic life (5/14/90): Federal Register, v. 55, no. 93, p. 19986–19992.

———1991a, Amendments to the water quality standards regulation to establish the numeric criteria for priority toxic pollutants necessary to bring all states into compliance with Section 303 (c)(2)(B)—Proposed Rule (11/19/91): Federal Register, v. 56, no. 223, p. 58420–58478.

———1991b, National study of chemical residues in fish, v. 1: Office of Science and Technology, EPA 823–R–92–008a [variously paged].

———1992a, Drinking water regulations and health advisories: Washington, D. C., Office of Water, 12 p.

———1992b, Integrated risk information system (IRIS) data base (as of 6/92): Cincinnati, Ohio, Office of Health and Environmental Assessment, Environmental Criteria and Assessment Office.

———1992c, National study of chemical residues in fish, v. I and II: Office of Science and Technology, EPA 823–R92–008a and EPA 823–R92–008b [variously paged].

———1992d, Water quality standards—Establishment of numeric criteria for priority toxic pollutants—States compliance, final rule (12/22/92): Federal Register, v. 57, no. 246, p. 60848–60923.

———1993, Guidance for assessing chemical contaminant data for use in fish advisories, v. 1—Fish sampling and analysis: Washington, D.C., Office of Water, EPA 823–R–002, 351 p.

U.S. Food and Drug Administration, 1992, Action levels for poisonous or deleterious substances in human food and animal feed: Washington, D.C., Center for Food Safety and Applied Nutrition, 14 p.

U.S. Geological Survey, 1986, Land use and land cover digital data from 1:250,000- and 1:100,000-scale maps: Data user's guide 4, 36 p.

U.S. Soil Conservation Service, U.S. Department of Agriculture Stabilization and Conservation Service, University of California Cooperative Extension, and West Stanislaus Resource Conservation District, 1992, West Stanislaus hydrologic unit area project: Stanislaus County 1992 progress report: U.S. Soil Conservation Service, p. 32.

Washington State Administrative Code, January 6, 1988, Water quality standards for surface waters of the State of Washington, chapters 173–201: Olympia, Washington State Administrative Code, p. 6.

Wauchope, R.D., Buttler, T.M., Hornsby, A.G., Augustijn-Beckers, P.W.M., and Burt, J.P., 1992, The SCS/ARS/CES pesticide properties database for environmental decision-making: Reviews of Environmental Contamination and Toxicology, v. 123, 164 p.

Weiping, L., Merril, D.E., and Haith, D.A., January/February 1990, Loading functions for pesticide runoff: Research Journal WPCF, v. 62, no.1, p. 16–26.

Wershaw, R.L., and Aiken, G.R., 1985, Molecular size and weight measurements of humic substances, *in* Aiken, G.R., McKnight, D.M., Wershaw, R.L., and MacCarthy, P., eds., Humic substances in soil, sediment and water: New York, John Wiley & Sons, p. 477–492.

Wershaw, R.L., and Mikita, M.A., 1987, NMR of humic substances and coal: Chelsea, Michigan, Lewis Publishers, Inc.

Wershaw, R.L., Fishman, M.J., Grabbe, R.R., and Lowe, L.E., eds, 1987, Methods for the determination of organic substances in water and fluvial sediments: U.S. Geological Survey Techniques of Water-Resources Investigations, Book 5, Chapter A3, 80 p.

Witkowski, P.J., Smith, J.A., Fusillo, T.V., and Chiou, C.T., 1987, A review of surface-water sediment fractions and their interactions with persistent man-made organic compounds: U.S. Geological Survey Circular 993, 39 p.

SUPPLEMENTAL DATA SECTION

Table 59. Estimate of mean daily pesticide use for major agricultural areas, Yakima River Basin, Washington, 1989

[Mean daily application is calculated by dividing the annual use by 365 days and is reported in grams per day; East side, the area east of the Yakima River from Moxee Valley downstream to Benton City near Kiona including the Moxee, Granger, Sunnyside and Whitstran areas; West side, the area west of the Yakima River downstream from Ahtanum Ridge to the city of Mabton including the Wapato, Toppenish, and Mabton areas; Other areas, small areas throughout the basin, for example, crops in the Wenas Creek Subbasin; Total basin, all of the Yakima River Basin downstream to Benton City near Kiona excluding the Kennewick Irrigation District; 2,4-D, (2,4-dichlorophenoxy) acetic acid; MCPA, (4-chloro-2-methylphenoxy) acetic acid; MSMA, sodium hydrogen methylarsonate; see text for source of information]

Pesticide	Mean daily application of pesticides							Major crop or use
	Kittitas area	Tieton area	East side	West side	Other areas	Total basin	Moxee Subbasin[1]	
Abamectin	10	90	200	35	15	350	9	Pears
Acephate	200	300	1,000	1,000	500	3,000	100	Urban, mint
Acrolein	700	600	1,500	800	400	4,000	150	Aquatic, urban
Alachlor	4,000	0	9,000	25,000	7,000	45,000	150	Corn
Amitraz	300	2,000	3,500	800	2,400	9,000	200	Pears
Amitrole	100	200	400	300	0	1,000	70	Urban
Atrazine	1,500	90	10,000	10,000	3,410	25,000	150	Corn
Avermectin	10	70	150	25	95	350	7	Pears
Azinphos-methyl	5,000	100,000	100,000	45,000	100,000	350,000	3,000	Apples, pears, cherries
BEE (bee attractant)	150	100	2,500	400	1,350	4,500	100	Pears
Bacillus thuringiensis	4	90	90	35	31	250	25	Apples
Bendiocarb	15	25	50	20	40	150	7	Urban
Benfluralin	1,000	400	5,000	3,500	100	10,000	400	Alfalfa, urban
Benomyl	90	450	2,500	500	1,460	5,000	150	Grapes, apples
Bentazone	50	0	2,000	2,000	450	4,500	10	Corn,
Boron	50	450	7,000	600	1,900	10,000	300	Cherries, urban
Bromacil	1,000	500	2,000	600	400	4,500	90	Roads
Bromoxynil	1,500	50	1,000	2,000	450	5,000	300	Wheat, alfalfa
Butylate	1,000	0	25,000	15,000	4,000	45,000	250	Corn
Byclobutanil	0	0	900	150	50	1,100	0	Grapes
Cacodylic acid	10	10	40	15	25	100	6	Urban
Calcium polysulfide	1,050	21,000	22,500	9,500	11,950	66,000	5,100	Apples, pears
Captan	1,000	5,000	40,000	15,000	9,000	70,000	1,500	Corn, apples, grapes
Carbaryl	6,000	10,000	80,000	40,000	14,000	150,000	3,500	Asparagus, apples, grapes
Carbofuran	100	20	6,000	1,500	1,380	9,000	30	Corn, grapes
Carboxin	3,000	0	6,000	7,000	0	16,000	800	Wheat, corn
Chinomethionat	100	2,500	2,500	1,000	1,900	8,000	600	Pears, apples
Chlorflurenol	10	10	25	15	30	90	5	Urban
Chlorophacinone	60	300	300	150	90	900	70	Apples
Chloropicrin	100	70	450	450	430	1,500	30	Potatoes, urban
Chlorpyrifos	3,000	45,000	80,000	40,000	32,000	200,000	10,000	Potatoes, urban
Chlorothalonil	400	100	600	600	300	2,000	60	Mint, potatoes
Chlorsulfuron	200	0	8,000	5,000	1,800	15,000	1,500	Roads, wheat
Clopyralid	25,000	1,500	10,000	10,000	3,500	50,000	500	Pasture, hay
Copper	2,500	8,000	140,000	50,000	9,500	210,000	40,000	Hops, cherries, pears

Table 59. Estimate of mean daily pesticide use for major agricultural areas, Yakima River Basin, Washington, 1989—Continued

Pesticide	Mean daily application of pesticides							Major crop or use
	Kittitas area	Tieton area	East side	West side	Other areas	Total basin	Moxee Subbasin[1]	
Cryolite	10	300	300	150	140	900	70	Apples
Cyanazine	600	0	9,500	4,500	400	15,000	200	Field corn, wheat
Cyfluthrin	20	150	350	60	120	700	15	Pears
Cytokinins	15	400	400	150	35	1,000	90	Apples
Dacthal (DCPA)	0	50	20	15	5	90	25	Roads, onions
Dalapon	300	0	10,000	5,000	700	16,000	100	Field corn
Demeton[2]	1	0	2	2	2	7	0	Barley
Diatomaceous earth (DE)	25	150	400	60	165	800	20	Pears
Diazinon	5,000	4,000	100,000	50,000	41,000	200,000	40,000	Hops, alfalfa, urban
Dicamba	40,000	3,000	25,000	20,000	12,000	100,000	5,000	Pasture, hay, asparagus
Dichlobenil	1,500	2,500	2,000	1,500	2,500	10,000	600	Apples, aquatics, roads, urban
2,4-D	250,000	40,000	150,000	100,000	160,000	700,000	35,000	Pasture, roads, apples
Dichloropropene	10,000	0	6,000	10,000	4,000	30,000	600	Potatoes
Dichlorvos	35	50	100	50	65	300	20	Urban
Diclofop	250	0	200	400	50	900	70	Wheat
Dicofol	20	300	10,000	10,000	4,680	25,000	3,000	Hops, mint
Difenzoquat	250	0	200	350	100	900	70	Wheat
Dimethoate	1,500	2,000	38,000	9,500	3,000	54,000	900	Grapes, wheat, apples
4,6-dinitro-*o*-cresol (DNOC)	1,000	25,000	25,000	10,000	9,000	70,000	5,000	Apples, pears
Dinocap	150	1,000	2,500	450	900	5,000	150	Pears, apples
Dinoseb[3]	35	50	100	50	65	300	20	Urban
Diphacinone	10	300	300	150	140	900	70	Apples
Diquat	250	150	200	150	150	900	70	Urban, aquatics
Disulfoton	10,000	90	35,000	35,000	910	81,000	4,000	Asparagus, wheat, hops
Diuron	9,000	35,000	70,000	40,000	6,000	160,000	10,000	Roads, apples, asparagus
Dodine	500	10,000	10,000	4,500	10,000	35,000	2,000	Apples, pears
S-ethyl dipropylthiocarbamate (EPTC)	4,500	900	24,000	15,000	600	45,000	1,000	Alfalfa, corn, potatoes
Endosulfan	2,000	30,000	45,000	15,000	8,000	100,000	7,000	Apples, pears, peaches
Endothal	250	100	10,000	6,000	3,650	20,000	4,500	Hops
Esfenvalerate	40	60	200	200	100	600	10	Corn, pears
Ethephon	150	3,500	3,500	1,500	1,350	10,000	800	Apples
Ethion	35	900	900	400	765	3,000	200	Apples
Ethoprop	1,000	0	700	1,300	0	3,000	60	Potatoes
Ethylan (Perthane)	9	60	150	20	11	250	6	Pears
Fenarimol	25	600	1,500	350	525	3,000	170	Apples, cherries
Fenbutatin oxide	50	500	1,000	250	700	2,500	90	Pears, apples
Fenoxaprop	15	0	9	20	6	50	5	Wheat
Fenvalerate	10	90	200	35	65	400	10	Pears
Fluazifop	0	0	600	350	50	1,000	0	Asparagus, grapes
Fluridone	25	25	6	25	19	100	2	Aquatics
Fonofos	450	0	7,000	5,000	2,550	15,000	20	Asparagus, corn

Table 59. Estimate of mean daily pesticide use for major agricultural areas, Yakima River Basin, Washington, 1989—Continued

Pesticide	Mean daily application of pesticides							Major crop or use
	Kittitas area	Tieton area	East side	West side	Other areas	Total basin	Moxee Subbasin[1]	
Formetanate	30	700	800	300	170	2,000	150	Apples, pears
Formulated hydrochloride	0	0	20	10	5	35	0	Peaches
Fosamine	20	10	10	25	25	90	2	Aquatics
Gibberellic acid (GA)	50	1,000	1,500	500	950	4,000	300	Apples, cherries
Glyphosate	40,000	80,000	200,000	80,000	0	400,000	15,000	Apples, corn, grapes
Hexazinone	6,000	300	3,500	2,500	2,700	15,000	450	Alfalfa, hay, oats
Imazalil	0	0	2	1	2	5	0	Barley
Imazamethabenz	80	0	100	150	20	350	20	Wheat, barley
Imazapyr	0	0	100	25	575	700	0	Firing range, roads
Iprodione	300	10	4,000	900	90	5,300	25	Hops, potatoes
Iron	0	0	7,000	1,000	1,000	9,000	0	Grapes, hops
Lactic acid	10	300	300	150	140	900	70	Apples
Lindane (γ-hexachlorocyclohexane)	6,000	50	4,000	9,000	950	20,000	2,000	Wheat, barley
Linuron	0	0	2,500	2,000	0	4,500	0	Asparagus
Malathion	90,000	45,000	200,000	80,000	35,000	450,000	25,000	Pasture, apples, grapes, cherries
Maleic hydrazide	500	0	300	600	100	1,500	25	Potatoes
Mancozeb	3,500	15,000	15,000	10,000	6,500	50,000	4,000	Asparagus, wheat
Maneb	200	0	600	700	0	1,500	10	Potatoes
Manganese	200	0	0	0	50	250	0	Oats
MCPA	15,000	700	6,000	5,000	3,300	30,000	1,500	Pasture, hay
Mecoprop (MCPP)	0	0	450	200	350	1,000	0	Urban
Metalaxyl	30	25	40,000	20,000	945	61,000	15,000	Hops
Metam	40,000	0	23,000	50,000	1,000	114,000	0	Potatoes
Methamidophos	1,000	0	600	1,000	400	3,000	0	Potatoes
Methidathion	45,000	4,000	9,000	6,000	6,000	70,000	900	Hay, apples
Methomyl	900	25	25,000	14,000	75	40,000	50	Grapes, asparagus
Methoxychlor	900	300	6,000	4,000	800	12,000	300	Alfalfa, asparagus
Methyl bromide	3,500	5,000	10,000	5,000	6,500	30,000	2,000	Urban
Methyl parathion	100	2,000	2,000	800	1,100	6,000	400	Apples, pears
Metiram	450	8,000	8,000	4,000	3,550	24,000	1,500	Asparagus
Metolachlor	400	0	4,500	3,500	600	9,000	45	Corn
Metribuzin	900	200	13,000	10,000	900	25,000	200	Asparagus, alfalfa
Metsulfuron	350	10	150	300	190	1,000	50	Oats, wheat
Mevinphos	35	300	500	250	215	1,300	80	Apples, alfalfa, corn
Monocarbamide dihydrogensulfate	40,000	0	23,000	50,000	2,000	115,000	0	Potatoes
MSMA	20	35	70	35	40	200	10	Urban
Myclobutanil	25	700	2,000	450	825	4,000	200	Apples, grapes
Naled	35	15	6,000	2,000	950	9,000	900	Hops, grapes alfalfa seed
1-Naphthylacetamide (NAD)	9	200	200	100	191	700	50	Apples

Table 59. Estimate of mean daily pesticide use for major agricultural areas, Yakima River Basin, Washington, 1989—Continued

Pesticide	Mean daily application of pesticides							Major crop or use
	Kittitas area	Tieton area	East side	West side	Other areas	Total basin	Moxee Subbasin[1]	
1-Naphthylacetic acid (NAA)	60	1,000	1,500	500	940	4,000	250	Apples, pears
Napropamide	15	25	13,000	6,000	960	20,000	7	Grapes, mint, asparagus
Nicosulfuron	0	0	20	9	6	35	0	Corn
Norflurazon	600	10,000	90,000	40,000	9,400	150,000	30,000	Hops, grapes, asparagus, pears
Oil	35,000	700,000	1,000,000	350,000	415,000	2,500,000	150,000	Apples, cherries, pears, peaches
Oryzalin	450	9,000	45,000	10,000	15,550	80,000	2,500	Grapes, apples, roads
Oxadiazon	15	25	35	20	5	100	7	Urban
Oxamyl	350	8,000	8,000	3,500	5,150	25,000	2,000	Apples, pears
Oxydemeton-methyl	15	300	500	400	285	1,500	70	Apples, mint
Oxyfluorfen	15	450	4,300	2,400	835	8,000	100	Grapes, mint, apples, cherries
Oxytetracycline (Terramycin)	100	1,000	2,000	400	1,000	4,500	150	Pears, apples
Paraquat	1,000	3,000	20,000	9,000	7,000	40,000	4,500	Hops, grapes, apples
Parathion-ethyl	3,000	20,000	80,000	25,000	22,000	150,000	15,000	Apples, cherries, grapes
Pendimethalin	350	300	6,000	3,000	350	10,000	100	Corn, grapes, apples
Permethrin	400	50	1,500	2,500	550	5,000	40	Corn, alfalfa, asparagus
Phorate	1,000	0	13,000	10,000	1,000	25,000	4,500	Hops, potatoes, corn
Phosalone	10	300	300	150	140	900	70	Apples
Phosmet	600	8,000	10,000	3,500	2,900	25,000	3,000	Apples, pears
Phosphamidon	1,500	35,000	35,000	15,000	13,500	100,000	8,000	Apples
Picloram	350	10	500	800	2,340	4,000	150	Firing range, wheat, pasture
Piperonyl butoxide	25	32	105	32	91	285	11	Urban, prunes
Primisulfuron	5	0	150	60	35	250	1	Corn
Prometon	90	100	200	100	110	600	70	Urban, roads
Pronamide	700	500	9,000	3,500	1,300	15,000	300	Alfalfa, grapes, apples
Propachlor	150	0	5,000	2,500	350	8,000	1,500	Corn
Propargite	400	5,000	70,000	40,000	14,600	130,000	3,500	Hops, mint, apples, grapes
Propetamphos	25	50	100	50	25	250	15	Urban
Propoxur	90	100	250	100	160	700	50	Urban
Propyzamide	60	300	300	150	190	1,000	90	Apples
Pyrethrins	35	350	350	150	115	1,000	70	Apples, urban
Quintozene (PCNB)	20	0	20	15	15	70	0	Barley, oats
Rotenone	60	1,500	2,000	800	640	5,000	300	Apples, cherries
Ryania	4	30	800	90	76	1,000	30	Cherries
Sethoxydim	15	5	900	400	180	1,500	5	Asparagus, grapes
Simazine	1,000	15,000	40,000	15,000	9,000	80,000	3,500	Apples, grapes, asparagus, prunes
Sodium trichloroacetate (TCA)	0	9	10	10	16	45	5	Roads
Streptomycin	25	400	500	150	225	1,300	80	Grapes, apples, prunes
Strychnine	10	300	350	150	190	1,000	80	Apples, cherries
Sulfometuron methyl	25	600	700	900	775	3,000	350	Roads
Sulfur	1,500	10,000	90,000	15,000	33,500	150,000	4,000	Cherries, peaches, hops, apples

Table 59. Estimate of mean daily pesticide use for major agricultural areas, Yakima River Basin, Washington, 1989—Continued

Pesticide	Mean daily application of pesticides							Major crop or use
	Kittitas area	Tieton area	East side	West side	Other areas	Total basin	Moxee Subbasin[1]	
Sulfuryl fluoride	1,500	2,000	5,000	2,000	4,500	15,000	800	Roads, urban
Tebuthiuron	1,000	70	600	500	330	2,500	150	Pasture
Terbacil	700	3,500	15,000	25,000	800	45,000	1,000	Mint, apples, alfalfa
Terbufos	60	0	2,000	900	40	3,000	20	Corn
Thiabendazole (TBZ)	25	0	20	35	20	100	9	Wheat
Thifensulfuron	450	0	300	600	150	1,500	150	Wheat
Thiophanate-methyl	350	0	900	1,000	250	2,500	20	Potatoes
Thiram	3,000	0	10,000	10,000	2,000	25,000	800	Corn, wheat
Triadimefon	90	350	800	300	460	2,000	100	Apples, grapes
Triadimenol	150	0	2,000	1,000	350	3,500	40	Corn, wheat
Tri-allate	150	0	200	350	0	700	50	Wheat, barley
Tribenuron	60	0	35	90	15	200	20	Wheat
Trichlorfon	25	10	900	400	165	1,500	9	Corn
Triclopyr	1,000	1,500	5,000	500	2,000	10,000	1,500	Roads, pasture
Trifluralin	600	100	20,000	15,000	4,300	40,000	4,000	Hops, asparagus
Triforine	0	50	90	20	40	200	10	Apples, pears
Trithion (carbophenothion)	6	90	100	45	59	300	50	Apples
Vinclozolin	10	300	300	150	140	900	70	Apples
Xylene	1,500	2,000	3,500	4,000	0	11,000	500	Aquatics
Zinc	700	4,500	15,000	2,500	7,300	30,000	700	Pears, cherries, grapes
Ziram	1,000	20,000	25,000	8,000	6,000	60,000	4,000	Apples, pears

[1]Application amounts included in tabulation of East side, listed in this table.
[2]Demeton-S and demeton-O were not sold in 1989.
[3]Dinoseb was not sold in 1989

Table 60. Concentrations and instantaneous loads of suspended sediment and selected organic compounds in water, Yakima River Basin, Washington, June 25–30, 1989

[Streamflow reported in cubic feet per second; concentrations reported in nanograms per liter, except for suspended sediment which is in milligrams per liter; loads reported in grams per day, except for suspended-sediment loads which are in kilograms per day; concentrations are from filtered-water samples except chlordane, chlorpyrifos, DDT+DDE+DDD (dichlorodiphenyltrichloroethane {DDT} plus dichlorodiphenydichloroethylene {DDE} plus dichlorodiphenyldichloroethane {DDD}), diazinon, dieldrin, endosulfan I, malathion, parathion, and phorate which are from unfiltered-water samples; for subtotal-loading calculations, less-than (<) loads were assumed to be equal to zero; all concentrations less than the minimum reporting level were assumed to be zero for these load calculations; trace concentrations are not identified in this listing; --, not applicable or not determined]

Station number	Station name	Streamflow	Concentration	Load
	SUSPENDED SEDIMENT			
	Background			
12478200	Cooper River at Salmon LaSac	229	1.6	900
12484550	Umtanum Creek near mouth	1	1.0	2
12496510	Pacific Power and Light Company Wasteway	--	4.4	--
12507594	Satus Creek above Wilson-Charley Canyon	12	2.3	68
			Subtotal	970
	Drinking water			
12496511	City of Yakima—Finish water	--	.2	--
461720120043200	Well 1, Sunnyside (09/22E-04P01)	--	--	--
462510120323900	Well 2, Harrah (11N/18E-22R02)	--	--	--
	Kittitas			
12484100	Wilson Creek above Cherry Creek	144	13	4,600
12484480	Cherry Creek at Thrall	126	120	37,000
			Subtotal	41,600
	Tieton			
12500445	Wide Hollow Creek near mouth	33	8	640
12502500	Ahtanum Creek at Union Gap	23	9	510
			Subtotal	1,150
	East-side tributaries			
12500430	Moxee Drain at Thorp Road	65	440	70,000
12505460	Granger Drain at mouth	37	640	58,000
12508850	Sulphur Creek Wasteway	259	240	150,000
12509710	Spring Creek at mouth	47	200	23,000
12509829	Snipes Creek at mouth	50	65	8,000
			Subtotal	309,000
	West-side tributaries			
12505350	East Toppenish Drain at Wilson Road	53	52	6,700
12505410	Sub 35 Drain at Parton Road	56	28	3,800
12505510	Marion Drain at Indian Church Road	71	29	5,000
12507508	Toppenish Creek at Indian Church Road	39	32	3,000
12508620	Satus Creek at gage at Satus	124	38	12,000
12508630	South Drain near Satus	82	140	28,000
			Subtotal	58,500
	Main stem and Naches River			
12479500	Yakima River at Cle Elum	3,780	6	55,000
12484500	Yakima River at Umtanum	3,630	25	220,000
12499000	Naches River near North Yakima	1,190	6	17,000
12500450	Yakima River above Ahtanum Creek	3,630	24	210,000
12507585	Yakima River at river mile 72	1,070	44	120,000
12509050	Yakima River at Euclid Bridge at river mile 55	1,560	47	180,000
12510500	Yakima River at Kiona	1,500	30	110,000
	ALACHLOR			
	Background			
12478200	Cooper River at Salmon LaSac	229	<.99	<.55
12484550	Umtanum Creek near mouth	1	<.42	<.01
12496510	Pacific Power and Light Company Wasteway	--	<.50	--
12507594	Satus Creek above Wilson-Charley Canyon	12	<.40	<.01

Table 60. Concentrations and instantaneous loads of suspended sediment and selected organic compounds in water, Yakima River Basin, Washington, June 25–30, 1989—Continued

Station number	Station name	Streamflow	Concentration	Load
	ALACHLOR—Continued			
	Drinking water			
12496511	City of Yakima—Finish water	--	<.64	--
461720120043200	Well 1, Sunnyside (09N/22E-04P01)	--	<.64	--
462510120323900	Well 2, Harrah (11N/18E-22R02)	--	<.64	--
	Kittitas			
12484100	Wilson Creek above Cherry Creek	144	<2.1	<.74
12484480	Cherry Creek at Thrall	126	<1.2	<.37
	Tieton			
12500445	Wide Hollow Creek near mouth	33	<2.0	<.16
12502500	Ahtanum Creek at Union Gap	23	<1.3	<.07
	East-side tributaries			
12500430	Moxee Drain at Thorp Road	65	<4.0	<.64
12505460	Granger Drain at mouth	37	33	3.0
12508850	Sulphur Creek Wasteway	259	33	21
12509710	Spring Creek at mouth	47	<2.0	<.23
12509829	Snipes Creek at mouth	50	<2.0	<.24
			Subtotal	24.0
	West-side tributaries			
12505350	East Toppenish Drain at Wilson Road	53	4.5	.58
12505410	Sub 35 Drain at Parton Road	56	98	13
12505510	Marion Drain at Indian Church Road	71	20	3.5
12507508	Toppenish Creek at Indian Church Road	39	16	1.5
12508620	Satus Creek at gage at Satus	124	13	3.9
12508630	South Drain near Satus	82	17	3.4
			Subtotal	25.9
	Main stem and Naches River			
12479500	Yakima River at Cle Elum	3,780	<.52	<4.8
12484500	Yakima River at Umtanum	3,630	<1.6	<14
12499000	Naches River near North Yakima	1,190	<.39	<1.1
12500450	Yakima River above Ahtanum Creek	3,630	<1.5	<13
12507585	Yakima River at river mile 72	1,070	11	29
12509050	Yakima River at Euclid Bridge at river mile 55	1,560	19	72
12510500	Yakima River at Kiona	1,500	12	44
	ATRAZINE			
	Background			
12478200	Cooper River at Salmon LaSac	229	<1.4	<.78
12484550	Umtanum Creek near mouth	1	<.77	<.01
12496510	Pacific Power and Light Company Wasteway	--	<.17	--
12507594	Satus Creek above Wilson-Charley Canyon	12	<.56	<.02
	Drinking water			
12496511	City of Yakima—Finish water	--	<.88	--
461720120043200	Well 1, Sunnyside (09N/22E-04P01)	--	18	--
462510120323900	Well 2, Harrah (11N/18E-22R02)	--	57	--
	Kittitas			
12484100	Wilson Creek above Cherry Creek	144	27	9.5
12484480	Cherry Creek at Thrall	126	71	22
			Subtotal	31.5
	Tieton			
12500445	Wide Hollow Creek near mouth	33	6.8	.55
12502500	Ahtanum Creek at Union Gap	23	<1.8	<.10
			Subtotal	.55

Table 60. Concentrations and instantaneous loads of suspended sediment and selected organic compounds in water, Yakima River Basin, Washington, June 25–30, 1989—Continued

Station number	Station name	Streamflow	Concentration	Load
	ATRAZINE—Continued			
	East-side tributaries			
12500430	Moxee Drain at Thorp Road	65	8.5	1.4
12505460	Granger Drain at mouth	37	48	4.3
12508850	Sulphur Creek Wasteway	259	49	31
12509710	Spring Creek at mouth	47	13	1.5
12509829	Snipes Creek at mouth	50	12	1.5
			Subtotal	39.7
	West-side tributaries			
12505350	East Toppenish Drain at Wilson Road	53	130	17
12505410	Sub 35 Drain at Parton Road	56	88	12
12505510	Marion Drain at Indian Church Road	71	34	5.9
12507508	Toppenish Creek at Indian Church Road	39	54	5.2
12508620	Satus Creek at gage at Satus	124	46	14
12508630	South Drain near Satus	82	130	27
			Subtotal	81.1
	Main stem and Naches River			
12479500	Yakima River at Cle Elum	3,780	.27	2.5
12484500	Yakima River at Umtanum	3,630	10	89
12499000	Naches River near North Yakima	1,190	.26	.76
12500450	Yakima River above Ahtanum Creek	3,630	5.6	50
12507585	Yakima River at river mile 72	1,070	26	68
12509050	Yakima River at Euclid Bridge at river mile 55	1,560	61	230
12510500	Yakima River at Kiona	1,500	32	120
	CHLORDANE, cis- and trans-			
	Background			
12478200	Cooper River at Salmon LaSac	229	.15	.08
12484550	Umtanum Creek near mouth	1	<.42	<.01
12496510	Pacific Power and Light Company Wasteway	--	<.44	--
12507594	Satus Creek above Wilson-Charley Canyon	12	<.84	<.02
			Subtotal	.08
	Drinking water			
12496511	City of Yakima—Finish water	--	<.35	--
461720120043200	Well 1, Sunnyside (09N/22E-04P01)	--	<.35	--
462510120323900	Well 2, Harrah (11N/18E-22R02)	--	<.35	--
	Kittitas			
12484100	Wilson Creek above Cherry Creek	144	<4.0	<1.4
12484480	Cherry Creek at Thrall	126	4.0	1.2
			Subtotal	1.2
	Tieton			
12500445	Wide Hollow Creek near mouth	33	.76	.06
12502500	Ahtanum Creek at Union Gap	23	.41	.02
			Subtotal	.08
	East-side tributaries			
12500430	Moxee Drain at Thorp Road	65	<3.5	<.55
12505460	Granger Drain at mouth	37	.14	.01
12508850	Sulphur Creek Wasteway	259	<8.5	<5.4
12509710	Spring Creek at mouth	47	1.4	.16
12509829	Snipes Creek at mouth	50	.37	.04
			Subtotal	.21
	West-side tributaries			
12505350	East Toppenish Drain at Wilson Road	53	2.6	.34
12505410	Sub 35 Drain at Parton Road	56	.45	.06

Table 60. Concentrations and instantaneous loads of suspended sediment and selected organic compounds in water, Yakima River Basin, Washington, June 25–30, 1989—Continued

Station number	Station name	Streamflow	Concentration	Load
	CHLORDANE, cis- and trans-—Continued			
	West-side tributaries—Continued			
12505510	Marion Drain at Indian Church Road	71	.38	.07
12507508	Toppenish Creek at Indian Church Road	39	3.0	.29
12508620	Satus Creek at gage at Satus	124	<4.0	<1.2
12508630	South Drain near Satus	82	<3.5	<.70
			Subtotal	.76
	Main stem and Naches River			
12479500	Yakima River at Cle Elum	3,780	<.46	<4.2
12484500	Yakima River at Umtanum	3,630	<1.4	<13
12499000	Naches River near North Yakima	1,190	<.38	<1.1
12500450	Yakima River above Ahtanum Creek	3,630	<1.4	<12
12507585	Yakima River at river mile 72	1,070	<2.5	<6.5
12509050	Yakima River at Euclid Bridge at river mile 55	1,560	--	--
12510500	Yakima River at Kiona	1,500	<1.4	<5.2
	CHLORPYRIFOS			
	Background			
12478200	Cooper River at Salmon LaSac	229	<1.6	<.90
12484550	Umtanum Creek near mouth	1	1.0	<.01
12496510	Pacific Power and Light Company Wasteway	--	<.75	--
12507594	Satus Creek above Wilson-Charley Canyon	12	<.69	<.02
	Drinking water			
12496511	City of Yakima—Finish water	--	<1.0	--
461720120043200	Well 1, Sunnyside (09N/22E-04P01)	--	<1.0	--
462510120323900	Well 2, Harrah (11N/18E-22R02)	--	<1.0	--
	Kittitas			
12484100	Wilson Creek above Cherry Creek	144	<3.5	<1.2
12484480	Cherry Creek at Thrall	126	.30	.09
			Subtotal	.09
	Tieton			
12500445	Wide Hollow Creek near mouth	33	<3.2	<.26
12502500	Ahtanum Creek at Union Gap	23	<2.3	<.13
	East-side tributaries			
12500430	Moxee Drain at Thorp Road	65	.48	.08
12505460	Granger Drain at mouth	37	<6.1	<.55
12508850	Sulphur Creek Wasteway	259	.16	.10
12509710	Spring Creek at mouth	47	3.0	.34
12509829	Snipes Creek at mouth	50	.20	.02
			Subtotal	.54
	West-side tributaries			
12505350	East Toppenish Drain at Wilson Road	53	3.7	.48
12505410	Sub 35 Drain at Parton Road	56	<3.2	<.44
12505510	Marion Drain at Indian Church Road	71	<3.2	<.56
12507508	Toppenish Creek at Indian Church Road	39	.10	.01
12508620	Satus Creek at gage at Satus	124	.20	.06
12508630	South Drain near Satus	82	<6.8	<1.4
			Subtotal	.55
	Main stem and Naches River			
12479500	Yakima River at Cle Elum	3,780	<.79	<7.3
12484500	Yakima River at Umtanum	3,630	<2.5	<22
12499000	Naches River near North Yakima	1,190	<.69	<2.0
12500450	Yakima River above Ahtanum Creek	3,630	<2.3	<21
12507585	Yakima River at river mile 72	1,070	<3.2	<8.4

Table 60. Concentrations and instantaneous loads of suspended sediment and selected organic compounds in water, Yakima River Basin, Washington, June 25–30, 1989—Continued

Station number	Station name	Streamflow	Concentration	Load
	CHLORPYRIFOS—Continued			
	Main stem and Naches River—Continued			
12509050	Yakima River at Euclid Bridge at river mile 55	1,560	<2.3	<8.8
12510500	Yakima River at Kiona	1,500	9.6	35
	(2,4-DICHLOROPHENOXY) ACETIC ACID [2,4-D]			
	Background			
12478200	Cooper River at Salmon LaSac	229	<10	<5.6
12484550	Umtanum Creek near mouth	1	<10	<.02
12496510	Pacific Power and Light Company Wasteway	--	<10	--
12507594	Satus Creek above Wilson-Charley Canyon	12	<10	<.29
	Drinking water			
12496511	City of Yakima—Finish water	--	<10	--
461720120043200	Well 1, Sunnyside (09N/22E-04P01)	--	<10	--
462510120323900	Well 2, Harrah (11N/18E-22R02)	--	<10	--
	Kittitas			
12484100	Wilson Creek above Cherry Creek	144	<10	<3.5
12484480	Cherry Creek at Thrall	126	290	89
			Subtotal	89
	Tieton			
12500445	Wide Hollow Creek near mouth	33	<10	<.81
12502500	Ahtanum Creek at Union Gap	23	--	--
	East-side tributaries			
12500430	Moxee Drain at Thorp Road	65	120	19
12505460	Granger Drain at mouth	37	140	13
12508850	Sulphur Creek Wasteway	259	80	51
12509710	Spring Creek at mouth	47	60	6.9
12509829	Snipes Creek at mouth	50	<10	<1.2
			Subtotal	89.9
	West-side tributaries			
12505350	East Toppenish Drain at Wilson Road	53	<10	<1.3
12505410	Sub 35 Drain at Parton Road	56	30	4.1
12505510	Marion Drain at Indian Church Road	71	140	24
12507508	Toppenish Creek at Indian Church Road	39	<10	<.95
12508620	Satus Creek at gage at Satus	124	10	3.0
12508630	South Drain near Satus	82	20	4.0
			Subtotal	35.1
	Main stem and Naches River			
12479500	Yakima River at Cle Elum	3,780	<10	<92
12484500	Yakima River at Umtanum	3,630	<10	<89
12499000	Naches River near North Yakima	1,190	<10	<29
12500450	Yakima River above Ahtanum Creek	3,630	70	620
12507585	Yakima River at river mile 72	1,070	<10	<26
12509050	Yakima River at Euclid Bridge at river mile 55	1,560	70	270
12510500	Yakima River at Kiona	1,500	90	330
	DDT+DDE+DDD			
	Background			
12478200	Cooper River at Salmon LaSac	229	.61	.34
12484550	Umtanum Creek near mouth	1	.50	<.01
12496510	Pacific Power and Light Company Wasteway	--	1.80	--
12507594	Satus Creek above Wilson-Charley Canyon	12	.61	.02
			Subtotal	.36

Table 60. Concentrations and instantaneous loads of suspended sediment and selected organic compounds in water, Yakima River Basin, Washington, June 25–30, 1989—Continued

Station number	Station name	Streamflow	Concentration	Load
	DDT+DDE+DDD—Continued			
	Drinking water			
12496511	City of Yakima—Finish water	--	.36	--
461720120043200	Well 1, Sunnyside (09N/22E-04P01)	--	.50	--
462510120323900	Well 2, Harrah (11N/18E-22R02)	--	<.50	--
	Kittitas			
12484100	Wilson Creek above Cherry Creek	144	.29	.10
12484480	Cherry Creek at Thrall	126	8.0	2.4
			Subtotal	2.50
	Tieton			
12500445	Wide Hollow Creek near mouth	33	2.7	.21
12502500	Ahtanum Creek at Union Gap	23	1.4	.08
			Subtotal	.30
	East-side tributaries			
12500430	Moxee Drain at Thorp Road	65	36	5.7
12505460	Granger Drain at mouth	37	83	7.5
12508850	Sulphur Creek Wasteway	259	57	36
12509710	Spring Creek at mouth	47	50	5.7
12509829	Snipes Creek at mouth	50	22	2.6
			Subtotal	57.5
	West-side tributaries			
12505350	East Toppenish Drain at Wilson Road	53	12	1.6
12505410	Sub 35 Drain at Parton Road	56	10	1.4
12505510	Marion Drain at Indian Church Road	71	4.6	.80
12507508	Toppenish Creek at Indian Church Road	39	4.3	.41
12508620	Satus Creek at gage at Satus	124	2.5	.76
12508630	South Drain near Satus	82	20	4.0
			Subtotal	8.97
	Main stem and Naches River			
12479500	Yakima River at Cle Elum	3,780	.50	4.6
12484500	Yakima River at Umtanum	3,630	.43	3.8
12499000	Naches River near North Yakima	1,190	.75	2.2
12500450	Yakima River above Ahtanum Creek	3,630	3.2	28
12507585	Yakima River at river mile 72	1,070	10	26
12509050	Yakima River at Euclid Bridge at river mile 55	1,560	12	46
12510500	Yakima River at Kiona	1,500	15	55
	DIAZINON			
	Background			
12478200	Cooper River at Salmon LaSac	229	<1.4	<.78
12484550	Umtanum Creek near mouth	1	<1.2	<.01
12496510	Pacific Power and Light Company Wasteway	--	<.66	--
12507594	Satus Creek above Wilson-Charley Canyon	12	<.68	<.02
	Drinking water			
12496511	City of Yakima—Finish water	--	.06	--
461720120043200	Well 1, Sunnyside (09N/22E-04P01)	--	<.86	--
462510120323900	Well 2, Harrah (11N/18E-22R02)	--	<.87	--
	Kittitas			
12484100	Wilson Creek above Cherry Creek	144	<2.9	<1.0
12484480	Cherry Creek at Thrall	126	<2.8	<.86
	Tieton			
12500445	Wide Hollow Creek near mouth	33	5.9	.48
12502500	Ahtanum Creek at Union Gap	23	3.3	.18
			Subtotal	.66

Table 60. Concentrations and instantaneous loads of suspended sediment and selected organic compounds in water, Yakima River Basin, Washington, June 25–30, 1989—Continued

Station number	Station name	Streamflow	Concentration	Load
	DIAZINON—Continued			
	East-side tributaries			
12500430	Moxee Drain at Thorp Road	65	410	65
12505460	Granger Drain at mouth	37	13	1.2
12508850	Sulphur Creek Wasteway	259	66	42
12509710	Spring Creek at mouth	47	72	8.3
12509829	Snipes Creek at mouth	50	32	4.0
			Subtotal	120
	West-side tributaries			
12505350	East Toppenish Drain at Wilson Road	53	28	3.6
12505410	Sub 35 Drain at Parton Road	56	25	3.4
12505510	Marion Drain at Indian Church Road	71	55	9.6
12507508	Toppenish Creek at Indian Church Road	39	31	3.0
12508620	Satus Creek at gage at Satus	124	22	6.7
12508630	South Drain near Satus	82	150	30
			Subtotal	56.3
	Main stem and Naches River			
12479500	Yakima River at Cle Elum	3,780	.67	6.2
12484500	Yakima River at Umtanum	3,630	<2.1	<19
12499000	Naches River near North Yakima	1,190	<.68	<2.0
12500450	Yakima River above Ahtanum Creek	3,630	7.9	72
12507585	Yakima River at river mile 72	1,070	35	92
12509050	Yakima River at Euclid Bridge at river mile 55	1,560	120	460
12510500	Yakima River at Kiona	1,500	39	140
	DICAMBA			
	Background			
12478200	Cooper River at Salmon LaSac	229	<10	<5.6
12484550	Umtanum Creek near mouth	1	<10	<.02
12496510	Pacific Power and Light Company Wasteway	--	<10	--
12507594	Satus Creek above Wilson-Charley Canyon	12	<10	<.29
	Drinking water			
12496511	City of Yakima—Finish water	--	<10	--
461720120043200	Well 1, Sunnyside (09N/22E-04P01)	--	<10	--
462510120323900	Well 2, Harrah (11N/18E-22R02)	--	<10	--
	Kittitas			
12484100	Wilson Creek above Cherry Creek	144	50	18
12484480	Cherry Creek at Thrall	126	100	31
			Subtotal	49
	Tieton			
12500445	Wide Hollow Creek near mouth	33	<10	<.81
12502500	Ahtanum Creek at Union Gap	23	--	--
	East-side tributaries			
12500430	Moxee Drain at Thorp Road	65	<10	<1.6
12505460	Granger Drain at mouth	37	50	4.5
12508850	Sulphur Creek Wasteway	259	10	6.3
12509710	Spring Creek at mouth	47	10	1.2
12509829	Snipes Creek at mouth	50	10	1.2
			Subtotal	13.2
	West-side tributaries			
12505350	East Toppenish Drain at Wilson Road	53	<10	<1.3
12505410	Sub 35 Drain at Parton Road	56	<10	<1.4
12505510	Marion Drain at Indian Church Road	71	<10	<1.7
12507508	Toppenish Creek at Indian Church Road	39	<10	<.95

Table 60. Concentrations and instantaneous loads of suspended sediment and selected organic compounds in water, Yakima River Basin, Washington, June 25–30, 1989—Continued

Station number	Station name	Streamflow	Concentration	Load
	DICAMBA—Continued			
	West-side tributaries—Continued			
12508620	Satus Creek at gage at Satus	124	<10	<3.0
12508630	South Drain near Satus	82	<10	<2.0
	Main stem and Naches River			
12479500	Yakima River at Cle Elum	3,780	<10	<92
12484500	Yakima River at Umtanum	3,630	10	89
12499000	Naches River near North Yakima	1,190	<10	<29
12500450	Yakima River above Ahtanum Creek	3,630	<10	<89
12507585	Yakima River at river mile 72	1,070	<10	<26
12509050	Yakima River at Euclid Bridge at river mile 55	1,560	<10	<38
12510500	Yakima River at Kiona	1,500	<10	<37
	DIELDRIN			
	Background			
12478200	Cooper River at Salmon LaSac	229	<.37	<.21
12484550	Umtanum Creek near mouth	1	<.12	<.01
12496510	Pacific Power and Light Company Wasteway	--	<.21	--
12507594	Satus Creek above Wilson-Charley Canyon	12	<.39	<.01
	Drinking water			
12496511	City of Yakima—Finish water	--	<.20	--
461720120043200	Well 1, Sunnyside (09N/22E-04P01)	--	<.20	--
462510120323900	Well 2, Harrah (11N/18E-22R02)	--	<.20	--
	Kittitas			
12484100	Wilson Creek above Cherry Creek	144	.01	<.01
12484480	Cherry Creek at Thrall	126	4.7	1.4
			Subtotal	1.4
	Tieton			
12500445	Wide Hollow Creek near mouth	33	1.0	.08
12502500	Ahtanum Creek at Union Gap	23	.003	<.01
			Subtotal	.08
	East-side tributaries			
12500430	Moxee Drain at Thorp Road	65	.86	.14
12505460	Granger Drain at mouth	37	3.6	.32
12508850	Sulphur Creek Wasteway	259	37	23
12509710	Spring Creek at mouth	47	7.9	.91
12509829	Snipes Creek at mouth	50	6.1	.75
			Subtotal	25.1
	West-side tributaries			
12505350	East Toppenish Drain at Wilson Road	53	.05	.01
12505410	Sub 35 Drain at Parton Road	56	1.2	.16
12505510	Marion Drain at Indian Church Road	71	1.4	.24
12507508	Toppenish Creek at Indian Church Road	39	.06	.01
12508620	Satus Creek at gage at Satus	124	1.7	.52
12508630	South Drain near Satus	82	.02	<.01
			Subtotal	.94
	Main stem and Naches River			
12479500	Yakima River at Cle Elum	3,780	<.23	<2.1
12484500	Yakima River at Umtanum	3,630	.97	8.7
12499000	Naches River near North Yakima	1,190	<.35	<1.0
12500450	Yakima River above Ahtanum Creek	3,630	.77	6.8
12507585	Yakima River at river mile 72	1,070	1.1	2.9
12509050	Yakima River at Euclid Bridge at river mile 55	1,560	≥5.7	22
12510500	Yakima River at Kiona	1,500	3.7	14

Table 60. Concentrations and instantaneous loads of suspended sediment and selected organic compounds in water, Yakima River Basin, Washington, June 25–30, 1989—Continued

Station number	Station name	Streamflow	Concentration	Load
	DIMETHOATE			
	Background			
12478200	Cooper River at Salmon LaSac	229	<2.2	<1.2
12484550	Umtanum Creek near mouth	1	<1.1	<.01
12496510	Pacific Power and Light Company Wasteway	--	<1.1	--
12507594	Satus Creek above Wilson-Charley Canyon	12	<1.3	<.04
	Drinking water			
12496511	City of Yakima—Finish water	--	<1.4	--
461720120043200	Well 1, Sunnyside (09N/22E-04P01)	--	<1.4	--
462510120323900	Well 2, Harrah (11N/18E-22R02)	--	<1.4	--
	Kittitas			
12484100	Wilson Creek above Cherry Creek	144	<4.7	<1.6
12484480	Cherry Creek at Thrall	126	<3.5	<1.1
	Tieton			
12500445	Wide Hollow Creek near mouth	33	5.7	.46
12502500	Ahtanum Creek at Union Gap	23	<4.2	<.24
			Subtotal	.46
	East-side tributaries			
12500430	Moxee Drain at Thorp Road	65	<13	<2.1
12505460	Granger Drain at mouth	37	<12	<1.1
12508850	Sulphur Creek Wasteway	259	330	210
12509710	Spring Creek at mouth	47	<4.6	<.53
12509829	Snipes Creek at mouth	50	<4.4	<.54
			Subtotal	210
	West-side tributaries			
12505350	East Toppenish Drain at Wilson Road	53	<4.5	<.58
12505410	Sub 35 Drain at Parton Road	56	<4.4	<.60
12505510	Marion Drain at Indian Church Road	71	4.6	.80
12507508	Toppenish Creek at Indian Church Road	39	<4.0	<.38
12508620	Satus Creek at gage at Satus	124	7.7	2.3
12508630	South Drain near Satus	82	<13	<2.6
			Subtotal	3.1
	Main stem and Naches River			
12479500	Yakima River at Cle Elum	3,780	<1.2	<11
12484500	Yakima River at Umtanum	3,630	<3.7	<33
12499000	Naches River near North Yakima	1,190	<1.3	<3.8
12500450	Yakima River above Ahtanum Creek	3,630	<3.5	<31
12507585	Yakima River at river mile 72	1,070	<3.5	<9.2
12509050	Yakima River at Euclid Bridge at river mile 55	1,560	7.3	28
12510500	Yakima River at Kiona	1,500	<3.7	<14
	ENDOSULFAN I			
	Background			
12478200	Cooper River at Salmon LaSac	229	<.34	<.19
12484550	Umtanum Creek near mouth	1	--	--
12496510	Pacific Power and Light Company Wasteway	--	8.0	--
12507594	Satus Creek above Wilson-Charley Canyon	12	<.21	<.01
	Drinking water			
12496511	City of Yakima—Finish water	--	1.3	--
461720120043200	Well 1, Sunnyside (09N/22E-04P01)	--	<.18	--
462510120323900	Well 2, Harrah (11N/18E-22R02)	--	<.18	--
	Kittitas			
12484100	Wilson Creek above Cherry Creek	144	<.83	<.29
12484480	Cherry Creek at Thrall	126	<.30	<.09

Table 60. Concentrations and instantaneous loads of suspended sediment and selected organic compounds in water, Yakima River Basin, Washington, June 25–30, 1989—Continued

Station number	Station name	Streamflow	Concentration	Load
	ENDOSULFAN I—Continued			
	Tieton			
12500445	Wide Hollow Creek near mouth	33	2.6	.21
12502500	Ahtanum Creek at Union Gap	23	2.3	.13
			Subtotal	.34
	East-side tributaries			
12500430	Moxee Drain at Thorp Road	65	12	1.9
12505460	Granger Drain at mouth	37	<1.4	<.13
12508850	Sulphur Creek Wasteway	259	6.8	4.3
12509710	Spring Creek at mouth	47	<.81	<.09
12509829	Snipes Creek at mouth	50	<.80	<.10
			Subtotal	6.2
	West-side tributaries			
12505350	East Toppenish Drain at Wilson Road	53	3.5	.45
12505410	Sub 35 Drain at Parton Road	56	<.80	<.11
12505510	Marion Drain at Indian Church Road	71	1.8	.31
12507508	Toppenish Creek at Indian Church Road	39	1.2	.11
12508620	Satus Creek at gage at Satus	124	<.80	<.24
12508630	South Drain near Satus	82	<1.6	<.32
			Subtotal	.87
	Main stem and Naches River			
12479500	Yakima River at Cle Elum	3,780	<.17	<1.6
12484500	Yakima River at Umtanum	3,630	.82	7.3
12499000	Naches River near North Yakima	1,190	.73	2.1
12500450	Yakima River above Ahtanum Creek	3,630	.92	8.2
12507585	Yakima River at river mile 72	1,070	<.10	<.26
12509050	Yakima River at Euclid Bridge at river mile 55	1,560	<.80	<3.1
12510500	Yakima River at Kiona	1,500	1.2	4.4
	EPTC (*S*-ethyl dipropylthiocarbamate)			
	Background			
12478200	Cooper River at Salmon LaSac	229	<.31	<.17
12484550	Umtanum Creek near mouth	1	<.22	<.01
12496510	Pacific Power and Light Company Wasteway	--	<.34	--
12507594	Satus Creek above Wilson-Charley Canyon	12	<.27	<.01
	Drinking water			
12496511	City of Yakima—Finish water	--	<.20	--
461720120043200	Well 1, Sunnyside (09/22E-04P01)	--	<.20	--
462510120323900	Well 2, Harrah (11N/18E-22R02)	--	1.0	--
	Kittitas			
12484100	Wilson Creek above Cherry Creek	144	1.6	.56
12484480	Cherry Creek at Thrall	126	37	11
			Subtotal	11.6
	Tieton			
12500445	Wide Hollow Creek near mouth	33	<.60	<.05
12502500	Ahtanum Creek at Union Gap	23	<.87	<.05
	East-side tributaries			
12500430	Moxee Drain at Thorp Road	65	<2.7	<.43
12505460	Granger Drain at mouth	37	3.1	.28
12508850	Sulphur Creek Wasteway	259	<5.0	<3.2
12509710	Spring Creek at mouth	47	3.3	.38
12509829	Snipes Creek at mouth	50	1.5	.18
			Subtotal	.84

Table 60. Concentrations and instantaneous loads of suspended sediment and selected organic compounds in water, Yakima River Basin, Washington, June 25–30, 1989—Continued

Station number	Station name	Streamflow	Concentration	Load
	EPTC (*S*-ethyl dipropylthiocarbamate)—Continued			
	West-side tributaries			
12505350	East Toppenish Drain at Wilson Road	53	1.8	.23
12505410	Sub 35 Drain at Parton Road	56	6.3	.86
12505510	Marion Drain at Indian Church Road	71	3.2	.56
12507508	Toppenish Creek at Indian Church Road	39	2.8	.27
12508620	Satus Creek at gage at Satus	124	2.9	.88
12508630	South Drain near Satus	82	3.4	.68
			Subtotal	3.48
	Main stem and Naches River			
12479500	Yakima River at Cle Elum	3,780	<.35	<3.2
12484500	Yakima River at Umtanum	3,630	1.6	14
12499000	Naches River near North Yakima	1,190	<.26	<.76
12500450	Yakima River above Ahtanum Creek	3,630	1.1	9.8
12507585	Yakima River at river mile 72	1,070	7.3	19
12509050	Yakima River at Euclid Bridge at river mile 55	1,560	3.8	14
12510500	Yakima River at Kiona	1,500	3.3	12
	MALATHION			
	Background			
12478200	Cooper River at Salmon LaSac	229	<2.8	<1.6
12484550	Umtanum Creek near mouth	1	<1.0	<.01
12496510	Pacific Power and Light Company Wasteway	--	5.6	--
12507594	Satus Creek above Wilson-Charley Canyon	12	<1.1	<.03
	Drinking water			
12496511	City of Yakima—Finish water	--	<1.8	--
461720120043200	Well 1, Sunnyside (09N/22E-04P01)	--	<1.7	--
462510120323900	Well 2, Harrah (11N/18E-22R02)	--	2.1	--
	Kittitas			
12484100	Wilson Creek above Cherry Creek	144	<5.9	<2.1
12484480	Cherry Creek at Thrall	126	<2.8	<.86
	Tieton			
12500445	Wide Hollow Creek near mouth	33	13	1.0
12502500	Ahtanum Creek at Union Gap	23	6.3	.35
			Subtotal	1.35
	East-side tributaries			
12500430	Moxee Drain at Thorp Road	65	<11	<1.8
12505460	Granger Drain at mouth	37	<10	<.91
12508850	Sulphur Creek Wasteway	259	<44	<28
12509710	Spring Creek at mouth	47	48	5.5
12509829	Snipes Creek at mouth	50	27	3.3
			Subtotal	8.8
	West-side tributaries			
12505350	East Toppenish Drain at Wilson Road	53	12	1.6
12505410	Sub 35 Drain at Parton Road	56	19	2.6
12505510	Marion Drain at Indian Church Road	71	31	5.4
12507508	Toppenish Creek at Indian Church Road	39	12	1.1
12508620	Satus Creek at gage at Satus	124	14	4.2
12508630	South Drain near Satus	82	<11	<2.2
			Subtotal	14.9
	Main stem and Naches River			
12479500	Yakima River at Cle Elum	3,780	1.5	14
12484500	Yakima River at Umtanum	3,630	<4.0	<36
12499000	Naches River near North Yakima	1,190	1.6	4.6

Table 60. Concentrations and instantaneous loads of suspended sediment and selected organic compounds in water, Yakima River Basin, Washington, June 25–30, 1989—Continued

Station number	Station name	Streamflow	Concentration	Load
	MALATHION—Continued			
	Main stem and Naches River—Continued			
12500450	Yakima River above Ahtanum Creek	3,630	25	220
12507585	Yakima River at river mile 72	1,070	14	37
12509050	Yakima River at Euclid Bridge at river mile 55	1,560	8.6	33
12510500	Yakima River at Kiona	1,500	70	260
	METOLACHLOR			
	Background			
12478200	Cooper River at Salmon LaSac	229	<.91	<.51
12484550	Umtanum Creek near mouth	1	<.25	<.01
12496510	Pacific Power and Light Company Wasteway	--	<.41	--
12507594	Satus Creek above Wilson-Charley Canyon	12	<.31	<.01
	Drinking water			
12496511	City of Yakima—Finish water	--	<.59	--
461720120043200	Well 1, Sunnyside (09N/22E-04P01)	--	<.59	--
462510120323900	Well 2, Harrah (11N/18E-22R02)	--	<.59	--
	Kittitas			
12484100	Wilson Creek above Cherry Creek	144	<1.9	<.67
12484480	Cherry Creek at Thrall	126	<.70	<.22
	Tieton			
12500445	Wide Hollow Creek near mouth	33	<1.8	<.14
12502500	Ahtanum Creek at Union Gap	23	<1.0	<.06
	East-side tributaries			
12500430	Moxee Drain at Thorp Road	65	<3.1	<.49
12505460	Granger Drain at mouth	37	<2.8	<.25
12508850	Sulphur Creek Wasteway	259	<15	<9.5
12509710	Spring Creek at mouth	47	<1.8	<.21
12509829	Snipes Creek at mouth	50	<1.8	<.22
	West-side tributaries			
12505350	East Toppenish Drain at Wilson Road	53	<1.8	<0.23
12505410	Sub 35 Drain at Parton Road	56	<1.8	<.25
12505510	Marion Drain at Indian Church Road	71	3.6	.62
12507508	Toppenish Creek at Indian Church Road	39	2.2	.21
12508620	Satus Creek at gage at Satus	124	4.1	1.2
12508630	South Drain near Satus	82	6.0	1.2
			Subtotal	3.23
	Main stem and Naches River			
12479500	Yakima River at Cle Elum	3,780	<.42	<3.9
12484500	Yakima River at Umtanum	3,630	<1.3	<12
12499000	Naches River near North Yakima	1,190	<.30	<.87
12500450	Yakima River above Ahtanum Creek	3,630	<1.2	<11
12507585	Yakima River at river mile 72	1,070	.93	2.4
12509050	Yakima River at Euclid Bridge at river mile 55	1,560	1.3	5.0
12510500	Yakima River at Kiona	1,500	1.9	7.0
	PARATHION			
	Background			
12478200	Cooper River at Salmon LaSac	229	<2.6	<1.4
12484550	Umtanum Creek near mouth	1	<.04	<.01
12496510	Pacific Power and Light Company Wasteway	--	<1.0	--
12507594	Satus Creek above Wilson-Charley Canyon	12	<1.4	<.04

Table 60. Concentrations and instantaneous loads of suspended sediment and selected organic compounds in water, Yakima River Basin, Washington, June 25–30, 1989—Continued

Station number	Station name	Streamflow	Concentration	Load
	PARATHION—Continued			
	Drinking water			
12496511	City of Yakima—Finish water	--	<1.7	--
461720120043200	Well 1, Sunnyside (09N/22E-04P01)	--	<1.7	--
462510120323900	Well 2, Harrah (11N/18E-22R02)	--	<1.7	--
	Kittitas			
12484100	Wilson Creek above Cherry Creek	144	<5.6	<2.0
12484480	Cherry Creek at Thrall	126	<3.1	<.96
	Tieton			
12500445	Wide Hollow Creek near mouth	33	<5.2	<.42
12502500	Ahtanum Creek at Union Gap	23	<4.6	<.26
	East-side tributaries			
12500430	Moxee Drain at Thorp Road	65	.14	.02
12505460	Granger Drain at mouth	37	<12	<1.1
12508850	Sulphur Creek Wasteway	259	.10	.06
12509710	Spring Creek at mouth	47	12	1.4
12509829	Snipes Creek at mouth	50	.20	.02
			Subtotal	1.50
	West-side tributaries			
12505350	East Toppenish Drain at Wilson Road	53	10	1.3
12505410	Sub 35 Drain at Parton Road	56	<5.2	<.71
12505510	Marion Drain at Indian Church Road	71	7.4	1.3
12507508	Toppenish Creek at Indian Church Road	39	16	1.5
12508620	Satus Creek at gage at Satus	124	180	55
12508630	South Drain near Satus	82	23	4.6
			Subtotal	63.7
	Main stem and Naches River			
12479500	Yakima River at Cle Elum	3,780	<1.0	<9.3
12484500	Yakima River at Umtanum	3,630	<3.3	<29
12499000	Naches River near North Yakima	1,190	<1.4	<3.9
12500450	Yakima River above Ahtanum Creek	3,630	<3.2	<28
12507585	Yakima River at river mile 72	1,070	11	29
12509050	Yakima River at Euclid Bridge at river mile 55	1,560	30	110
12510500	Yakima River at Kiona	1,500	<3.3	<12
	PHORATE			
	Background			
12478200	Cooper River at Salmon LaSac	229	<2.4	<1.3
12484550	Umtanum Creek near mouth	1	<.96	<.01
12496510	Pacific Power and Light Company Wasteway	--	<.88	--
12507594	Satus Creek above Wilson-Charley Canyon	12	<.92	<.03
	Drinking water			
12496511	City of Yakima—Finish water	--	<1.6	--
461720120043200	Well 1, Sunnyside (09N/22E-04P01)	--	<1.6	--
462510120323900	Well 2, Harrah (11N/18E-22R02)	--	<1.6	--
	Kittitas			
12484100	Wilson Creek above Cherry Creek	144	<5.3	<1.9
12484480	Cherry Creek at Thrall	126	5.5	1.7
			Subtotal	1.7
	Tieton			
12500445	Wide Hollow Creek near mouth	33	<4.9	<.40
12502500	Ahtanum Creek at Union Gap	23	<3.0	<.17

Table 60. Concentrations and instantaneous loads of suspended sediment and selected organic compounds in water, Yakima River Basin, Washington, June 25–30, 1989—Continued

Station number	Station name	Streamflow	Concentration	Load
	PHORATE—Continued			
	East-side tributaries			
12500430	Moxee Drain at Thorp Road	65	130	21
12505460	Granger Drain at mouth	37	<8.1	<.73
12508850	Sulphur Creek Wasteway	259	<40	<25
12509710	Spring Creek at mouth	47	<5.0	<.57
12509829	Snipes Creek at mouth	50	<4.9	<.60
			Subtotal	21
	West-side tributaries			
12505350	East Toppenish Drain at Wilson Road	53	10	1.3
12505410	Sub 35 Drain at Parton Road	56	<4.9	<.67
12505510	Marion Drain at Indian Church Road	71	19	3.3
12507508	Toppenish Creek at Indian Church Road	39	<2.8	<.27
12508620	Satus Creek at gage at Satus	124	<4.9	<1.5
12508630	South Drain near Satus	82	<9.1	<1.8
			Subtotal	4.6
	Main stem and Naches River			
12479500	Yakima River at Cle Elum	3,780	<.92	<8.5
12484500	Yakima River at Umtanum	3,630	<2.9	<26
12499000	Naches River near North Yakima	1,190	<.91	<2.6
12500450	Yakima River above Ahtanum Creek	3,630	3.6	32
12507585	Yakima River at river mile 72	1,070	<3.5	<9.2
12509050	Yakima River at Euclid Bridge at river mile 55	1,560	<3.0	<11
12510500	Yakima River at Kiona	1,500	<2.9	<11
	PHOSPHAMIDON			
	Background			
12478200	Cooper River at Salmon LaSac	229	<2.2	<1.2
12484550	Umtanum Creek near mouth	1	<.86	<.01
12496510	Pacific Power and Light Company Wasteway	--	<2.0	--
12507594	Satus Creek above Wilson-Charley Canyon	12	<1.0	<.03
	Drinking water			
12496511	City of Yakima—Finish water	--	<1.5	--
461720120043200	Well 1, Sunnyside (09N/22E-04P01)	--	<1.4	--
462510120323900	Well 2, Harrah (11N/18E-22R02)	--	<1.5	--
	Kittitas			
12484100	Wilson Creek above Cherry Creek	144	<4.7	<1.6
12484480	Cherry Creek at Thrall	126	<2.4	<.74
	Tieton			
12500445	Wide Hollow Creek near mouth	33	8.1	.65
12502500	Ahtanum Creek at Union Gap	23	5.3	.30
			Subtotal	.95
	East-side tributaries			
12500430	Moxee Drain at Thorp Road	65	43	6.8
12505460	Granger Drain at mouth	37	56	5.1
12508850	Sulphur Creek Wasteway	259	<37	<23
12509710	Spring Creek at mouth	47	6.7	.77
12509829	Snipes Creek at mouth	50	28	3.4
			Subtotal	16.1
	West-side tributaries			
12505350	East Toppenish Drain at Wilson Road	53	52	6.7
12505410	Sub 35 Drain at Parton Road	56	16	2.2
12505510	Marion Drain at Indian Church Road	71	10	1.7
12507508	Toppenish Creek at Indian Church Road	39	<3.2	<.30

Table 60. Concentrations and instantaneous loads of suspended sediment and selected organic compounds in water, Yakima River Basin, Washington, June 25–30, 1989—Continued

Station number	Station name	Streamflow	Concentration	Load
	PHOSPHAMIDON—Continued			
	West-side tributaries—Continued			
12508620	Satus Creek at gage at Satus	124	<4.5	<1.4
12508630	South Drain near Satus	82	<11	<2.2
			Subtotal	10.6
	Main stem and Naches River			
12479500	Yakima River at Cle Elum	3,780	<2.1	<19
12484500	Yakima River at Umtanum	3,630	<6.5	<58
12499000	Naches River near North Yakima	1,190	<1.0	<2.9
12500450	Yakima River above Ahtanum Creek	3,630	<6.1	<54
12507585	Yakima River at river mile 72	1,070	<2.4	<6.3
12509050	Yakima River at Euclid Bridge at river mile 55	1,560	22	84
12510500	Yakima River at Kiona	1,500	5.3	19
	PROMETON			
	Background			
12478200	Cooper River at Salmon LaSac	229	<1.2	<.67
12484550	Umtanum Creek near mouth	1	<.58	<.01
12496510	Pacific Power and Light Company Wasteway	--	<.26	--
12507594	Satus Creek above Wilson-Charley Canyon	12	<.27	<.01
	Drinking water			
12496511	City of Yakima—Finish water	--	<.81	--
461720120043200	Well 1, Sunnyside (09N/22E-04P01)	--	<.80	--
462510120323900	Well 2, Harrah (11N/18E-22R02)	--	<.81	--
	Kittitas			
12484100	Wilson Creek above Cherry Creek	144	7.8	2.8
12484480	Cherry Creek at Thrall	126	1.4	.43
			Subtotal	3.23
	Tieton			
12500445	Wide Hollow Creek near mouth	33	22	1.8
12502500	Ahtanum Creek at Union Gap	23	3.0	.17
			Subtotal	1.97
	East-side tributaries			
12500430	Moxee Drain at Thorp Road	65	3.8	.60
12505460	Granger Drain at mouth	37	32	2.9
12508850	Sulphur Creek Wasteway	259	26	16
12509710	Spring Creek at mouth	47	<2.5	<.29
12509829	Snipes Creek at mouth	50	<2.5	<.30
			Subtotal	19.5
	West-side tributaries			
12505350	East Toppenish Drain at Wilson Road	53	<2.5	<.32
12505410	Sub 35 Drain at Parton Road	56	<2.5	<.34
12505510	Marion Drain at Indian Church Road	71	<2.5	<.43
12507508	Toppenish Creek at Indian Church Road	39	2.7	.26
12508620	Satus Creek at gage at Satus	124	<2.5	<.76
12508630	South Drain near Satus	82	<2.7	<.54
			Subtotal	.26
	Main stem and Naches River			
12479500	Yakima River at Cle Elum	3,780	<.27	<2.5
12484500	Yakima River at Umtanum	3,630	1.1	9.8
12499000	Naches River near North Yakima	1,190	<.26	<.76
12500450	Yakima River above Ahtanum Creek	3,630	1.4	12
12507585	Yakima River at river mile 72	1,070	2.4	6.3
12509050	Yakima River at Euclid Bridge at river mile 55	1,560	5.2	20
12510500	Yakima River at Kiona	1,500	3.0	11

Table 60. Concentrations and instantaneous loads of suspended sediment and selected organic compounds in water, Yakima River Basin, Washington, June 25–30, 1989—Continued

Station number	Station name	Streamflow	Concentration	Load
	PROPARGITE			
	Background			
12478200	Cooper River at Salmon LaSac	229	<1.8	<1.0
12484550	Umtanum Creek near mouth	1	<.61	<.01
12496510	Pacific Power and Light Company Wasteway	--	<.66	--
12507594	Satus Creek above Wilson-Charley Canyon	12	<.64	<.02
	Drinking water			
12496511	City of Yakima—Finish water	--	<1.2	--
461720120043200	Well 1, Sunnyside (09/22E-04P01)	--	<1.2	--
462510120323900	Well 2, Harrah (11N/18E-22R02)	--	<1.2	--
	Kittitas			
12484100	Wilson Creek above Cherry Creek	144	<3.8	<1.3
12484480	Cherry Creek at Thrall	126	<4.6	<1.4
	Tieton			
12500445	Wide Hollow Creek near mouth	33	<3.6	<.29
12502500	Ahtanum Creek at Union Gap	23	<2.1	<.12
	East-side tributaries			
12500430	Moxee Drain at Thorp Road	65	210	33
12505460	Granger Drain at mouth	37	40	3.6
12508850	Sulphur Creek Wasteway	259	260	160
12509710	Spring Creek at mouth	47	3.0	.34
12509829	Snipes Creek at mouth	50	45	5.5
			Subtotal	207
	West-side tributaries			
12505350	East Toppenish Drain at Wilson Road	53	19	2.5
12505410	Sub 35 Drain at Parton Road	56	11	1.5
12505510	Marion Drain at Indian Church Road	71	30	5.2
12507508	Toppenish Creek at Indian Church Road	39	27	2.6
12508620	Satus Creek at gage at Satus	124	14	4.2
12508630	South Drain near Satus	82	160	32
			Subtotal	48.0
	Main stem and Naches River			
12479500	Yakima River at Cle Elum	3,780	<.68	<6.3
12484500	Yakima River at Umtanum	3,630	<2.1	<19
12499000	Naches River near North Yakima	1,190	.57	1.6
12500450	Yakima River above Ahtanum Creek	3,630	3.0	27
12507585	Yakima River at river mile 72	1,070	17	44
12509050	Yakima River at Euclid Bridge at river mile 55	1,560	44	170
12510500	Yakima River at Kiona	1,500	7.2	26
	SIMAZINE			
	Background			
12478200	Cooper River at Salmon LaSac	229	<1.4	<.78
12484550	Umtanum Creek near mouth	1	<.67	<.01
12496510	Pacific Power and Light Company Wasteway	--	<.19	--
12507594	Satus Creek above Wilson-Charley Canyon	12	<.36	<.01
	Drinking water			
12496511	City of Yakima—Finish water	--	<.88	--
461720120043200	Well 1, Sunnyside (09N/22E-04P01)	--	1.9	--
462510120323900	Well 2, Harrah (11N/18E-22R02)	--	6.4	--
	Kittitas			
12484100	Wilson Creek above Cherry Creek	144	4.2	1.5
12484480	Cherry Creek at Thrall	126	11	3.4
			Subtotal	4.9

Table 60. Concentrations and instantaneous loads of suspended sediment and selected organic compounds in water, Yakima River Basin, Washington, June 25–30, 1989—Continued

Station number	Station name	Streamflow	Concentration	Load
	SIMAZINE—Continued			
	Tieton			
12500445	Wide Hollow Creek near mouth	33	7.1	0.57
12502500	Ahtanum Creek at Union Gap	23	6.7	.38
			Subtotal	.95
	East-side tributaries			
12500430	Moxee Drain at Thorp Road	65	<3.6	<.57
12505460	Granger Drain at mouth	37	130	12
12508850	Sulphur Creek Wasteway	259	81	51
12509710	Spring Creek at mouth	47	10	1.2
12509829	Snipes Creek at mouth	50	6.4	.78
			Subtotal	65.0
	West-side tributaries			
12505350	East Toppenish Drain at Wilson Road	53	14	1.8
12505410	Sub 35 Drain at Parton Road	56	17	2.3
12505510	Marion Drain at Indian Church Road	71	21	3.6
12507508	Toppenish Creek at Indian Church Road	39	43	4.1
12508620	Satus Creek at gage at Satus	124	27	8.2
12508630	South Drain near Satus	82	<3.6	<.72
			Subtotal	20.0
	Main stem and Naches River			
12479500	Yakima River at Cle Elum	3,780	.25	2.3
12484500	Yakima River at Umtanum	3,630	2.2	20
12499000	Naches River near North Yakima	1,190	.24	.70
12500450	Yakima River above Ahtanum Creek	3,630	3.9	35
12507585	Yakima River at river mile 72	1,070	16	42
12509050	Yakima River at Euclid Bridge at river mile 55	1,560	33	130
12510500	Yakima River at Kiona	1,500	18	66

Table 61. Physical, chemical, and biological properties of selected organic compounds analyzed, Yakima River Basin, Washington, 1988–91

[Water solubility reported in milligrams per liter; K_{oc} values reported in milliliters per gram; C, degrees Celsius; DDD, dichlorodiphenyldichloroethane; DDE,dichlorodiphenyldichloroethylene; DDT, dichlorodiphenyltrichloroethane; HCH, hexachlorocyclohexane; DEF, *S,S,S*-tributyl phosphorotrithioate; 2,4-D, (2,4-dichlorophenoxy) acetic acid; 2,4-DP, (2,4-dichlorophenoxy) propionic acid; 2,4,5-T, (2,4,5-trichlorophenoxy) acetic acid; EPTC, *S*-ethyl dipropylthiocarbamate; ~, approximately; --, no data found; >, greater than; E, estimate; <, less than; data sources for References are:

(1) = U.S. Environmental Protection Agency, 1991b;
(2)= Mercer and others, 1990;
(3)= U.S. Environmental Protection Agency, 1982;
(4)= Kidd and James, 1991;
(5) = Wauchope and others, 1992;
(6) = Becker and others, 1989;
(7) = Custom Chemicides [n.d.];
(8) = Callahan and others, 1979;
(9) = Smith and others, 1988;
(10) = Weiping and others, 1990;
(11) = Kenaga, 1980]

Compound	Water solubility	Temperature	Reference	Sediment-water sorption coefficient (K_{oc})	Reference	Half-life in soil	Reference
Organochlorine compounds							
Aldrin	0.18	(25 C)	(3)	96,000	(2)	Rapidly converted to dieldrin	(8)
Chlordane	.056–1.85		(1)	140,000	(1)	[1]~ 1 year	(1)
	.56		(2)	38,000	(10)	~9.6 years	(10)
	.056	(25 C)	(3)				
Chlorothalonil	.6	(25 C)	(4)	1,380	(5)	1.5– 3 months	(4)
						1 month	(5)
DDD	.1		(2)	770,000	(2)	--	
DDE	.04	(20 C)	(1)	4,400,000	(1)	Biodegraded very slowly	(1)
DDT	.005		(2)	243,000	(2)	[1]Photolysis slow: > 150 years	(8)
	.006	(25 C)	(3)	3,900,000	(3)	[1]Hydrolysis: 0.2 to 12 years	(8)
				238,000	(8)	~ 10 years	(10)
				240,000	(10)		
Dacthal	.5	(20 C)	(6)	5,000	(6)	100 days	(6)
Dieldrin	.186	(20 C)	(1)	1,700	(1)	[1]Hydrolysis: 10.5 years at pH 7	(1)
	.195	(25 C)	(3)			[1]Photolysis: about 2 months	(1)
						~ 2.4 years	(10)
Endosulfan	.32	(22 C)	(4)	12,400	(5)	50 days	(5)
				2,040	(6)	120 days	(6)
Endosulfan I	.16		(2)	--		--	
	.53	(25 C)	(3)				
Endosulfan II	.07		(2)	--		--	
	.28	(25 C)	(3)				
Endosulfan sulfate	.16		(2)	--		--	
	.22		(3)				
Endrin	.25	(25 C)	(1)	1,700	(1)	[1]Hydrolysis: > 4 years	(1)
	.024		(2)				
Endrin aldehyde	50	(25 C)	(3)	670	(3)	--	
Hexachlorobenzene (HCB)	.006	(25 C)	(1)	3,900	(1)	[1]Photolysis: ~ 90 days	(1)
	.11	(24 C)	(1)			Highly persistent	(8)
α-HCH	1.63	(25 C)	(3)	3,800	(1)	Hydrolysis, photolysis, and oxidation: not important	(1)
						Isomers not sorbed on sediment	(8)

Table 61. Physical, chemical, and biological properties of selected organic compounds analyzed, Yakima River Basin, Washington, 1988–91—Continued

Compound	Water solubility	Temperature	Reference	Sediment-water sorption coefficient (K_{oc})	Reference	Half-life in soil	Reference
				Organochlorine compounds—Continued			
β-HCH	0.24	(25 C)	(3)	3,800	(1)	Isomers not extensively sorbed onto sediment in aquatic environment	(8)
δ-HCH	31.4	(25 C)	(3)	6,600	(2)	Isomers not extensively sorbed onto sediment in aquatic environment	(8)
Heptachlor	.056–.18	(25 C)	(1)	220	(2)	[1]1–3 days	(1)
	.18	(25 C)	(3)	12,000	(3)	Persists in aquatic systems once it sorbs onto sediment	(1)
				24,000	(10)	~ 5.5 years	(10)
Heptachlor epoxide	.35	(25 C)	(1)	220	(1)	Several years	(1)
						Not susceptible to hydrolysis, oxidation, or photolysis	(1)
Lindane (γ-HCH)	7.8	(25 C)	(3)	3,800	(1)	400 days	(5)
	7		(5)	1,080	(2)	266 days	(10)
Methoxychlor	.1		(1)	80,000	(1)	[1]Hydrolysis: ~ 1 year at 27 C	(1)
	.003		(2)			120 days	(5)
Mirex	.2	(24 C)	(1)	24,000,000	(1)	Photolysis: slow rate	(1)
	.6		(2)			No significant decrease in concentration over 130 days	(1)
Nonachlor	.06 E		(1)	73,000	(1)	--	
Perthane	.1		(1)	1,100,000	(1)	Not as persistent as DDT	(1)
						Undergoes photolysis and biodegradation	(1)
Toxaphene	.5	(25 C)	(3)	964	(2)	Sorption important	(8)
	3		(5)	100,000	(5)	9 days	(5)
	.74		(9)	21,000	(10)	~ 2.5 years	(10)
				Organophosphorus compounds			
Azinphos-methyl	30	(25 C)	(4)	1,000	(5)	10 days	(5)
	29		(5)			40 days	(6)
Chlorpyrifos	2	(25 C)	(1)	13,600	(1)	7–141 days	(1)
	.3		(2)	6,070	(5)	~ 60–120 days	(4)
	.4		(5)			30 days	(5)
DEF	2.3	(20 C)	(4)	--		--	
Demeton-S-methyl	3,300	(20 C)	(4)	1 E	(6)	10 days E	(6)
	1,000,000	(20 C)	(6)			7 hours at pH 6	(7)
Diazinon	40	(20 C)	(5)	85	(2)	40 days	(5)
	60		(5)	1,000 E	(5)	32 days	(10)
				500 E	(6)		

Table 61. Physical, chemical, and biological properties of selected organic compounds analyzed, Yakima River Basin, Washington, 1988–91—Continued

Compound	Water solubility	Temperature	Reference	Sediment-water sorption coefficient (K_{oc})	Reference	Half-life in soil	Reference
Organophosphorus compounds—Continued							
Dimethoate	25,000	(21 C)	(4)	20	2(5)	7 days	(5)
	39,800		(5)	8	(6)		
				17 E	(11)		
Disulfoton	25		(2)	1,600	(2)	30 days E	(5)
	12	(22 C)	(4)	600 E	(5)	5 days	(6)
Ethion	2		(2)	15,400	(2)	150 days	(5)
	1.1		(5)	10,000	(5)		
Fonofos	13	(20 C)	(6)	870	(5)	3–16 weeks	(4)
	16.9		(5)	532	(6)	40 days	(5)
	68		(10)			45 days	(6)
						60 days	(10)
Isofenphos	23.8	(20 C)	(4)	600	(5)	150 days E	(5)
	24		(5)				
Malathion	145	(25 C)	(4)	1,800	(2)	1 day	(5)
	130		(5)	1,796	(10)		
Methamidophos	2,000,000		(4)	5	(5)	6 days	(5)
	1,000,000 E		(5)	1	(6)		
	1,000,000	(20 C)	(6)				
Methidathion	240	(20 C)	(4)	400	(6)	7 days	(5)
	220		(5)			21 days	(6)
Methyl-parathion	60		(2)	5,100	(2)	5 days	(5)
	55-60	(25 C)	(4)	9,800	(11)	15 days	(10)
Mevinphos	600,000		(5)	44	(5)	3 days	(5)
				1	(6)		
Parathion	24	(25 C)	(4)	10,700	(2)	14 days	(5)
	5,000 E		(5)			18 days	(10)
	11,000		(10)				
Phorate	50	(25 C)	(4)	3,260	(2)	2–14 days	(4)
	22		(5)	1,000 E	(5)	60 days E	(5)
				2,000	(6)	90 days	(6)
				660	(10)	82 days	(10)
Phosphamidon	1,000,000 E		(5)	7	(5)	17 days	(5)
Terbufos	4.5	(27 C)	(4)	500	(5)	9 –27 days	(4)
	5		(5)	3,000	(6)	5 days	(5)
				1,100	(11)		

Table 61. Physical, chemical, and biological properties of selected organic compounds analyzed, Yakima River Basin, Washington, 1988–91—Continued

Compound	Water solubility	Temperature	Reference	Sediment-water sorption coefficient (K_{oc})	Reference	Half-life in soil	Reference
Chlorophenoxy acid herbicides							
2,4-D	620	(25 C)	(4)	20	(2)	< 7 days	(4)
	890 (acid)		(5)			10 days (acid)	(5)
						7 days	(10)
2,4-DP (dichlorprop)	350	(20 C)	(4)	1,000 E (ester)	(5)	10 days (ester)	(5)
	50 E (ester)		(5)			10 days	(6)
Fenoprop (silvex)	140		(2)	2,600	(2)	--	
2,4,5-T	238		(2)	80 (amine salt)	(5)	33 days	(10)
	278		(9)			24 days (amine salt)	(5)
	500,000 E (amine salt)		(5)	53 E	(11)		
Triazine compounds							
Ametryn	185	(20 C)	(4)	388	(2)	70–120 days	(4)
	300		(5)			60 days	(5)
Atrazine	33		(2)	163	(2)	~ 6 –10 weeks	(4)
	28	(20 C)	(4)	100	(5)	60 days	(5)
						64 days	(10)
Cyanazine	171	(25 C)	(4)	190	(5)	~ 2 weeks	(4)
	170		(5)	168	(10)	14 days	(5)
Hexazinone	33,000	(25 C)	(4)	54	(5)	~ 1–6 months	(4)
						90 days	(5)
Metribuzin	1,050	(20 C)	(4)	60 E	(5)	~ 1–2 months	(4)
	1,220		(5)	41	(6)	40 days	(5)
				24	(10)	30 days	(6)
						37 days	(10)
Prometon	750	(20 C)	(4)	150	(5)	500 days	(5)
	720		(5)	300	(6)	120 days	(6)
				350 E	(11)		
Prometryn	48	(20 C)	(4)	614	(2)	40–70 days	(4)
	33		(5)	400	(5)	60 days	(5)
				610	(10)		
Propazine	8.6		(2)	153	(2)	135 days	(5)
	5	(20 C)	(4)	154	(5)		
Simazine	3.5		(2)	138	(2)	60 days	(5)
	5	(20 C)	(4)	130	(5)	75 days	(6)
	6.2		(5)				
Simetryn	450	(20 C)	(4)	--		--	

Table 61. Physical, chemical, and biological properties of selected organic compounds analyzed, Yakima River Basin, Washington, 1988–91—Continued

Compound	Water solubility	Temperature	Reference	Sediment-water sorption coefficient (K_{oc})	Reference	Half-life in soil	Reference
				Carbamate compounds			
Aldicarb	7,800						
	6,000	(25 C)	(4)	36	(10)	70 days	(10)
Carbaryl	40	(30 C)	(4)	230	(2)	7–14 days in sandy loam	(4)
	120		(5)	300	(5)	14–28 days in clay loam	(4)
				200	(6)	10 days	(5)
				229	(10)	22 days	(10)
Carbofuran	415		(2)	29	(2)	30–60 days	(4)
	320	(25 C)	(4)	22	(5)	50 days	(5)
	351		(5)	28	(10)	40 days	(10)
EPTC	375	(25 C)	(4)	200	(5)	6 days	(5)
	344		(5)	280	(6)	30 days	(6)
				240 E	(11)		
Methiocarb	27	(20 C)	(4)	300 E	(5)	--	
	24		(5)				
Methomyl	57,900	(25 C)	(4)	72	(5)	30 days	(5)
	58,000		(5)	160 E	(11)	33 days	(6)
Oxamyl	280,000	(25 C)	(4)	25	(5)	~ 7 days	(4)
	282,000		(5)	6	(10)	4 days	(5)
						6 days	(10)
Propham	250	(20 C)	(4)	200 E	(5)	15 days at 16 C	(4)
				51 E	(11)	5 days at 29 C	(4)
				210 E	(11)	10 days	(5)
Propoxur	2,000 E	(20 C)	(4)	30	(5)	30 days	(5)
	1,800		(5)	67 E	(11)		
				Thiocarbamate compounds			
Butylate	45	(22 C)	(4)	400	(5)	13 days	(5)
	44		(5)	126	(6)	12 days	(6)
	46	(20 C)	(6)				
Cycloate	75	(20 C)	(4)	430	(5)	~ 4–8 weeks	(4)
	95		(5)	400	(6)	30 days	(5)
	75	(20 C)	(6)	345	(11)		
Vernolate	90	(20 C)	(4)	260	(5)	8–16 days (27 C)	(4)
	108		(5)	330	(6)	> 64 days (4 C)	(4)
	95	(20 C)	(6)			12 days	(5)

Table 61. Physical, chemical, and biological properties of selected organic compounds analyzed, Yakima River Basin, Washington, 1988–91—Continued

Compound	Water solubility	Temperature	Reference	Sediment-water sorption coefficient (K_{oc})	Reference	Half-life in soil	Reference
Acetamide compounds							
Alachlor	242	(25 C)	(4)	190	(2)	15 days	(5)
	240		(5)	170	(5)	18 days	(10)
				120	(10)		
Diphenamid	260	(27 C)	(4)	210	(5)	30 days	(5)
Metolachlor	530	(20 C)	(4)	200	(5)	~ 30 days	(4)
				181	(10)	90 days	(5)
						20 days	(6)
						42 days	(10)
Anilide compounds							
Carboxin	170	(25 C)	(4)	260	(5)	3 days	(5)
	195		(5)	264	(6)	7 days	(6)
	170	(20 C)	(6)				
Benzoic compounds							
Dicamba	4,500		(2)	2	(2)	< 14 days	(4)
	6,500	(25 C)	(4)	2	(6)	14 days (salt)	(5)
	500,000	(20 C)	(6)	2	(10)		
				0.4 E	(11)		
Chloroacetanilide compounds							
Butachlor	20	(20 C)	(4)	--		--	
Propachlor	580		(2)	265	(2)	6.3 days	(5)
	613	(25 C)	(4)	80	(5)	6 days	(6)
				420	(10)	7 days	(10)
Conazole compounds							
Triadimefon	260	(20 C)	(4)	300	(5)	~ 18 days in sandy loam	(4)
	71.5		(5)	273	(6)	~ 6 days in loam	(4)
						26 days	(5)
Pyrethroid compounds							
Flucythrinate	.5	(21 C)	(4)	100,000 E	(5)	~ 2 months	(4)
	.06		(5)			21 days	(5)
Permethrin	.2	(20 C)	(4)	100,000	(5)	< 38 days	(4)
	.006		(5)	86,600	(6)	30 days	(5)
						32 days	(6)

Table 61. Physical, chemical, and biological properties of selected organic compounds analyzed, Yakima River Basin, Washington, 1988–91—Continued

Compound	Water solubility	Temperature	Reference	Sediment-water sorption coefficient (K_{oc})	Reference	Half-life in soil	Reference
Pyridine compounds							
Picloram	430	(25 C)	(4)	26	(2)	90 days	(6)
				16	(5)	90 days (salt)	(5)
				48	(10)	100 days	(10)
Sulfite compounds							
Propargite	.5	(25 C)	(4)	4,000 E	(5)	2–4 weeks in moist soils	(4)
						~ 18 weeks in sand/loam	(4)
						56 days	(5)
Trifluoromethyl compounds							
Trifluralin	.3		(1)	13,700	(1)	3 to > 27 weeks	(1)
	4	(27 C)	(1)	8,000	(5)	21–35 days irrigated	(1)
	.6		(2)	7,300	(10)	126 to > 190 days in sandy loam	(1)
						60 days	(5)
						132 days	(10)
Uracil compounds							
Bromacil	820		(2)	72	(2)	60 days (acid)	(5)
	815	(25 C)	(4)	32 (lithium salt)	(5)	60 days (lithium salt)	(5)
	700 (acid)		(5)	32 (acid)	(5)	350 days	(10)
Terbacil	710	(25 C)	(4)	41	(2)	120 days	(5)
				55	(5)	50 days	(10)
Urea compounds							
Linuron	75		(2)	863	(2)	~ 2–5 months in field conditions	(4)
	81	(25 C)	(4)	400	(5)	60 days	(5)
				370	(6)	75 days	(10)

[1]Half-life in water.

Table 62. Selected water-quality criteria, regulations, and health advisories for organic compounds detected in the Yakima River Basin, Washington, 1987–90

[Concentrations reported in nanograms per liter (ηg/L); -, not available; a, not to be exceeded at any time; b, 24-hour average; c, 4-day average not to be exceeded more than once every 3 years; d, 1-hour average not to be exceeded more than once every 3 years; DDT+DDE+DDD, dichlorodiphenyltrichloroethane (DDT) plus dichlorodiphenyldichloroethylene (DDE) plus dichlorodiphenyldichloroethane (DDD)]

Compound	Water-quality criteria for freshwater aquatic life		EPA drinking-water regulations[1]			Health advisory for cancer at a risk level of 1 in 1,000,000		
	Acute	Chronic	Maximum contaminant level goal	Maximum contaminant level	Lifetime health advisory[2]	Drinking water[3]	Water and fish ingestion[4]	Proposed water and fish ingestion[5]
Organochlorine compounds								
Aldrin	3,000a[6]	--	--	--	--	2	0.074	0.13
Aldrin/dieldrin	2,500a[7,8]	1.9b[7,8]	--	--	--	--	--	--
Chlordane	2,400a[6]	4.3b[6]	0	2,000	--	30	.46	.57
Chlorothalonil	--	--	--	--	--	1,500	--	--
DDT+DDE+DDD	1,100a[6]	1.0b[6]	--	--	--	[9]100	.024	--
DDT	--	--	--	--	--	--	--	.59
DDE	1,050,000a[6,10]	--	--	--	--	--	--	.59
DDD	600a[6,10]	--	--	--	--	--	--	.83
Dieldrin	2,500a[6]	1.9b[6]	--	--	--	2	.071	.14
Endosulfan	220a[6]	56b[6]	--	--	--	--	--	930 for α-isomer 930 for β-isomer
Endrin	180a[6]	2.3b[6]	2,000	2,000	2,000	--	--	760
HCB (hexachlorobenzene)	6,000d[11,12]	3,680c[11,12]	0	1,000	--	20	--	--
Heptachlor	520a[6]	3.8b[6]	0	400	--	8	.278	.21
Heptachlor epoxide	520a[5]	3.8b[5]	0	200	--	4	--	.1
hexachlorocyclohexane (HCH) [mixture of isomers]	100,000a[6,12]	--	--	--	--	--	9 for α-isomer 16 for β-isomer	3.9 14
Lindane (γ-HCH)	2,000a[6]	80b[6]	200	200	200	--	18.6	19
Methoxychlor	--	[6]30	40,000	40,000	40,000	--	--	--
Mirex	--	[6]1.0	--	--	--	--	--	--
Polychlorinated biphenyls (PCB)	2,000a[6]	14b[6]	0	500	--	5	.079	.044
Toxaphene	730d[6]	.2c[6]	0	3,000	--	30	.71	.73
Organophosphorus compounds								
Azinphos-methyl	--	[6]10	--	--	--	--	--	--
Chlorpyrifos	83d[6]	41c[6]	--	--	20,000	--	--	--
Demeton	--	[6]100	--	--	--	--	--	--
Diazinon	9[13]	--	--	--	600	--	--	--
Disulfoton	50[13]	--	--	--	300	--	--	--
Ethion	20[13]	--	--	--	--	--	--	--
Fonophos	--	--	--	--	10,000	--	--	--

Table 62. Selected water-quality criteria, regulations, and health advisories for organic compounds detected in the Yakima River Basin, Washington, 1987–90—Continued

Compound	Water-quality criteria for freshwater aquatic life		EPA drinking-water regulations[1]			Health advisory for cancer at a risk level of 1 in 1,000,000		
	Acute	Chronic	Maximum contaminant level goal	Maximum contaminant level	Lifetime health advisory[2]	Drinking water[3]	Water and fish ingestion[4]	Proposed water and fish ingestion[5]
Organophosphorus compounds—Continued								
Malathion	--	100[6]	--	--	200,000	--	--	--
Methyl parathion	--	--	--	--	2,000	--	--	--
Mevinphos	[13]2	--	--	--	--	--	--	--
Parathion	65d[6]	13c[6]	--	--	--	--	--	--
Phosphamidon	[13]30	--	--	--	--	--	--	--
Terbufos	--	--	--	--	900	--	--	--
Carbamate and thiocarbamate compounds								
Butylate	--	--	--	--	350,000	--	--	--
Carbaryl	[13]20	--	--	--	700,000	--	--	--
Carbofuran	--	--	40,000	40,000	40,000	--	--	--
Chlorophenoxy-acid herbicides plus dicamba								
Dicamba	[13]200,000	--	--	--	200,000	--	--	--
(2,4-dichlorophenoxy) acetic acid [2,4-D]	[13]3,000	--	70,000	70,000	70,000	--	[14]100,000	--
Fenoprop (silvex)	[13]1,400	--	50,000	50,000	50,000	--	[14]10,000	--
Triazine compounds								
Ametryne	--	--	--	--	60,000	--	--	--
Atrazine	--	--	3,000	3,000	3,000	--	--	--
Cyanazine	--	--	--	--	1,000	--	--	--
Metribuzin	--	--	--	--	200,000	--	--	--
Prometon	--	--	--	--	100,000	--	--	--
Propazine	--	--	--	--	10,000	--	--	--
Simazine	[13]10,000	--	4,000	4,000	4,000	--	--	--
Uracil compound								
Bromacil	--	--	--	--	90,000	--	--	--
Acetamide compound								
Metolachlor	--	--	--	--	100,000	--	--	--
Chloroacetanilide compounds								
Alachlor	--	--	0	2,000	--	400	--	--
Propachlor	--	--	--	--	90,000	--	--	--

Table 62. Selected water-quality criteria, regulations, and health advisories for organic compounds detected in the Yakima River Basin, Washington, 1987–90—Continued

Compound	Water-quality criteria for freshwater aquatic life		EPA drinking-water regulations[1]		Lifetime health advisory[2]	Health advisory for cancer at a risk level of 1 in 1,000,000		
	Acute	Chronic	Maximum contaminant level goal	Maximum contaminant level		Drinking water[3]	Water and fish ingestion[4]	Proposed water and fish ingestion[5]
Trifluoromethyl compound								
Trifluralin	--	--	--	--	5,000	--	--	--
Volatile compound								
Chloroform	[6]28,900,000	[6]1,240,000	--	100,000	--	6,000	.19	--
Auxin compound								
Picloram	--	--	500,000	500,000	500,000	--	--	--

[1]U.S. Environmental Protection Agency (EPA), 1992a.
[2]EPA health advisory for which the specified concentration appears to be without appreciable risk (U.S. Environmental Protection Agency, 1992a).
[3]EPA risk-assessment guideline for carcinogenity, assuming the consumption of 2 liters per day (U.S. Environmental Protection Agency, 1992a, unless noted otherwise).
[4]Water-quality criteria for human health, assuming the consumption of water at 2 liters per day and aquatic organisms at 6.5 grams per day (U.S. Environmental Protection Agency, 1986).
[5]U.S. Environmental Protection Agency, 1991a.
[6]U.S. Environmental Protection Agency, 1986.
[7]Washington State Administrative Code, 1988.
[8]Proposed water-quality criteria are: Acute = 359.5 ηg/L, and Chronic = 114.7 ηg/L (U.S. Environmental Protection Agency, 1991c).
[9]U.S. Environmental Protection Agency, 1992b.
[10]Value shown is not a criterion, but a lowest-observed adverse-effect level.
[11]U.S. Environmental Protection Agency, 1990.
[12]Value shown is a proposed criterion, not final.
[13]Recommended maximum concentration sampled at any time and any place (National Academy of Sciences-National Academy of Engineering Committee on Water Quality Criteria, 1973).
[14]Value is not based on carcinogenicity but on toxicity.

☆ U.S. GOVERNMENT PRINTING OFFICE: 1999 — 773-460 / 21003 Region No. 8